Biopesticides Handbook

Biopesticides have a great influence in sustainable agriculture, and their use in commercial farming ensures environmental protection, qualitative products, and effective use of resources. The second edition of *Biopesticides Handbook* is fully updated and includes five new chapters on microbial, biochemical, and RNAi pesticides. It details the benefits of biopesticides along the food chain, offering a full spectrum of the range of organisms and organic products that may be used in the biological control of pests. It discusses the uses and abuses of biopesticides, their positive and negative consequences, as well as more recent advances and the best mode of action to improve environmental safety.

FEATURES

- Thoroughly updated, this edition explores not only the benefits but also all aspects of biopesticides
- Includes new chapters on the uses of biochemical and microbial pesticides and plant-incorporated protectants
- Discusses the new field of RNAi pesticides
- Provides information on insect growth regulators and allelochemicals
- Introduces a new chapter on the uses of biopesticides in food and medicinal crops

This book is intended for professionals, researchers, academics, and postgraduate students with experience in fields such as chemistry, biochemistry, environmental sciences, ecology, and agriculture, as well as those invested in the supply chain of agricultural products, such as farmers, growers, and other stakeholders.

Biopesticides Handbook

Second Edition

Edited by
Leo M.L. Nollet
Showkat Rasool Mir

CRC Press
Taylor & Francis Group
Boca Raton London New York

CRC Press is an imprint of the
Taylor & Francis Group, an **informa** business

Designed cover image: © TBC

Second edition published 2024
by CRC Press
2385 NW Executive Center Drive, Suite 320, Boca Raton FL 33431

and by CRC Press
4 Park Square, Milton Park, Abingdon, Oxon, OX14 4RN

CRC Press is an imprint of Taylor & Francis Group, LLC

© 2024 selection and editorial matter, Leo M.L. Nollet and Showkat Rasool Mir; individual chapters, the contributors

First edition published by CRC Press 2015

ISBN: 978-1-032-20628-8 (hbk)
ISBN: 978-1-032-20767-4 (pbk)
ISBN: 978-1-003-26513-9 (ebk)

DOI: 10.1201/9781003265139

Typeset in Times
by Apex CoVantage, LLC

We would like to thank all the authors of the different chapters. Without their commitment and perseverance, this book would never have existed. We also have a word of thanks to the editorial team at CRC Press.

Leo M.L. Nollet and Showkat Rasool Mir

I dedicate this work to my teacher and guide, Dr. Mohd Ali, Retired Professor, Faculty of Pharmacy, Jamia Hamdard, New Delhi.

Showkat Rasool Mir

Contents

SECTION 3 Classes of Biopesticides

SECTION 4   Uses of Biopesticides

Preface

Biopesticides are naturally occurring chemicals or substances that are derived from animals, plants, and microorganisms used to control agricultural pests and pathogens. The use of biopesticides is increasing day by day due to safety and regulatory issues. This makes the concept all the more relevant. *Biopesticides Handbook* in its second edition intends to add to the knowledge base of farmers, growers, and other stakeholders related to the supply chain of agricultural products.

The book has been divided into four sections.

Section 1 deals with the introduction of biopesticides, their types, metabolism, and mechanisms of action. Recent trends in their regulatory status across the globe have also been highlighted in this section.

Section 2 deals with the details of residues of biopesticides once they are used. Their effect on foodstuffs, water, and soil have been particularly discussed at length.

Section 3 has been dedicated to elaboration on biochemical, microbial, plant-based, animal-based and fungal biopesticides. The chapter on allelochemicals highlights how the compounds synthesized by the plants prevent/control the growth of unwanted organisms in and around the plants. Products derived from industrial wastes and genetic materials with pesticidal activities are also discussed.

Section 4 gives a sneak peek into the use of biopesticides with special reference to their use against pests in food and medicinal plants.

> Genius is one percent inspiration, ninety-nine percent perspiration

Editors

Leo M.L. Nollet earned an MS (1973) and a PhD (1978) in biology at the Katholieke Universiteit, Leuven, Belgium. He is an editor and associate editor of numerous books. He edited for M. Dekker, New York—now CRC Press of the Taylor & Francis Publishing Group—the first, second, and third editions of *Food Analysis by HPLC* and *Handbook of Food Analysis*. The last edition is a two-volume book. Dr. Nollet also edited the *Handbook of Water Analysis* (first, second, and third editions) and *Chromatographic Analysis of the Environment* (third and fourth editions, CRC Press). With F. Toldrá, he coedited two books published in 2006, 2007, and 2017: *Advanced Technologies for Meat Processing* (CRC Press) and *Advances in Food Diagnostics* (Blackwell Publishing—now Wiley). With M. Poschl, he coedited the book *Radionuclide Concentrations in Foods and the Environment* in 2006 (CRC Press). Dr. Nollet has also coedited with Y. H. Hui and other colleagues several books: *Handbook of Food Product Manufacturing* (Wiley, 2007), *Handbook of Food Science, Technology, and Engineering* (CRC Press, 2005), *Food Biochemistry and Food Processing* (first and second editions, Blackwell Publishing—now Wiley, 2006 and 2012), and the *Handbook of Fruits and Vegetable Flavors* (Wiley, 2010). In addition, he edited the *Handbook of Meat, Poultry, and Seafood Quality* (first and second editions, Blackwell Publishing—now Wiley, 2007 and 2012). From 2008 to 2011, he published five volumes of animal product–related books with F. Toldrá: *Handbook of Muscle Foods Analysis*, *Handbook of Processed Meats and Poultry Analysis*, *Handbook of Seafood and Seafood Products Analysis*, *Handbook of Dairy Foods Analysis* (second edition, 2021), and *Handbook of Analysis of Edible Animal By-Products*. Also, in 2011, with F. Toldrá, he coedited two volumes for CRC Press: *Safety Analysis of Foods of Animal Origin* and *Sensory Analysis of Foods of Animal Origin*. In 2012, they published the *Handbook of Analysis of Active Compounds in Functional Foods* and, in a co-edition with Hamir Rathore, *Handbook of Pesticides: Methods of Pesticides Residues Analysis* (2009); *Pesticides: Evaluation of Environmental Pollution* (2012); *Biopesticides Handbook* (2015); and *Green Pesticides Handbook: Essential Oils for Pest Control* (2017). Other finished book projects include *Food Allergens: Analysis, Instrumentation, and Methods* (with A. van Hengel; CRC Press, 2011) and *Analysis of Endocrine Compounds in Food* (Wiley-Blackwell, 2011). Dr. Nollet's recent projects include *Proteomics in Foods* with F. Toldrá (Springer, 2013) and *Transformation Products of Emerging Contaminants in the Environment: Analysis, Processes, Occurrence, Effects, and Risks* with D. Lambropoulou (Wiley, 2014). In the series Food Analysis and Properties, he edited (with C. Ruiz-Capillas) *Flow Injection Analysis of Food Additives* (CRC Press, 2015), and *Marine Microorganisms: Extraction and Analysis of Bioactive Compounds* (CRC Press, 2016). With A.S. Franca, he coedited *Spectroscopic Methods in Food Analysis* (CRC Press, 2017), and with Horacio Heinzen and Amadeo R. Fernandez-Alba, he coedited *Multiresidue Methods for the Analysis*

of Pesticide Residues in Food (CRC Press, 2017). Further volumes in the series Food Analysis and Properties are *Phenolic Compounds in Food: Characterization and Analysis* (with Janet Alejandra Gutierrez-Uribe, 2018), *Testing and Analysis of GMO-Containing Foods and Feed* (with Salah E. O. Mahgoub, 2018), *Fingerprinting Techniques in Food Authentication and Traceability* (with K. S. Siddiqi, 2018), *Hyperspectral Imaging Analysis and Applications for Food Quality* (with N.C. Basantia, Leo M.L. Nollet, and Mohammed Kamruzzaman, 2018), *Ambient Mass Spectroscopy Techniques in Food and the Environment* (with Basil K. Munjanja, 2019), *Food Aroma Evolution: During Food Processing, Cooking, and Aging* (with M. Bordiga, 2019), *Mass Spectrometry Imaging in Food Analysis* (2020), *Proteomics in Food Authentication* (with S. Ötleş, 2020), *Analysis of Nanoplastics and Microplastics in Food* (with K.S. Siddiqi, 2020), *Chiral Organic Pollutants, Monitoring and Characterization in Food and the Environment* (with Edmond Sanganyado and Basil K. Munjanja, 2020), *Sequencing Technologies in Microbial Food Safety and Quality* (with Devarajan Thangardurai, Saher Islam, and Jeyabalan Sangeetha, 2021), *Nanoemulsions in Food Technology: Development, Characterization, and Applications* (with Javed Ahmad, 2021), *Mass Spectrometry in Food Analysis* (with Robert Winkler, 2022), *Bioactive Peptides from Food: Sources, Analysis, and Functions* (with Semih Ötles, 2022), and *Nutriomics: Well-Being through Nutrition* (with Devarajan Thangadurai, Saher Islam, and Juliana Bunmi Adetunji, 2022). In 2023 he published with Javed Ahmad *Analysis of Naturally Occurring Food Toxins of Plant Origin.*

Showkat Rasool Mir is currently working as an associate professor at Jamia Hamdardard, New Delhi. He started his career in the Faculty of Pharmacy, Jamia Hamdard, New Delhi, in 2002 after completing an MPharm at the same institute. He earned a PhD in 2006. He has more than 20 years of teaching experience. He teaches pharmacognosy and phytochemistry at the undergraduate and postgraduate levels. He has authored three books and several book chapters.

Dr. Mir has more than 15 years of experience in the field of natural product research. He has more than 80 research papers in national and international journals to his credit. His cumulative impact factor is more than 80, and his h-index is 28. He has supervised ten PhD and twenty PG scholars while four PhD and two PG scholars are currently pursuing research projects under his supervision. He has successfully completed four funded research projects. His field of expertise is the spectroscopic identification and characterization of natural products. He studies the role of plants and their products as drugs and health foods.

Contributors

Adil Ahamad
Department of Pharmacognosy and
  Phytochemistry
School of Pharmaceutical Education
  and Research
Jamia Hamdard, New Delhi, India

Javed Ahamad
Department of Pharmacognosy
Faculty of Pharmacy
Tishk International University
Erbil, Kurdistan Region, Iraq

Javed Ahmad
Department of Pharmaceutics
College of Pharmacy
Najran University
Najran, Saudi Arabia

Faraat Ali
Department of Inspection and Enforcement
Laboratory Services
Botswana Medicines Regulatory Authority
Gaborone, Botswana

Hasan Ali
Department of Pharmacy
Meerut Institute of Technology
Meerut, India

Antonia Garrido Frenich
Department of Chemistry and Physics
Research Centre for Mediterranean
  Intensive Agrosystems and Agri-
  Food Biotechnology (CIAMBITAL)
University of Almería
Agri-Food Campus of International
  Excellence
Almería, Spain

Vinod Kumar Gauttam
Department of Pharmacy
IES Institute of Pharmacy
Bhopal, India

Travis R. Glare
Faculty of Agriculture and Life
  Sciences
Lincoln University
Lincoln, New Zealand

Sumeet Gupta
MM College of Pharmacy
Maharishi Markandeshwar (Deemed to
  Be University)
Mullana, Haryana, India

Hyda Haroon
Department of Botany
Jamia Hamdard
New Delhi, India

Irzam Haroon
Department of Biotechnology
Jamia Hamdard
New Delhi, India

Zareeka Haroon
Department of Chemistry
National Institute of Technology
Srinagar, Jammu and Kashmir,
  India

Haseeb Ahmad Khan
Department of Biochemistry
College of Science
King Saud University
Riyadh, Saudi Arabia

Beatriz Martín-García
Department of Chemistry and Physics
Research Centre for Mediterranean
  Intensive Agrosystems and Agri-
  Food Biotechnology (CIAMBITAL)
University of Almería
Agri-Food Campus of International
  Excellence
Almería, Spain

José L. Martínez-Vidal
Department of Chemistry and Physics
Research Centre for Mediterranean
    Intensive Agrosystems and
    Agri-Food Biotechnology
    (CIAMBITAL)
University of Almería
Agri-Food Campus of International
    Excellence
Almería, Spain

Showkat Rasool Mir
Department of Pharmacognosy and
    Phytochemistry
School of Pharmaceutical Education
    and Research
Jamia Hamdard, New Delhi, India

Awanish Mishra
Department of Pharmacology and
    Toxicology
National Institute of Pharmaceutical
    Education and Research
    (NIPER)
Changsari, Assam, India

Kavita Munjal
Department of Pharmacy
Amity Institute of Pharmacy
Amity University
Noida, Uttar Pradesh, India

Mohd Javed Naim
Department of Pharmaceutical
    Chemistry
Faculty of Pharmacy
Tishk International University
Erbil, Kurdistan, Iraq

Nikita Nain
Department of Pharmacology
M.M. College of Pharmacy
M.M. (Deemed to Be University)
Mullana, Haryana, India

Kumari Neha
Department of Pharmaceutical
    Chemistry
Delhi Institute of Pharmaceutical
    Sciences and Research (DIPSAR)
DPSR University
New Delhi, India

Leo M.L. Nollet
University College Ghent
Ghent, Belgium

Debarati Rakshit
Department of Pharmacology and
    Toxicology
National Institute of Pharmaceutical
    Education and Research (NIPER)
Changsari, Assam, India

Alba Reyes-Ávila
Department of Chemistry and Physics
Research Centre for Mediterranean
    Intensive Agrosystems and Agri-
    Food Biotechnology (CIAMBITAL)
University of Almería
Agri-Food Campus of International
    Excellence
Almería, Spain

Roberto Romero-González
Department of Chemistry and Physics
Research Centre for Mediterranean
    Intensive Agrosystems and
    Agri-Food Biotechnology
    (CIAMBITAL)
University of Almería
Agri-Food Campus of International
    Excellence
Almería, Spain

Yawar Sadiq
Department of Botany
Aligarh Muslim University
Aligarh, India

Asfia Shabbir
Department of Biosciences
Jamia Millia Islamia
New Delhi, India

Arvind Kumar Sharma
Department of Pharmacognosy
Faculty of Pharmacy
Tishk International University
Erbil, Kurdistan, Iraq

Shariq Ibrahim Sherwani
Department of Communication
Utah Tech University
St. George, Utah, USA

Neelam Singh
I.T.S. College of Pharmacy
Dr. A.P.J. Abdul Kalam Technical
    University
Lucknow, Uttar Pradesh, India

Manisha Trivedi
Indian Pharmacopoeia Commission
Ministry of Health and Family Welfare
Government of India
Ghaziabad, India

Section 1

Biopesticides

1 Biopesticides
An Introduction

Leo M.L. Nollet

CONTENTS

Chemical crop protection is profit-induced poisoning of environment. It is very commonly used in the largest part of the agricultural world.

Now it has realized that the need to feed an ever-growing global population combined with increasing demand for sustainable agricultural practices has generated a significant rise in demand for biopesticides. Biopesticides offer unique benefits all along the food value chain, providing additional options for growers, buyers, dealers, consultants, and retailers. Some significant benefits to growers are resistance management; enhanced crop quality; partnering with traditional chemicals; labor and harvest flexibility; maintaining beneficial insects; environmental safety/no residue issues; strong return on investment; sustainable technology; and added value at the grower, distributor, and retailer levels.

Biological pest control (BPC) emerged as a scientific discipline in the last decade of the 19th century. The first complete and successful control of a major pest, cottony cushion scale, (*Icerya purchasi*) in California was by the introduction of the coccinellid predator *Rodalla cardinalis*. A widespread interest developed in its use among agricultural entomologists and served to begin a series of attempts at biological control. Harold Scott Smith in 1919 was first to designate and define the term *biological control* [1]. He emphasized the use of natural enemies to control insect pests. Since then, the control approach has been in vogue, with certain types of pesticides derived from such natural materials as plants, animals, bacteria, and certain minerals. For example, canola oil and baking soda have pesticidal applications and are considered biopesticides. Biopesticides fall into three major classes: (1) microbial pesticides, (2) plant-incorporated protectants (PIPs), and (3) biochemical pesticides.

DOI: 10.1201/9781003265139-2

In the early years of biopesticides development, some products promised but did not gave results. The US Environmental Protection Agency (EPA) states that in April 2016, there were 299 registered biopesticide active ingredients and 1,401 active biopesticide product registrations [2]. Now the annual sales of microbial pesticides are reported to be US $750 million globally, amounting to only 2.5% of the chemical market. The global market for synthetic and biopesticides should reach $79.3 billion by 2022 from $61.2 billion in 2017 at a compound annual growth rate (CAGR) of 5.3% [3]. Thus, commercially viable biopesticides found success in the market, and still more biopesticides technologies have been developed that give growers more targeted and effective pest management options. Regulatory changes, consumer demand for low residues, and the need for even more productive farming practices are inescapable market forces; biopesticides offer solutions all these areas. The result is increasing acceptance of biopesticides as an effective partner in crop protection programs.

1.1 KEY BENEFITS OF BIOPESTICIDES

Potential problems associated with continued long-term use of toxic insecticide include pest resistance and negative impacts on natural enemies. In addition, increasing documentation of the negative environmental and health impacts of toxic chemical pesticides and increasingly stringent environmental regulations of pesticides have resulted in renewed interest in the development and use of biopesticide pest management products.

Similarly, the pesticide residues in beverages like tea and coffee have affected their marketing. There is a considerable export market for cotton fabric and garments devoid of pesticide residues in Japan and Western countries as even traces of pesticidal chemicals can cause skin ailments in human beings. India has a very good export market for Basmati-type scented rice. All these products are highly susceptible to insect pests, so the use of biopesticides and integrated pest management (IPM) is a must for enhancing their export.

However, biochemical insect control methods produce rather subtle effects compared to conventional chemical insecticides, which are lethal upon contact. It is also clear that a single biochemical control method can rarely replace chemical insecticide treatments. Therefore, it has been realized that the available biochemical tools should be developed as components of an integral crop management program, rather than as sole agents.

Thus there is the need for a book that can put all the known biopesticides and their application techniques, including their advantages and limitations, on one canvas.

1.2 LAYMAN'S NEED FOR BIOPESTICIDES

This book will give a full spectrum of the range of organisms and some organic products that may be used in biological control of insects. It will discuss the uses and abuses of biopesticides that have been around for a long time, as well as the recent advances in this area. It will describe their metabolism and mode of action in order to prove the environmental safety of biopesticides. The book will provide the present status of biopesticides residues in foodstuffs, soil, and water to show that biopesticides can enhance crop quality. Some chapters will describe the degradation and dynamics of biopesticides to reflect the environmental safety as well as easy residue management with biopesticides. Other chapters detail findings about new technologies developed for use in biopesticide crop protection and biopesticide regulations and legislation.

1.3 SOCIETY'S NEED FOR BIOPESTICIDE EDUCATION

Stanford University researchers identified 237 of the most relevant research papers published to date, including 17 studies of populations consuming organic and conventional diets and 233 studies that compared the nutrient levels or the bacterial, fungal, or pesticide contamination of foods grown organically and conventionally. The duration ranged from two days to two years [4]. Smith-Spangler, an instructor in the school's division of general medical discipline did the comprehensive meta-analysis. "Some believe that organic food is always healthier and more nutritious," said Smith-Spangler. "We were a little surprised that we did not find that." Researchers said their aim was to educate people, not to discourage them from making organic food purchases.

Managing pests in ways that leave little or no toxic residue impact on non-target organisms that are not prone to pest resistance has always been a challenge in the modern agricultural system. Consumers are becoming more aware of environmental concerns and asking for chemical-free crops. These challenges are met by using the biopesticides in agriculture.

Biopesticide knowledge is essential for organic food growers; biopesticides buyers, dealers, consultants, retailers, and users; environmentalists; bureaucrats; policy makers; plant protection students and teachers; and technologists at institutes, universities, and research laboratories.

1.3.1 PUBLICATIONS IN THE AREA

A literature survey through chemical abstract shows that most publications on the subject are from China, followed by the US, Japan, India, Spain, and Germany.

Many general books are available on pesticides, environmental pollution, biochemistry, anatomy, etc. Only a few books deal with some aspects of biopesticides but not the whole picture. Here we have a specific and detailed book on the emerging field of biopesticides.

1.4 CONCLUSION

Biopesticides provide a good alternative to chemical crop protection. By responding concurrently to the interests of farming, forestry, and the industrial sector, biopesticides offer a considerable potential for utilization in sustainable agriculture. When all the features are added, the advantages of biopesticides in crop protection suggest that utilization of this class of pesticides can be a highly attractive proposition.

In view of the potential environmental problems associated with chemical crop protection, one can consider establishing centers for the production of biopesticides on a large scale and for the training of users (farmers) and suppliers.

Though biopesticides have proved a promising class of pesticides, a lot of research is yet required to prove their multiple uses for mankind. This book will be a stepping stone in this area.

REFERENCES

1. Smith, H.S. On some phases of insect control by the biological method. Journal of Economic Entomology, 1919, 12 (4), 288–292. https://doi.org/10.1093/jee/12.4.288
2. www.epa.gov/ingredients-used-pesticide-products/what-are-biopesticides
3. www.bccresearch.com/market-research/chemicals/biopesticides-global-markets-report.html
4. Smith-Spangler, C., Brandeau, M.L., Hunter, G.H., Bavinger, C., Pearson, M., Eschbach, P.J., Sundaram, V., Liu, H., Schirmer, P., Stave, C., Olkin, I., Bravata, D.M. Are organic foods safer or healthier than conventional alternatives?: A systematic review. Annals of Internal Medicine, 2012, 157, 348–366. https://doi:org/10.7326/0003-4819-157-5-201209040-00007

2 Types of Biopesticides

Travis R. Glare and Leo M.L. Nollet

CONTENTS

2.1 INTRODUCTION

Pests, diseases, and weeds are major issues throughout the world. In the area of crop production, for example, there are thought to be around 67,000 species that can be considered pests, across microbes, plants (weeds), invertebrates, and vertebrates (Oerke et al. 1994). Estimates of crop yield losses around the world are around 40% (IPPC Secretariat 2021, Oerke et al. 1994). To combat these threats, humans have developed and used many methods. The most successful approaches have been through the application of substances to reduce the populations of pests, disease, and weeds. These have been termed *pesticides*. The advent of chemical pesticides was a major underpinning factor in the green revolution of the 1940s through the 1960s, when crop yields were increased dramatically. The world is again in need of an increase in production similar to the green revolution just to feed the ever-growing population, at a time when environmental and health concerns are leading to a reduction in the arsenal of pesticides available. Biopesticides are touted as an alternate to synthetic pesticides. But what are biopesticides?

DOI: 10.1201/9781003265139-3

7

2.2 WHAT ARE BIOPESTICIDES?

The term *biopesticides* has been used to cover a wide variety of formulated products that are used for control of pests, diseases, and weeds. Simply searching "biopesticide(s)" using the Google Scholar search engine produced 94,000 hits. While many active agents are used in biopesticides, and the targets can be diverse and include vertebrates, invertebrates, plants and microbes, the main rationale for the term is to recognise that the active agent is of biological rather than synthetic origin. In the broadest sense, a biopesticide is simply a formulation based on the activity of an agent of natural origin that has pesticidal action. But given the range of science disciplines and target pests, diseases, and weeds that can be involved, many researchers use the term *biopesticide* in a more restricted sense, making a specific definition difficult.

In some cases, only microbial based biopesticides are considered (e.g., Kiewnick 2007); in others, semiochemicals and pheromones are also forms of biopesticide (Copping and Menn 2000). The US Environmental Protection Agency (EPA) has defined biopesticides as naturally occurring substances that control pests (biochemical pesticides), microorganisms that control pests (microbial pesticides), and pesticidal substances produced by plants containing added genetic material (plant-incorporated protectants) or PIPs (www.epa.gov/pesticides/biopesticides#what).

Biopesticides are derived from such natural materials as animals, plants, bacteria, and certain minerals. For example, canola oil and baking soda have pesticidal applications and are considered biopesticides. As of August 31, 2020, there were 390 registered biopesticide active ingredients (www.epa.gov/ingredients-used-pesticide-products/what-are-biopesticides). The EPA have classified biopesticides into three categories for registrations purposes: (1) microbial pesticides, in which the active ingredient is a microorganism (e.g., a bacterium, fungus, virus, or protozoan); (2) plant-incorporated protectants (PIPs), in which genetic material has been added to a plant to produce pesticidal substances (such as the well-known "Bt maize"); and (3) biochemical pesticides, defined as naturally occurring substances that control pests by non-toxic mechanisms, such as pheromones and some plant extracts.

Similar categories were used by Copping and Menn (2000), who listed biopesticides as including microbial (viral, bacterial, and fungal) organisms; entomophagous nematodes; plant-derived pesticides (botanicals); secondary metabolites from micro-organisms (antibiotics); insect pheromones applied for mating disruption, monitoring, or lure-and-kill strategies; and genes used to transform crops to express resistance to insect, fungal, and viral attacks or to render them tolerant of herbicide application. Tijjani et al. (2016) included microbial pesticides, biochemical pesticides, PIPs, and semiochemicals.

The EU European Environment Agency defines *biopesticide* as a "pesticide made from biological sources, that is from toxins which occur naturally. . . . [N]aturally occurring biological agents used to kill pests by causing specific biological effects rather than by inducing chemical poisoning" (www.eea.europa.eu/help/glossary/chm-biodiversity/biopesticide, accessed December 2022). The definition specifies biological effects rather than chemical poisons. In the past, the EU did not specific separate biopesticides from synthetics, and they were generally considered under the same legislation as pesticides (Villaverde et al. 2014); however, that is now changing (see later in this chapter).

Thakore (2006) also defined biopesticides as "living organisms (plants, microscopic animals such as nematodes, and microorganisms, including bacteria, viruses, and fungi) or natural products derived from these organisms, that are used to suppress pest populations." This author listed three major categories of biopesticides: microbial pesticides, plant-incorporated protectants (genetically modified organisms [GMOs]), and "other," although, confusingly, later in the article, GMOs are discussed as being in competition with biopesticides for market share. Gupta and Dikshit (2010) listed three categories: microbial pesticides (microbial pesticides containing whole microorganisms), plant pesticides (GM plants), and biochemical pesticides (naturally occurring substances that control pests by non-toxic mechanisms). The latter category included substances that interfere with growth or mating, such as plant-growth regulators, and substances that repel or attract pests, such as pheromones. Gupta and Dikshit (2010) included a discussion of plant extracts and insect parasites as examples, without assigning them to a specific category. Commonly, authors define biopesticides based on their own area of interest, such as those based on living microbes (often including nematodes).

Glare et al. (2012) added another class of biopesticides by including beneficial endophytes. This was done on the basis that endophytes—microbes that live within plants without having detrimental effects on them—can have pesticidal or pest-deterrent properties. These endophytes can be introduced through seed or during propagation, and as the plants can be used as the delivery system, the combination seems to fit the definition of a biopesticide. In a sense, GM plants are like endophytes as a form of biopesticide, offering a product that delivers naturally occurring anti-pest compounds produced within the plant. More recent publications have also suggested bacterial endophytes as biopesticides (e.g., Maksimov et al. 2018, Rong et al. 2020).

As noted by the EPA, it can be difficult to determine what substances or even organisms meet the criteria to be considered as agents for use as biopesticides, and there are many grey areas in these definitions. Some authors and regulators have not included pheromones and plant extracts as biopesticides. There is also debate about the inclusion of some other compounds as biopesticides, especially when the mode of action is based on toxic effects.

Some biopesticides are based on microbial and plant extracts, without any live organisms in the final product, acting through toxins. One approach to defining biopesticides has been to use the mode of action, limiting biopesticides to those agents that have a non-toxic mode of action, such as infection. However, this approach is difficult to apply if plant and microbial extracts are included as biopesticide active agents. Use of the term *non-toxic mode of action* in a definition of biochemical biopesticides is problematic as many non-infectious modes of action currently accepted in biopesticides act through toxins. For example, the bacterium *Bacillus thuringiensis* (Bt) kills insects via toxins, rather than infection. In some cases, a toxic mode of action is defined or implied as neurotoxic, but as the mode of action of many naturally occurring toxins produced as part of an infection process are unknown, separating biopesticides from other pesticides based on the type of toxin produced is difficult. Another example is spinosad, a metabolite group first identified from a soil actinomycete *Saccharopolyspora spinosa* and widely used as an insecticide. Copping and Menn (2000) and Santos and Pereira (2020) list spinosad as a biopesticide, whereas most others do not, due to its neurotoxic (nicotinic acetylcholine receptor) nature. Villaverde et al. (2014) suggest classification of biopesticides based on their mode of action, with those based on living organisms acting by exploitation, competition, antibiosis, and lysis and natural products by contact, ingestion, systemic action, suffocation and/or attraction/repulsion, and/or induced resistance.

Some authors include beneficial predators and parasitoids in the group of biopesticides, while most do not. As many companies supply agents, such as the parasitoids *Trichogramma* spp., as products (van Lenteren 2011), their use as inundative pest control products does suggest a biopesticide. Sundh and Goettel (2013) separated non-indigenous invertebrate biocontrol agents (IBCAs) from microbial biocontrol agents (MBCAs), pointing out that the former are largely unregulated and so are treated differently to the MBCAs, which are treated more as chemicals. As outlined by Köhl et al. (2019), of importance in classification of biopesticides and potential for risk to humans and environment is the mode of action. The risk of microbial metabolites is nuanced, depending on when they are produced, and current regulations do not account for this.

The main aim of defining pesticides as biopesticides is to indicate a reduced risk of non-target impacts, other environmental issues, and increased mammalian safety and to indicate the biological origin of the control approach. It can cover products used against vertebrates or invertebrates, plants, or microbial-caused diseases. The term *biopesticide* acknowledges the use of a naturally occurring active agent and distinguishes it from a chemical synthetic pesticide. As discussed earlier, the definition of biopesticides can be limited to those using live agents or be as broad as to include genetically modified plants. In reality, agents used in pesticides fall along a continuum from whole live organism through to near-synthetic molecules. The definition of *biopesticide* is, therefore, to an extent an arbitrary limitation along this continuum. Future developments, such as overexpression of metabolites through directed production, will further stretch these definition issues.

The term *biopesticide* implies the need for repeated application, a pesticide approach. The assumption is that after application, the active agent will not persistent long term and will not replicate enough to control the target pest, weed, or disease without repeat application. Delivery is typically spray or drench, but increasingly, seed coat and, in the case of endophytic microbes, within seed are becoming more common. Given these methods of application, formulation is often inherent in a biopesticide, or at least packaging for distribution.

The use of the term *biopesticide* may, in fact, create a perception of chemical equivalence (Sundh and Goettel 2013), given that it includes the term *pesticide*. Unlike parasitoids and predators, which Sundh and Goettel (2013) referred to as IBCAs (often called natural enemies or beneficial organisms), use of the term *biopesticides* suggests they are still pesticides. The line between biopesticide and synthetic pesticide can be blurred by the use of naturally derived bioactives as templates for synthetic production. Examples are the use of pyrethroids, synthetic versions of pyrethrins (original extracted from the plant *Chrysanthemum*) that have improved UV tolerance, and spinosad. Spinosad is a mixture of compounds originally isolated from the soil actinomycete *Saccharopolyspora spinosa* (Sparks et al. 1995). This family of unique metabolites was identified and developed by Dow AgroSciences as a selective, environmentally friendly insecticide composed of a mixture of spinosyn A and spinosyn D. Spinosads are very effective against a range of insects and have low mammalian toxicity (O'Callaghan and Glare 2001). Despite being derived from naturally occurring microbes, spinosyns have been treated as chemical insecticides by regulators (Chandler et al. 2011), and many synthetic versions have been produced (Sparks et al. 1995). Abamectin is another common insecticidal compound. It is usually a mixture of avermectins, macrocyclic lactone compounds produced by the soil bacterium *Streptomyces avermitilis* (Khalil 2013). As abamectin is produced from the fermentation of the bacterium, it can be considered a biopesticide (Chandler et al. 2011), but the EPA has classified it as a pesticide because of the toxicity level against some mammals, such as rats. Abamectin can be synthesised. One reason these compounds have not been universally treated as biopesticides is that they pose risks to non-target organisms as they have broad toxicity against a range of hosts. That diminishes one of the benefits of biopesticides: lack of non-target toxicity.

There a number of strategies that are similar to biopesticidical approaches that can't be considered biopesticides. Biostimulants, biofertilizers, and some bioinoculants have actions which are not directly pesticidal, although they may promote an effect indistinguishable from a biopesticide, such as plant growth. For example, biofertilizers may contain microbes that promote plant growth through increased nitrogen or phosphorus availability, and healthier plants are more resistant to disease, but the action is not directly on the disease-causing organism. In some cases, the use of invertebrate parasitoids or predators is inundative and may fit a broad definition of a biopesticide, but they are rarely considered as such.

The term *biorationals* has also been used to refer to biopesticides and related approaches (e.g., Rosell et al. 2008). Other names used which fall under the auspices of biopesticides include bionematicides, bioherbicides and biofungicides. Products and prototypes in these categories contain agents that act directly against the pest, whether a nematode, weed, or disease.

Seiber et al. (2014) discussed the differences in biopesticide definition and also provided a useful list of desired qualities: naturally occurring or their derivatives, reduced toxicity to non-target organisms and low persistence in the environment, usable in organic agriculture, low mammalian toxicity and safe for workers and residents, "green" technology, and non-restricted use.

2.3 HISTORY OF BIOPESTICIDES

The earliest pesticides may have been compounds such as elemental sulphur dusting, which was used around 4,500 years ago. Reference to the use of plant derivatives or botanical-based pesticides can be traced back at least 2,000 years to Egypt, Greece, China, and India, with the oldest records referring to the use of the neem tree extracts as a pesticide 4,000 years ago (El-Wakeil 2013).

While botanical-based pesticides are likely to have been used throughout our history, chemical toxins such as arsenic, lead, and mercury were known to be used from around the 15th century. There are records of the use of biopesticides from the 17th century, with the use of some plant extracts to control pests. Nictone sulphate was used to control plum beetles as early as the 17th century, and in the 19th century, more natural substances like pyrethrum and rotenone, which is derived from the roots of tropical vegetables, became commonly used (El-Wakeil 2013).

The first biological introduced in Europe was probably a rodenticide based on *Salmonella enterica* (Ratin) used in Sweden and other European countries in 1904 (Sundh and Goettel 2013). The discovery that microbes cause disease, attributed to Agostino Bassi in 1835, led to the idea of using microbes to control insect pests and disease. First proposed by Louis Pasteur, it was in Russia in the 1890s that the first efforts were made to use fungi against a wheat cockchafer (Zimmermann et al. 1995). A commercially available biopesticide based on the insecticidal bacterium *Bacillus thuringiensis* was sold as Sporéine in France in 1938.

In the 1950s, the development of DDT as an insecticide and other chemical control methods had a detrimental effect on further development and use of biologically based pest and disease control options around the world. These chemical pesticides were cheap and effective control options for many of the world's pest problems. It was not until the emergence of evidence of negative environmental impacts and mammalian toxicity issues that biological options again became of interest. Despite the increasingly public antipathy to chemical pesticides, they still dominate the current pest, weed, and disease control landscape.

There is a resurgence of interest in biopesticides as the pressure on chemical control approaches increases. Many standard synthetic chemical pesticides have been withdrawn by regulators, and there are fewer new synthetics appearing on the market (Glare et al. 2016). Biopesticides are also gaining market share due to applications in areas that have no current synthetic crop protection (Olson 2015). While the number of biopesticides competing with current synthetic pesticides is relatively small, the trend is strongly for this to increase, a trend supported by most commenters. Demand is also being driven by efforts by government and others to increase the use of ecofriendly agri-inputs, such as the EU Farm to Fork strategy.

2.4 PERCEIVED ADVANTAGES AND DISADVANTAGES

The term *biopesticide* has become associated with many perceived attributes, both positive and negative, which currently influence market attitudes. The perceived advantages include low environmental and mammalian risk, including non-target safety through more specificity; lower resistance development; and fewer residue issues. As many biopesticide active agents are based on multigene systems for the mortality of the target (such as in infection), it is less likely that host resistance will develop. The low persistence of many biopesticides can be a benefit where residues are an issue, such as when spraying on a food crop, but can be a negative when there is little persistence of effect after application.

Other perceived negative factors include expense, low efficacy, slow to kill, limited shelf life, and variability in performance. We term these "perceived" issues because no issue is universal to all biopesticides. There is also an issue that many biopesticides target only niche markets due to the specificity of the bioactive agent, which results in registration costs being too high for the size of the market, limiting commercialisation possibilities.

There have been significant advances in developing biopesticides, especially around formulation, which have overcome some of the issues with early biopesticides. However, the perception remains in the minds of many and still restricts uptake of new products. Biopesticides have been promoted as suitable for inclusion in integrated pest management programmes. Generally, a biopesticide will not harm distantly related organisms, so can be used where, for example, parasitoids and predators may be active on insect pests. Further information on the advantages and disadvantages of biopesticides may be found in Pavela (2014), Rajamani and Negi (2021), Glare et al. (2016), and Hassan and Gökçe (2014).

2.5 AGENTS USED IN BIOPESTICIDES

The registration of new products in the US by the EPA is increasing slowly, with more than 390 registered biopesticide active ingredients and over 1,250 actively registered biopesticide products. The area is growing rapidly, with most of the leading pesticide companies around the world acquiring or developing biopesticide products and capabilities (Beer 2012). There have also been significant advances made in the area of the application of "omics" sciences, which are likely to result in many new products (Köberl et al. 2012) and in formulation and application technologies. As discussed earlier, the definition of *biopesticides* can be restrictive, such as microbial-based biopesticides, or inclusive of all pesticides based on naturally produced actives. Some examples follow.

2.5.1 MICROBES AND MICROBIAL EXTRACTS

Microbes and nematodes cause many diseases in other organisms, and it is this ability that has attracted researchers to the idea of developing microbial-based biopesticides. Microorganisms are often used as the active agent in biopesticides, making use of natural actions to infect other organisms (Koul 2011, Sundh and Goettel 2013, Glare et al. 2012). Common microbes used in biopesticides include bacteria, notably *Bacillus* spp., fungi, and viruses. Traditionally, nematodes with biocontrol potential have also been included under the heading of microorganisms but have also been included as augmentative biopesticides (van Lenteren 2011). Bacteriophages have also been formulated as biopesticides, for use against bacterial disease (e.g., Roach et al. 2008). According to Koul (2011) microbial-based biopesticides account for around 90% of all biopesticides. Several market forecasters have found that microbial-based biopesticides are the fastest-growing sector of biopesticides in 2022.

Live microbes have been used in biopesticides to target invertebrate and vertebrate pests, plants (as herbicides), and diseases caused by other microbes. The mode of action can vary from competitive exclusion and infection through to toxicity, where multiple toxins are often produced by the live microbe that contributes to target mortality.

A book by Kabaluk et al. (2010) contains reviews of microbial-based biopesticides available in 15 countries, and the EU reported 547 products (24 bacteriocides,

167 fungicides, 8 herbicides, 302 insecticides, 46 nematicides), although these products were based on only 247 species of nematodes, bacteria, fungi, and virus used as the source of the bioactives.

Biopesticides derived from fungi, bacteria, algae, viruses, nematodes, and protozoa, as well as some other compounds produced directly from these microbes, such as metabolites, are the main microbial pest control agents. Up until now, more than 3,000 species of microbes have been reported to cause diseases in insects.

Over 100 bacteria have been identified as insect pathogens, among which *Bacillus thuringiensis* Berliner (Bt) has the maximum importance as a microbial control agent. So far, more than 1,000 insect species viruses have been isolated, such as nuclear polyhedrosis virus (NPV), which infested 525 insects worldwide. Over 800 species of entomopathogenic fungi and 1,000 species of protozoa pathogenic have been described and identified. The two major groups of entomopathogenic nematodes are Steinernema (55 species) and Heterorhabditis (12 species) (Nawaz et al. 2016, Ruiu 2018, Thakur et al. 2020).

2.5.2 PLANT-BASED EXTRACTS

Plants have been a source of bioactives for use against pests, disease, and weeds since early agricultural times. Some plant extracts are toxic to the target, some are repellent, and other work as attractants. Despite the large number of botanical derivatives that are known to have useful activity, few have been developed into commercially available biopesticides (see review by Mathew 2016, Khursheed et al. 2022, Singh 2014). In the first half of the 20th century, four main groups of compounds dominated plant extract–based biopesticides: nicotine and alkaloids, rotenone and rotenoids, pyrethrum and pyrethrins, and vegetable oils (El-Wakeil 2013). Pyrethrins and neem are now well established as bioactives in multiple biopesticides for arthropod control, but few others appear to be widespread in use. Pyrethrins, fast-acting insecticidal compounds, are produced by *Tanacetum* (*Chrysanthemum*) *cinerariifolium* (Asteraceae) or related species. Naturally produced pyrethrins do not persist well, leading to the development of synthetic pyrethroids (Chandler et al. 2011). Neem oil, from the seeds of *Azadirachta indica*, is a widely used botanical biopesticide for insect control (Chandler et al. 2011). There are over 6,000 other potential plant extracts known to have insecticidal activity that might be used as biopesticides (Nawaz et al. 2016). Joseph and Sujatha (2012) report over 200 plants that have been demonstrated to have activity against defoliating pests in the laboratory. Plants produce many secondary compounds to deter the feeding of herbivores.

Essential oils from plants are also used as biopesticides, especially as insecticides (López et al. 2021). Some have direct toxicity; others can act as feeding deterrents, repellents, or other forms of antagonists. Examples include citronella oil for mosquito repellency, cinnamon oil for aphids and mites, and eugenol from basil or cloves for a range of insects (Rosell et al. 2008).

Plant extracts have been promoted, like other biopesticides, on the basis of being less toxic to non-target organisms and mammals. While this is often true, it is not universal as some of the most toxic substances and carcinogens known are plant derived (El-Wakeil 2013).

Phytochemicals are relegated as either primary or secondary metabolites, and the majority of the plant species still remain unexploited and untapped for pesticidally active principles. Among the plants studied for pesticidal properties, only a negligible number have achieved commercial status. Currently, the whole world is before the development and promotion of ecofriendly biopesticides which only attack the target pest and are harmless to beneficial biota. Plants have inbuilt genetical mechanisms like repellency and insecticidal action to protect from pest attacks. Likewise, botanicals are considered as an effective substitute for chemical pesticides as these botanicals and their derivatives have good potentiality to regulate and control harmful pests.

2.5.3 Microbes in Plants (Endophytes)

A growing area of interest in biological control is microbes associated with plants. Many microbes have been found to promote plant growth (see Compant et al. 2010; Partida-Martínez and Heil 2011). Microbes that colonise the rhizosphere or phyllosphere can contribute to a plant's health. These microbes have been seen as sources of new agents, genes, and bioactive compounds for use in biopesticides due to the extent of specificity within the plant microbiome. In particular, endophytes, microbes (fungi and bacteria usually) that live within plants without causing symptoms of disease, have become of interest, especially as many have been shown to have effects against pests and pathogens (Lodewyckx et al. 2002, Sikora et al. 2008, Backman and Sikora 2008; Ownley et al. 2008, Porras-Alfaro and Bayman 2011). Recent publications (Maksimov et al. 2018, Rong et al. 2020, Glare et al. 2012) have included endophytes as a class of biopesticide, on the basis that the microbe-plant combination with demonstrated plant protection benefits can be propagated and distributed as a product. The classic example is the grass endophytes, *Neotyphodium* spp.: fungi which produce bioactive compounds within the plant and have insecticidal and other activities. Selection of specific endophyte grass combinations has resulted in successful commercialisation in many temperate grassland areas in New Zealand, Australia, the US, and South America (Easton 2007). The possibilities for exploitation of endophytes seems extensive. In some cases, microbes with direct biocontrol potential, such as the entomopathogenic fungi *Beauveria bassiana* and *Metarhizium anisopliae,* can also occur as plant endophytes and, in addition to showing insecticidal activity, can be antagonists of plant diseases, rhizosphere colonizers, and plant growth promoters (Ownley et al. 2008). There is increasing interest in the range of endophytes protecting plants from pests and diseases (Xia et al. 2022, Kaddes et al. 2019), which will no doubt lead to increased commercial use.

2.5.4 Genetically Modified Plants

Genetically modified (GM) plants with pest- and disease-resistant capabilities are a growing and controversial area in plant protection. The EPA includes plant-incorporated protectants (PIPs) among the classes of biopesticides. It defines PIPs as plants that produce pesticidal substances from genetic material that has been added to the plant, such as those plants expressing pesticidal proteins from the bacterium *Bacillus thuringiensis*. The EPA states that the "protein and its genetic material, but not the plant itself, are regulated by EPA" (website). The number of transgenic plants, those expressing foreign genes, is increasing, but the technology remains controversial in many countries. Interestingly, although most current GM plants sold commercially are based on herbicide resistance (for weed control) or insecticidal protein expression, the new prototypes under development cover much broader traits, including pharmaceuticals and human health diet supplements such as vitamin A, oleic acid, and omega-3 fatty acid (Chen and Lin 2013, Ahmad et al. 2012). The global area of GM crops in 2012 was estimated at 170M ha (James 2012), with herbicide-tolerant soybean accounting for 47%.

Plant-incorporated protectants (PIPs) are biopesticides expressed in genetically modified (GM) crops and are typically macromolecular in nature. First-generation insecticidal PIPs were Cry proteins expressed in GM crops containing transgenes from the soil bacterium *Bacillus thuringiensis*; next-generation double-stranded ribonucleic acid (dsRNA) PIPs have been recently approved. Like conventional synthetic pesticides, the use of either Cry protein or dsRNA PIPs results in their release into receiving environments. However, as opposed to conventional low molecular weight pesticides, the environmental fate of macromolecular PIPs remains less studied and is poorly understood (Parker and Sanders 2017, Schiemann et al. 2019, Nelson 2014).

2.5.5 INVERTEBRATES

Classical biological control involves the point introduction of live organisms antagonistic to a pest, weed, or disease, which are then left to spread and multiply in the pest population. This is not a biopesticide. When the agents are mass produced, packaged, and used inundatively with no expectation of spread and long-term maintenance of control without further release, these could be considered as biopesticides. In some cases, natural enemies, such as parasitoids or predators, are produced in bulk, packaged, and distributed for periodic inundative release as what van Lenteren (2011) calls augmentative biocontrol. According to van Lenteren (2011), there are 230 commercially available species, covering insects (especially Hymenoptera), mites, and nematodes. An example is the production of natural enemies for some glasshouse insect pests around the world with parasitoids, such as *Encarsia formosa* for whitefly control, available in over 20 countries.

2.5.6 SEMIOCHEMICALS AND OTHER ACTIVES

Semiochemicals, informative molecules used in insect-insect or plant-insect interaction (Heuskin et al. 2011), can be classed as biopesticides, depending on use. The most common approaches using semiochemicals are those that are derived from and/or used against insect pests and include pheromones, kairomones, allomones, and other classes of behaviourally active compounds. Semiochemicals can be used in mating disruption and lure-and-kill approaches or simply for monitoring insect populations. The use for monitoring does not, in my opinion, fit with a biopesticide definition, as there is no direct effect on populations, but it can be a useful component of an integrated pest management strategy. Copping (2009) described 74 semiochemicals used in mating disruption, lure-and-kill, or insect-monitoring strategies. Semiochemicals are considered an environmentally safe option as they are usually very specific in target, very low quantities are required, and, as volatiles, they have low persistence (Rosell et al. 2008). Mating disruption has been used successfully against a number of insects. Through the release of large amounts of pheromones, the search for mates is confused, leading to reductions in successful breeding. Examples include pheromone-based products for control of spruce budworm, *Choristoneura fumiferana* (Rhainds et al. 2012) and other lepidopteran, coleopteran, hemipteran, and heteropteran species (Rosell et al. 2008).

Mass trapping using semiochemicals can reduce pest damage using traps baited with pheromone lures. Mass trapping has been successful in some situations with a number of insects, such as the codling moth, *Cydia pomonella*; pink bollworm, *Pectinophora gossypiella*; bark beetles; palm weevils; corn rootworms (*Diabrotica* spp.); gypsy moth, *Lymantria dispar*; and boll weevil, *Anthonomus grandis grandis* (El-Sayed et al. 2006). The other tactic that uses semiochemical approaches is termed "lure and kill," in which semiochemical lures are used to bring insects to a pesticide-baited trap. This may not involve a biopesticide.

Some of the oldest known biopesticide actives come not from plants or microbes but from the earth. Copper, diatomaceous earth, kaolin clay, hydrogen peroxide, potassium bicarbonate, salts, and soaps are considered biopesticides in some countries, including the US and Canada (Bailey et al. 2010). Other products potentially classified as biopesticides come from biological sources, such as the byproducts of organic processing systems (corn gluten meal, acetic acid) (Bailey et al. 2010). Baking soda and canola oil have also been used as biopesticides, and in one report, cow urine was termed a potential biopesticide (Gahukar 2013).

2.5.6.1 RNAi-Based Technology

RNA interference (RNAi), a eukaryotic process in which transcript expression is reduced in a sequence-specific manner, can be co-opted for the control of plant pests and pathogens in a topical application system. Double-stranded RNA (dsRNA), the key trigger molecule of RNAi, has been shown to provide protection without the need for integration of dsRNA-expressing constructs as transgenes. Consequently, development of RNA-based biopesticides is gaining momentum as a narrow-spectrum alternative to chemical-based control measures, with pests and pathogens targeted with accuracy and specificity. Limitations for a commercially viable product to overcome include stable delivery of the topically applied dsRNA and extension of the duration of protection (Fletcher et al. 2020).

Double- stranded RNA (dsRNA) molecules regulate gene expression by targeting specific endogenous mRNA molecules in a sequence-specific manner. By exploiting this sequence-dependent mode of action, RNAi-based products with higher selectivity and better safety profiles (less mobile through the soil, less persistent, and less toxic) than contentious chemical pesticides are being developed. RNAi-based control has several unique features that offer additional opportunities compared with contentious chemical pesticides. The dsRNA active molecules can be designed to target the expression of different genes without the need to change the sequence-dependent mode of action, and, depending on the gene targeted in the pest, various outcomes ranging from sublethal to lethal effects can be achieved (Taning et al. 2021).

2.6 MARKET FOR BIOPESTICIDES

Estimating the total market share of biopesticides is dependent on what is included as a biopesticide. Most figures are based on a restricted definition: for example, excluding GM plants and augmentative natural enemies. Until recently, biopesticides' (microbial and plant-extract based) commercial sales made around 1% of the estimated more than US$30 billion world pesticide market. In the 1990s, one biopesticide active (*B. thuringiensis*) made up over 90% of all biopesticides sales (Rosell et al. 2008); however, this situation has changed. Biopesticide sales in 2010 were estimated at 4.5% of the world pesticide market (Bailey et al. 2010). This figure is continuing to rise, and sales of synthetic pesticides are falling. There are a range of companies which regularly evaluate and forecast the biopesticide market (e.g., Market and Markets, Fortune Business Insights, Allied Market Research), all of which have repeatedly reported that biopesticide sales are increasing year on year.

The development of the interest in biopesticides is being driven by a convergence of factors: withdrawal of pesticides, environmental and health concerns about chemical pesticides, and improvements in products. This has led to all major pesticide companies showing interest in acquiring biopesticide products and/or companies.

The future of the biopesticide market looks promising, with opportunities in seed treatment, soil treatment, and postharvest modes of applications. The global biopesticide market is expected to grow with a compound annual growth rate (CAGR) of 15% from 2019 to 2024 (Marketsandmarkets 2022), although there was less growth during the COVID-impacted years of 2020–2021 (Fortune Business Insights 2022). The major drivers for this market are increasing demand for organic food, lower cost of raw material, and faster regulatory approval. Bioinsecticides continue to experience relatively high demand as compared to other products such as bioherbicides, biofungicides, and bionematicides. As bioinsecticides account for a ~70% share of the market, manufacturers are particularly focused on this category, offering enhanced pest control and ensuring food security. Further, to tackle difficulties associated with finding bioactivity against insects, key manufacturers are targeting to increase the number of microbes tested against insects and develop effective products.

In the biopesticides market, product innovations and development remain influenced by the increased calls for the adoption of sustainable and cost-effective measures. While biopesticides are used as pest-control materials in the cultivation of various agricultural products, it is the demographic shift in the preference for organic fruits and vegetables which has sprung the demand in recent times.

With the increased stringency of chemical residue–related regulations, high incorporation of IPM programs in developing countries, and easier registration of biopesticides than chemical pesticides, manufacturers are likely to capitalize on the new-found opportunities. Innovations have also been forthcoming for liquid formulation of biopesticides for foliar application, as the category continues to witness higher demand (Persistence Market Research 2022).

2.7 PRODUCTION, FORMULATION, AND APPLICATION

As mentioned earlier, the term *biopesticide* suggests a product which will usually involve production, formulation, and application. Production is specific to the organism or bioactive and can involve solid or liquid fermentation or simple culturing and propagation. For extracts such as metabolites or semiochemicals, isolation and/or extraction processes are required.

Formulation is the most complex area for biopesticides. Biopesticides in the broadest sense can range from those that have no true formulation step, such as endophytes or augmentative biocontrols, through to actives that rely on formulations to provide stability and efficacy in the field (Hynes and Boyetchko 2006). The science of formulation is rarely available in the research literature, as much of the information is held by companies as trade secrets. Burges (1998) produced an edited treatise on microbial biopesticide formulation which provided a summary of the state of the art at that time. Gašić and Tanović (2013) review biopesticide formulations currently marketed and trends for the future. They listed dry formulations such as dusts, seed dressing formulations, water dispersible granules, and wettable powders. Among the liquid formulations were emulsions, suspension concentrates, oil dispersions, suspo-emulsions, capsule suspensions, and ultra-low volume formulations. López et al. (2021) recently reviewed solid and liquid formulations of essential oils, while other recent information can be found in Kala et al. (2020) and Lengai and Muthomi (2018).

Semiochemicals, dependent on volatile molecules for effect, are particularly difficult to formulate. Slow-release devices have been developed to ensure the agents are able to act over time (Heuskin et al. 2011). These compounds, like almost all bioactives and agents used in biopesticides, need to be protected from UV and other environmentally caused degradation. Sensitivity to UV is a characteristic shared by almost all biopesticides.

In most cases, biopesticides are applied using standard pesticide equipment, such as seed drills and spray apparatus. However, the requirement for many biopesticides to keep organisms alive can mean modified methods are used. In general, biopesticides which require specialised equipment for application are less likely to achieve market penetration, as this adds to the cost of application.

2.8 REGULATORS' RECOGNITION OF BIOPESTICIDES

Defining what is a biopesticide is more than a etymological discussion. Regulators are grappling with the issue in many countries. Reviews of registration procedures for biopesticides have been proliferating (Kabaluk et al. 2010), as many authorities have identified product registration as a bottleneck in the development of new biopesticides. Keswani et al. (2016) wrote a synthesis of policy support, quality control, and regulatory management of biopesticides. Few countries recognise biopesticides as a separate classification under any pesticide regulations, resulting in inappropriate evaluation methods in many cases (Sundh and Goettel 2013), with the EPA being at the forefront of developing separate processes. As mentioned earlier, the EPA recognises three classes of biopesticides: microbial pesticides, plant-incorporated protectants, and biochemical pesticides. The definition of at least some groups that can be classed as biopesticides allows for appropriate and simplified registration procedures. It has been the bane of many biopesticide developers that registration processes based on chemical pesticides can lead to requirements that are inappropriate, unnecessary, and expensive. For example, risk assessment required under most regimes are cost intensive, time consuming, and often confusing. The lack of recognition of live organisms as opposed to bioactive compounds and the extended registration process have also been cited as issues for biopesticides (Chandler et al. 2011).

There has been progress in some regions. Canada has been moving to recognise differences between biopesticides and synthetic pesticides (Bailey et al. 2010), and the European Union have been considering the issue for many years. The European Union is moving to recognise biopesticides as a separate category, after years of attempting to develop tailored regulations that were expected to favour low-risk products (EC 2009a, 2009b) but did not specifically address biopesticides and failed to reduce the complexity of registration (Helepciuc and Todor 2022). The new approach, targeted for introduction late in 2022, will be based on the biology and ecology of each microbe, which is hoped to increase the number of biopesticides registered. There is a need for alternatives, given the EU's "Farm to Fork" strategy, which requires a 50% reduction in pesticide use by 2030 (Helepciuc and Todor 2022).

Clearly, a concerted effort to modify and align the regulations governing categories of biopesticides will be crucial to the development of novel biopesticides (Glare et al. 2012).

2.9 CONCLUSIONS

Many pest control products have been called biopesticides or fall under definitions used for biopesticides. The use of the term can be restricted, limited to live organisms, expanded to include extracts and metabolites directly from those organisms, or used for any naturally occurring compound that has a pesticidal ability. The main classes of biopesticides are

1. Live microbe–based products
2. Microbial-sourced biochemicals
3. Plant extracts
4. GM plants
5. Live arthropods
6. Semiochemicals
7. Endophytes
8. Other organically sourced materials

We are sure different authors could argue with this list as it is very broad. The term *biopesticide* will need to be defined for each purpose and may be too broad to be useful in tackling the major issues around regulation. Part of the drive for use of more naturally occurring products is the consumer's wish to have less risk from unwanted toxins in food. The same concerns that are driving the withdrawal of synthetic pesticides will also result in increased scrutiny of any new protection products, including biopesticides. It is naïve to think biopesticides as an entire group can be judged as low risk. Clearly, defining categories of biopesticides and making risk assessments appropriate for that category would greatly increase the availability of new agents.

The science around biopesticides is likely to advance rapidly in the next ten years in response to a growing crisis in pest, disease, and weed control as synthetic pesticides become unavailable. The exploitation of knowledge gained through genomic and associated technologies, breakthroughs in formulation approaches, and continued discovery of new agents or novel activities will expand the commercialisation of biopesticides to become a mainstream pest control tool.

REFERENCES

Ahmad, P., Ashraf, M., Younis, M., et al. 2012. Role of transgenic plants in agriculture and biopharming. *Biotechn Adv* 30: 524–540.

Backman, P.A. and Sikora, R.A. 2008. Endophytes: An emerging tool for biological control. *Biol Cont* 46: 1–3.

Bailey, K.L., Boyetchko, S.M. and Längle, T. 2010. Social and economic drivers shaping the future of biological control: A Canadian perspective on the factors affecting the development and use of microbial biopesticides. *Biol Cont* 52: 221–229.

Beer, A. 2012. Company mergers, acquisitions and crop protection deals in 2012. *Agrow Annu Rev* xi–xiii.

Burges, H.D. (ed.). 1998. *Formulation of microbial biopesticides: Beneficial microorganisms, nematodes and seed treatments.* Kluwer Academic, 396 pp.

Chandler, D., Bailey A.S., Tatchell G.M., et al. 2011. The development, regulation and use of biopesticides for integrated pest management. *Phil Trans R Soc B* 366: 1987–1998.

Chen, H. and Lin, Y. 2013. Promise and issues of genetically modified crops. *Curr Opinion Plant Biol* 16: 255–260.

Compant, S., Clément, C. and Sessitsch, A. 2010. Plant growth-promoting bacteria in the rhizo- and endosphere of plants: Their role, colonization, mechanisms involved and prospects for utilization. *Soil Biol Biochem* 42: 669–678.

Copping, L.G. (ed). 2009. *The manual of biocontrol agents: A World compendium.* British Crop Production Council, 851 pp.

Copping, L.G. and Menn, J.J. 2000. Biopesticides: A review of their action, applications and efficacy. *Pest Manag Sci* 56: 651–676.

Easton, H.S. 2007. Grasses and *Neotyphodium* endophytes: Co-adaptation and adaptive breeding. *Euphytica* 154: 295–306.

EC 2009a. Regulation (EC) No 1107/2009 of the European Parliament and of the Council of 21 October 2009 concerning the placing of plant protection products on the market and repealing Council Directives 79/117/EEC and 91/414/EEC. *Off J Europ Union* 309(52): 1–50.

EC 2009b. Directive 2009/128/EC of the European Parliament and the council establishing a framework for community action to achieve the sustainable use of pesticides. *Off J Europ Union* 309(52): 71–85.

El-Sayed, A.M., Suckling, D.M., Wearing, C.H. and Byers, J.A. 2006. Potential of mass trapping for long-term pest management and eradication of invasive species. *J Econ Entomol* 99: 1550–1564.

El-Wakeil, N.E. 2013. Botanical pesticides and their mode of action. *Gesunde Pflanzen* 65: 125–149.

Fletcher, S.J., Reeves, P.T., Hoang, B.T. and Mitter, N. 2020. A perspective on RNAi-based biopesticides. *Frontiers Plant Sci* 11: 51.

Fortune Business Insights. 2022. www.fortunebusinessinsights.com/industry-reports/biopesticides-market-100073 (accessed November 2022).

Gahukar, R.T. 2013. Cow urine: A potential biopesticide. *Indian J Entomol* 75: 212–216.

Gašić, S. and Tanović, B. 2013. Biopesticide formulations, possibility of application and future trends. *Pestic Phytomed (Belgrade)* 28: 97–102.

Glare, T.R., Caradus, J., Gelernter, W.D., et al. 2012. Have biopesticides come of age? *Trends Biotechn* 30: 250–258.

Glare, T.R., Gwynn, R.L. and Moran-Diez, M.E. 2016. Development of biopesticides and future opportunities. *Microb Biopest* 211–221.

Gupta, S. and Dikshit, A.K. 2010. Biopesticides: An ecofriendly approach for pest control. *J Biopesticides* 3: 186–188.

Hassan, E. and Gökçe, A. 2014. Production and consumption of biopesticides. In *Advances in plant biopesticides* (pp. 361–379). Springer.

Helepciuc, F.E., and Todor, A. 2022. EU microbial pest control: A revolution in waiting. *Pest Manag Sci* 78(4): 1314–1325.

Heuskin, S., Verheggen, F.J., Haubruge, E., Wathelet, J.-P. and Lognay, G. 2011. The use of semiochemical slow-release devices in integrated pest management strategies. *Biotechnol Agron Soc Environ* 15: 459–470.

Hynes, R.K. and Boyetchko, S.M. 2006. Research initiatives in the art and science of biopesticide formulations. *Soil Biol Biochem* 38: 845–849.

IPPC Secretariat. 2021. *Scientific review of the impact of climate change on plant pests—A global challenge to prevent and mitigate plant pest risks in agriculture, forestry and ecosystems.* FAO on behalf of the IPPC Secretariat. https://doi.org/10.4060/cb4769en

James, C. 2012. *Global status of commercialized biotech/GM crops: ISAAA Briefs No. 44.* ISAAA.

Joseph, B. and Sujatha, S. 2012. Insight of botanical based biopesticides against economically important pest. *Intern J Pharm Life Sci* 3: 2138–2148.

Kabaluk, J.T., Svircev A.M., Goettel M.S. and Woo S.G. (eds.) 2010. *The use and regulation of microbial pesticides in representative jurisdictions worldwide.* IOBC Global, 99 pp.

Kaddes, A., Fauconnier, M.L., Sassi, K., Nasraoui, B. and Jijakli, M.H. (2019). Endophytic fungal volatile compounds as solution for sustainable agriculture. *Molecules* 24: 1065.

Kala, S., Sogan, N., Agarwal, A., Naik, S.N., Patanjali, P.K. and Kumar, J. 2020. Biopesticides: Formulations and delivery techniques. In *Natural remedies for pest, disease and weed control* (pp. 209–220). Academic Press.

Keswani, C., Sarma, B.K. and Singh, H.B. 2016. Synthesis of policy support, quality control, and regulatory management of biopesticides in sustainable agriculture. In *Agriculturally important microorganisms* (pp. 3–12). Springer.

Khalil, M.S. 2013. Abamectin and Azadirachtin as eco-friendly promising biorational tools in integrated nematodes management programs. *J Plant Pathol Microb* 4: 4.

Khursheed, A., Rather, M.A., Jain, V., Rasool, S., Nazir, R., Malik, N.A. and Majid, S.A. 2022. Plant based natural products as potential ecofriendly and safer biopesticides: A comprehensive overview of their advantages over conventional pesticides, limitations and regulatory aspects. *Microb Pathog* 105854.

Kiewnick, S. 2007. Practicalities of developing and registering microbial biological control agents. *CAB Rev Persp Agric Vet Sci Nutrit Nat Res* 2: 013.

Köberl, M., Ramadan, E.M., Roßmann, B., et al. 2012. Using ecological knowledge and molecular tools to develop effective and safe biocontrol strategies. In *Pesticides in the modern world.* IntechOpen.

Köhl, J., Kolnaar, R. and Ravensberg, W.J. 2019. Mode of action of microbial biological control agents against plant diseases: Relevance beyond efficacy. *Front Plant Sci* 10: 845.

Koul, O. 2011. Microbial biopesticides: Opportunities and challenges. *CAB Rev Persp Agric Vet Sci Nutrit Nat Res* 6: 056.

Lengai, G.M. and Muthomi, J.W. 2018. Biopesticides and their role in sustainable agricultural production. *J Biosci Med* 6: 7.

Lodewyckx, C., Vangronsveld, J., Porteous, F., et al. 2002. Endophytic bacteria and their potential applications. *Crit Rev Plant Sci* 21: 583–606.

López, M.D., Cantó-Tejero, M. and Pascual-Villalobos, M.J. 2021. New insights into biopesticides: Solid and liquid formulations of essential oils and derivatives. *Front Agron* 3.

Maksimov, I.V., Maksimova, T.I., Sarvarova, E.R., Blagova, D.K. and Popov, V.O. 2018. Endophytic bacteria as effective agents of new-generation biopesticides. *Appl Biochem Microbiol* 54: 128–140.

Marketsandmarkets. 2022. www.marketsandmarkets.com/Market-Reports/biopesticides-267. html?gclid=CjwKCAiApvebBhAvEiwAe7mHSN3_EhSLSk-yHO7GrpZM8NDOwVcuWF_hC0jninwAbJtar2SUJ30mvRoCS1AQAvD_BwE (accessed November 2022).

Nawaz, M., Mabubu, J.I. and Hua, H. 2016. Current status and advancement of biopesticides: Microbial and botanical pesticides. *J Entomol Zool Stud* 4: 241–246.

Nelson, M.E. and Alves, A.P. 2014. Plant incorporated protectants and insect resistance. In *Insect resistance management* (pp. 99–147). Academic Press.

O'Callaghan, M. and Glare, T.R. 2001. *Environmental and health impacts of Spinosad*. Report for the New Zealand Ministry of Agriculture and Forestry, 61 pp.

Oerke, E.C., Dehne, H.W., Schoenbeck, F. and Weber, A. 1994. *Crop production and crop protection: Estimated losses in major food and cash crops*. Elsevier Science Publishers B.V.

Olson, S. 2015. An analysis of the biopesticide market now and where it is going. *Outlooks Pest Manag* 26: 203–206.

Ownley, B.H., Griffin, M.R., Klingeman, W.E., et al. 2008. *Beauveria bassiana*: Endophytic colonization and plant disease control. *J Invertebr Pathol* 98: 267–270.

Parker, K.M. and Sander, M. 2017. Environmental fate of insecticidal plant-incorporated protectants from genetically modified crops: Knowledge gaps and research opportunities. *Environ Sci Technol* 51(21): 12049–12057.

Partida-Martínez, L.P. and Heil, M. 2011. The microbe-free plant: Fact or artefact? *Front Plant Sci* 2: 100 (16 pp).

Pavela, R. 2014. Limitation of plant biopesticides. In *Advances in plant biopesticides* (pp. 347–359). Springer.

Persistence Market Research. 2022. www.persistencemarketresearch.com/market-research/biopesticides-market.asp (accessed November 2022).

Porras-Alfaro, A. and Bayman, P. 2011. Hidden fungi, emergent properties: Endophytes and microbiomes. *Annu Rev Phytopathol* 49: 291–315.

Rajamani, M. and Negi, A. 2021. Biopesticides for pest management. In *Sustainable bioeconomy* (pp. 239–266). Springer.

Rhainds, M., Kettela, E.G. and Silk, P.J. 2012. Thirty-five years of pheromone-based mating disruption studies with *Choristoneura fumiferana* (Clemens) (Lepidoptera: Tortricidae). *Can Entomol* 144: 379–395.

Roach, D.R., Castle, A.J., Svircev, A.M. and Tumini, F.A. 2008. Phage-based biopesticides: Characterization of phage resistance and host range for sustainability. *Acta Hort* 793: 397–401.

Rong, S., Xu, H., Li, L., Chen, R., Gao, X. and Xu, Z. 2020. Antifungal activity of endophytic *Bacillus safensis* B21 and its potential application as a biopesticide to control rice blast. *Pesticide Biochem Physiol* 162: 69–77.

Rosell, G., Quero, C., Coll, J. and Guerrero, A. 2008. Biorational insecticides in pest management. *J Pestic Sci* 33: 103–121.

Ruiu, L. 2018. Microbial biopesticides in agroecosystems. *Agronomy* 8: 235.

Santos, V.S.V. and Pereira, B.B. 2020. Properties, toxicity and current applications of the biolarvicide spinosad. *J Toxicol Environ Health Part B* 23: 13–26.

Schiemann, J., Dietz-Pfeilstetter, A., Hartung, F., Kohl, C., Romeis, J. and Sprink, T. 2019. Risk assessment and regulation of plants modified by modern biotechniques: Current status and future challenges. *Annu Rev Plant Biol* 70: 10–1146.

Seiber, J.N., Coats, J., Duke, S.O. and Gross, A.D. 2014. Biopesticides: State of the art and future opportunities. *J Agric Food Chem* 62: 11613–11619.

Sikora, R.A., Pocasangre, L., Zum Felde, A., et al. 2008. Mutualistic endophytic fungi and *in-planta* suppressiveness to plant parasitic nematodes. *Biol Cont* 46: 15–23.

Singh, D. (ed.). (2014). *Advances in plant biopesticides.* Springer.

Sparks, T.C., Thompson, G.D., Larson, L.L., et al. 1995. Biological characteristics of the spinosyns: A new naturally derived insect control agents. *Proc Beltwide Cotton Conf* 2: 903–907.

Sundh, I. and Goettel, M.S. 2013. Regulating biocontrol agents: A historical perspective and a critical examination comparing microbial and macrobial agents. *BioControl* 58: 575–593.

Taning, C.N.T., Mezzetti, B., Kleter, G., Smagghe, G. and Baraldi, E. 2021. Does RNAi-based technology fit within EU sustainability goals? *Trends Biotechnol* 39: 644–647.

Thakore, Y. 2006. The biopesticide market for global agricultural use. *Industrial Biotechnol* 2: 194–208.

Thakur, N., Kaur, S., Tomar, P., Thakur, S. and Yadav, A.N. 2020. Microbial biopesticides: Current status and advancement for sustainable agriculture and environment. In *New and future developments in microbial biotechnology and bioengineering* (pp. 243–282). Elsevier.

Tijjani, A., Bashir, K.A., Mohammed, I., Muhammad, A., Gambo, A. and Musa, H. 2016. Biopesticides for pests control: A review. *J Biopest Agric* 3: 6–13.

van Lenteren, J.C. 2011. The state of commercial augmentative biological control: Plenty of natural enemies, but a frustrating lack of uptake. *BioControl* 57: 1–20.

Villaverde, J.J., Sevilla-Moran, B., Sandin-Espana, P., Lopez-Goti, C. and Alonso-Prados, J.L. 2014. Biopesticides in the framework of the European Pesticide Regulation (EC) No. 1107/2009. *Pest Manag Sci* 70: 2–5.

Xia, Y., Liu, J., Chen, C., et al. 2022. The multifunctions and future prospects of endophytes and their metabolites in plant disease management. *Microorganisms* 10: 1072.

Zimmermann, G., Papierok, B. and Glare, T. 1995. Elias Metschnikoff, Elie Metchnikoff or Ilya Ilich Mechnikov (1845–1916): A pioneer in insect pathology, the first describer of the entomopathogenic fungus *Metarhizium anisopliae* and how to translate a Russian name. *Biocon Sci Technol* 5: 527–530.

3 Metabolism of Biopesticides

Leo M.L. Nollet

CONTENTS

DOI: 10.1201/9781003265139-4

3.1 INTRODUCTION

Biopesticides are pesticides that are derived from natural substances such as minerals, plants, animals, and bacteria. The main types of biopesticides are microbial, biochemical, and plant-incorporated protectants [1]. Other types of biopesticides are detailed in Section 3 of this book. Their main advantage over conventional pesticides is that they have less persistence in the environment, due to the absence of unnatural aromaticity and halogen components in their structures [2]. As a result, people's views about them have since changed worldwide, and they are now considered to be a breakthrough in the agrochemical industry. As a result of this, the global biopesticides market was valued at US$5.2 billion in 2021 and is projected to generate revenue of US$12.2 billion by 2028 at a compound annual growth rate (CAGR) of 15.3% during the forecast period, 2022–2028 [3]. Scientists nowadays are focusing their research more on understanding the mode of action of these biopesticides in different plants and animals, with the objective of improving their efficacy as pest control agents and also understanding their toxicity. **Table 3.1** summarizes the general properties of a number of biopesticides, their specific mode of action, and their merits and drawbacks.

This chapter discusses the metabolic pathways of the different biopesticides, paying special attention to the types of reactions involved and the outcomes of the metabolism in both plants and animals.

3.2 XENOBIOTIC METABOLISM

Metabolism refers to the chemical reactions that take place in living organisms. The reactions involve conversion of one molecule into another molecule via a defined pathway. Metabolism can be catabolic, in which a compound is broken down, usually by enzymes, to produce energy and other essential molecules required for biological processes. It can also be anabolic, in which smaller molecules react to create larger molecules. It is important to note that these two processes, anabolism and catabolism, are interdependent as the products of catabolism are used as intermediates in anabolism. Thus, the major importance of metabolism is that it provides the energy and raw materials needed for different biological processes [4].

Xenobiotic metabolic processes involve two main pathways referred to as phase I and phase II reactions. A phase I reaction involves the introduction of a polar group onto lipophilic toxicant molecules. Therefore, the product of a phase I reaction is usually more water soluble than the parent species. Examples of phase I reactions include hydroxylation, N-oxidation, and deamination. **Table 3.2** summarizes some of the phase I and phase II reactions in the metabolism of xenobiotics, the enzymes involved, and the typical reactions. Phase II reactions involve conjugation reactions

TABLE 3.1
Mode of Action of Biopesticides

Example	Mode of Action	Advantages	Disadvantages
	Microbial Protectants		
1. Bacteria, e.g., *Bacillus thuringiensis* and *Pseudomonas putida*	*Antibiosis*: insecticidal proteins kill insect larvae *Competition*: competes with the pests for nutrients from habitat or exuded by plant	Little effect on other organism Considered more environment friendly than synthetic pesticide	May require ingestion to be effective Low field persistence Slow killing
2. Fungi, e.g., *Trichoderma atroviridae*	*Mycoparasitism/predation*: lives and feeds internally or externally on the host	They do not need to be eaten to be effective	Require a narrow range of conditions including moist soil and cool temperatures to proliferate
3. Virus, e.g., granulosis virus (GV)	*Antibiosis*: infects digestive cells in larvae gut	Not known to infect plants or vertebrate Have narrow host insect range No residues associated with applications	Must be ingested by the host to produce an infection High cost of production
	Biochemical Protectants		
1. Plant extracts and oils, e.g., neem limonoids and lemongrass oil	Suffocation Repellents (deterrents) Disruptor of insect development	Not very specific as compared to other biopesticides; non-toxic to birds and animals; noncarcinogenic	May require ingestion to be effective

(Continued)

TABLE 3.1 (Continued)

Example	Mode of Action	Advantages	Disadvantages
	Microbial Protectants		
1. Insect growth regulators, e.g., precocenes	Alter the growth and development of insects	Effective when applied in very minute quantities	Not species specific, hence have huge impact on non-target species Must be absorbed by plant tissue in order to be effective
2. Plant growth regulators, e.g., gibberellins and auxins	Enhance crop yield, crop shelf life, and the appearance of the crop	Low concentration required on the order of ppm or ppb	
3. Insect sex pheromones; usually a combination of molecules	Mating disruptants, pest repellents attract pests to traps	High species or strain specificity Relatively low toxicity	Some are known to be human carcinogens and endocrine disrupters They often must be used in combination with other pest management strategies to achieve the desired efficacy
	Plant-Incorporated Protectants		
1. Bt corn	Same as parent gene carrying the microbial	Reduced risk of mycotoxin contamination Effective pest control Lower applications of insecticides	Pest resistance to Cry protein Transfer of gene to other plants

TABLE 3.2
Phase I and Phase II Reactions of Xenobiotic Metabolism

Enzymes	Reaction Responsible	Example of the Reaction
Oxidation	Cytochrome P450 monooxygenases	$RH + NADPH + H^+ + O_2 \rightarrow ROH + NADP + H_2O$
Reduction	Aldehydes and ketone reductases	$RCHO + NADPH + H^+ \rightarrow RCH_2OH + NADP$
Conjugation	UDP glucoronyl transeferases	$ROH + UDP$ glucuronic acid $\rightarrow ROH$ glucuronide + UTP

such as the addition of glucuronic acid, amino acids, sulfate, or acetate. Phase I and phase II reactions lessen the toxicity of toxic substances, at the same time enabling excretion of the water-soluble metabolites through the normal excretion channels of the organism. However, some toxicants inhibit the enzymes that carry out phase I and phase II reactions, thereby increasing the toxicity of these substances [4]. In summary, metabolism may result in increased or decreased toxicity of the product, depending on the metabolic pathway followed.

To have a better comprehension of metabolism of biopesticides in plants or animals, the studies have to be planned carefully. In most cases, the process begins with radiolabeling of the active ingredient using the ^{14}C isotope. However, other isotopes such as ^3H may also be used. The parent radiolabeled compounds and the metabolites are extracted from the matrix of interest and analyzed using high-pressure liquid chromatography to separate the extracts [5], hyphenated (coupled) to radio detectors, to detect the fraction containing the radiolabeled parent or its metabolites, and mass spectrometry to characterize the metabolites in a particular fraction by mass [6]. The trend experienced nowadays is a shift from the traditional methods of extraction, which consume more resources, to the latest methods, which consume fewer resources. Furthermore a significant development is the shift from use of selective detectors such as a fluorescence detector to mass spectrometers, which provide better details about the metabolite structures.

3.3 MICROBIAL BIOPESTICIDES

Microbial biopesticides are pesticides derived from microorganisms such as fungi, bacteria, and viruses [7].

3.3.1 ENTOMOPATHOGENIC BACTERIA

3.3.1.1 *Bacillus thuringiensis*

Bacillus thuringiensis (*Bt*) is a gram positive bacterium that can form spores. During the formation of spores, *Bacillus thuringiensis* forms crystal proteins that have insecticidal properties. The crystal proteins, Crystal (Cry) and Cytolytic (Cyt), are also known as δ-endotoxins. Crystal proteins produced by different strains of *Bacillus thuringiensis* have very significant toxicity to the target organism. Moreover, they have high specificity towards the target organisms, which include Lepidoptera,

Coleoptera, Diptera, and some nematodes [8]. This makes them effective pest control agents. To date, there are over 200 registered products based on *Bacillus thuringiensis* in the US. The major setback of this type of biopesticides is resistance to the crystal proteins after continuous usage.

The mode of action of the Cry toxins is thought to occur by producing spores in the epithelium cells of the midgut in the insect. This causes the cells to break down, and the structure of the midgut epithelium is disturbed [9]. Moreover, their mode of action makes them very effective against insect species because (1) the toxins bind irreversibly to the membrane of the larvae; (2) after insertion, the toxins mix with the membrane; and (3) the toxins mix with their three-domain structure with high specificity to receptors [10–12].

At least four different non–structurally related families of proteins form the Cry group of toxins. The expression of certain Cry toxins in transgenic crops has contributed to an efficient control of insect pests, resulting in a significant reduction in chemical insecticide use. The mode of action of the three-domain Cry toxin family involves sequential interaction of these toxins with several insect midgut proteins, facilitating the formation of a pre-pore oligomer structure and subsequent membrane insertion that leads to the killing of midgut insect cells by osmotic shock. Interestingly, similar Cry-binding proteins have been identified in the three insect orders as cadherin, aminopeptidase-N, and alkaline phosphatase, suggesting a conserved mode of action [13].

However, it was also proposed that, in contradiction to this mechanism, *Bt* toxins function by activating certain intracellular signaling pathways, which leads to the necrotic death of their target cells without the need for pore formation. Because work in this field has largely focused, for several years, on the elaboration and promotion of these two models, Vachon et al. [14] examine in detail the experimental evidence on which they are based. It is concluded that the presently available information still supports the notion that *Bt* Cry toxins act by forming pores, but most events leading to their formation, following binding of the activated toxins to their receptors, remain relatively poorly understood.

3.3.1.1.1 Thuringiensin Metabolism

Thuringiensin is an insecticide produced from *Bacillus thuringiensis*. It inhibits the production of RNA polymerase, thus inhibiting the production of ribosomal RNA. For the insecticide to be effective against the target insect, it has to be absorbed by the plant first, then translocated to the parts that are eaten by the insect [15, 16].

In a study by Mersie and Singh, thuringiensin in corn was not observed to degrade after treatment [17]. In a separate study by the same research group, the metabolism of radiolabeled thuringiensin was investigated in potato under controlled conditions with no observable metabolite being detected, as such no metabolism was observed [15]. To confirm the results, a similar study was performed on snap beans, and it was observed that the rate of absorption and, hence, metabolism was minimal. This further explains the high persistence of thuringiensin in plants such as cotton [18]. However, in animals such as the tobacco budworm, thuringiensin metabolism occurs by a dephosphorylation process that is facilitated by the phosphatase enzyme under prevailing acidic conditions that are not present in plant systems [19].

3.3.2 ENTOMOPATHOGENIC FUNGI

The use of fungal-based biopesticides is as old as the history of agriculture itself. Fungal-based biopesticides are mainly used to control diseases caused by microorganisms such as fungi or bacteria. In this category, the main biopesticides are *Trichoderma* spp. and *Beauveria bassiana*. They are commonly used in agricultural setups to control pests in horticultural, floricultural, and even forest plantations. What is common between these biopesticides is their mode of action, which is based on metabolite production, together with combined attachment and penetration of the host, thereby killing them [20, 21]. This phenomenon is called mycoparasitism.

The pathogenicity of entomopathogenic fungi depends on the ability of their enzymatic equipment, consisting of lipases, proteases, and chitinases, which degrade the insect's integument. Additionally, the researchers studied the content of β-galactosidase, -glutaminase, and catalase within entomopathogenic fungi [22].

The mode of action of entomopathogenic fungi against insects involves attaching a spore to the insect cuticle, followed by germination, penetration of the cuticle, and dissemination inside the insect. Strains of entomopathogenic fungi are concentrated in the following order: Hypocreales (various genera), Onygenales (*Ascosphaera* genus), Entomophthorales, and Neozygitales (Entomophthoromycota) [23].

3.3.2.1 *Trichoderma*

Trichoderma spp–based biopesticides have numerous applications in agriculture. The main strains used as pesticides are *Trichoderma virens*, *Trichoderma viride*, and *Trichoderma harzianum*. These are mainly used against root rot in dry land crops such as groundnuts. The strains have various modes of action such as (1) mycoparasitism, (2) production of antibiotics, (3) breakdown of cell walls using enzymes such as chitinase, and even (4) competition for nutrients and space [24–26]. For instance *Trichoderma harzianum's* activity against the fungi *Gaeumannomyces graminis* has been proposed to take place by the production of antibodies [27].

However, in a separate study, the action of two strains of *Trichoderma harzianum* (T22 and T39) against three phytopathogenic fungi (*Rhizoctonia solani*, *Pythium ultimum*, and *Gaeumannomyces graminis var tritici*) was observed to take place by the production of secondary metabolites, such as T22azaphilone and T39butenolide. The secondary metabolites formed were active against plant pathogens to varying degrees. Therefore, further research is still needed on their mode of action [28].

3.3.2.1.1 Metabolism of *Trichoderma* Strains

While considerable work has been done on the metabolism of *Trichoderma* strains, significant progress has been achieved in the secondary metabolites of the strains. The secondary metabolites have been classified according to the pathway. Good examples of these are secondary metabolites not related to acetate and those from the tricarboxylic acid cycle. The rest are grouped according to their chemicals classes, such as fatty acids, polyketides, and oxygen heterocyclic compounds, among others. For instance, harziandone, a diterpene, was isolated from *Trichoderma harzianum*. Generally, most *Trichoderma* strains are derived using the tricarboxylic acid cycle. Furthermore, depending on the metabolic pathway taken, different metabolites will have different biological activity [29]. However, because *Trichoderma* strains are easily metabolized even though by different mechanisms, this may suggest low toxicity.

3.4 MICROBIAL METABOLITES

3.4.1 SPINOSYNS

The spinosyns are a class of new insecticides that are produced from the aerobic fermentation of *Saccharopolyspora spinosa*. The major insecticide is spinosad, which is made up of two major components, spinosyn A and spinosyn D, in the ratio 85:15 [30]. They are very effective against many insects that damage crops, such as Lepidoptera, Coleoptera, and Orthoptera, and some parasites that affect livestock and humans, such as ticks. Their mode of action involves the disruption of the nicotinic acetylcholine receptors [31]. Moreover, the ultimate effect of spinosyns as a result of continuous use is paralysis in the insects. This paralysis is attributed to fatigue in the nervous system [32–34].

The anti-insecticidal activity against several commercially relevant pests is based on the allosteric, agonistic binding at the nicotinic acetylcholine receptor and an antagonistic effect on the γ-aminobutyric acid receptor [35]. Spinosyns result from the polyketide pathway. The biosynthesis of the tetracyclic lactone comprises an enzymatic Diels-Alder and a Rauhut-Currier-like reaction, followed by glycosylation with rhamnose and forosamine. The spinosyns, constituting the insecticidal active ingredients in Spinosad and Spinetoram from Dow AgroSciences, are manufactured by fermentation. However, in recent years, facilitated by computer-aided molecular modeling and bioactivity-directed synthesis, novel highly potent, structurally simplified spinosyn analogues were discovered, which might be accessible in the future by total synthesis on an industrial scale.

3.4.1.1 Metabolism in Plants

The metabolism of spinosyns was studied in different plant such as apples, tomatoes, cotton, and grapes. Several factors such as photolysis and translocation were investigated. It was observed that in apples, Spinosyn A and D were mainly degraded by photolysis, and translocation did not play a major role in the metabolism of spinosad. Studies on cotton seed, cabbages, and grapes showed that spinosyn A and D degraded in the sun to give highly polar metabolites.

The metabolic pathway in plants includes fast breakdown of the compound in light, followed by dealkylation on the sugars, change of the forosamine structure, and lastly the entire loss of the 2,3,4-tri-O- methyl rhamnose. The spinosyns undergo macrolide cleavage to form small fragments that become part of the plant constituents [36].

3.4.1.2 Metabolism in Animals

Metabolism studies were carried out in different animals such as rats, goats, and hens. The highest concentrations of spinetoram were found in the gastrointestinal tract, liver, and fat. For spinosad, the highest concentrations were found in the liver,

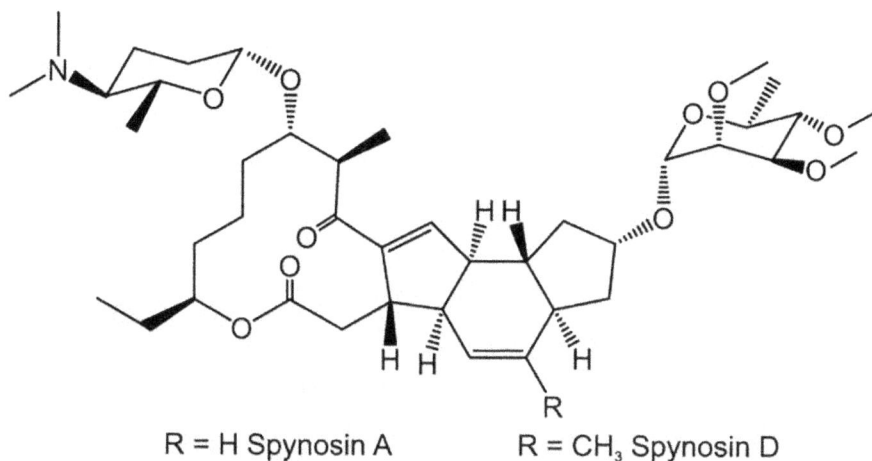

R = H Spynosin A   R = CH₃ Spynosin D

FIGURE 3.1   Spinosyn A and spinosyn D.

kidneys, and thyroid. The metabolic pathways included both phase I and phase II reactions. In rats, the pathway included glutathione conjugation and cysteine conjugation of the parent compound [37]. In livestock animals, research suggested that for both spinosad and spinetoram, the pathway was overall dealkylation.

Several studies were conducted on lactating goats, one dosed with spinosyn A and the other with spinosyn D. The milk and excreta were collected, and the goats sacrificed within a day. Tissue samples were analyzed. Both spinosyn A and spinosyn D were predominant in the tissue and milk, and some metabolites were identified but were not characterized. The metabolic pathway was by hydroxylation of the macrolide ring. Figure 3.1 illustrates the molecular structures of spinosyn A and spinosyn D.

In hens, residues were determined in tissues, eggs, and feces. The birds were sacrificed within a day, and analysis of the samples was carried out. The greatest amounts of residues were found in fat, of which the parent compound was predominant. Some of the compounds were present in muscle and eggs. Metabolism takes place in the liver by N-de-methylation, O- demethylation, and loss of furosamine sugar moiety.

3.4.2 STROBILURIN FUNGICIDES

Strobilurins are natural biopesticides identified and isolated from the fungus *Strobilurus tenacellus*. The pioneer in this field was Anke, who isolated strobilurin A from the liquid cultures of the mushroom *Strobilurus tenacellus*. As a result of his study, many other compounds related to strobilurin were discovered [38, 39]. These are known as strobilurin B, C, D, and so on. Furthermore, several other synthetic fungicides have been developed such as azoxystrobin. Generally, the strobilurin fungicides have protective, curative, and eradicant action. However, it is during the germination of spores that they are highly effective.

Their basic mode of action is binding to the quinol oxidation site in the mitochondria, thus stopping the electron transfer process between cytochrome B and cytochrome C, which further stops the oxidation of NADH and the synthesis of ATP [40]. By so doing, the fungus runs short of energy and dies. Their main advantage is that they are highly specific and have swift action. However, because they have only one site of action, their major disadvantage is resistance.

3.4.2.1 Metabolism

There is limited literature on the metabolism of the strobilurin fungicides. Many examples of metabolism dwell on azoxystrobin, kresoxim methyl, and trifloxystrobin. This calls for more research into the metabolic pathways of the other strobilurin fungicides, especially the natural ones. Their most important metabolic pathway is methyl ester hydrolysis. However, other reaction pathways such as hydroxylation of the aromatic ring may occur. These are illustrated in Figure 3.2

3.4.2.2 Metabolism in Plants

In plants, azoxystrobin was metabolized both biotically and abiotically to give 17 metabolites, the major one being cyanophenoxypyrimidinol, formed by ether bond cleavage. This product was readily converted as an N-glucoside conjugate [36, 38]. Figure 3.3 shows a detailed outline of the pathway taken during the metabolism of azoxystrobin in plants. In a more recent study, metabolism of the three synthetic strobilurins was investigated in wheat cell suspension cultures. The metabolic pathway of trifloxystrobin took place by demethylation followed by hydroxylation, while that for kresoxim-methyl largely took place by demethylation. The metabolic rates depended on the amounts of the compounds and cells added to the media. Moreover, it was observed that the metabolic rates of trifloxystrobin and kresoxim methyl were faster than that of azoxystrobin [39].

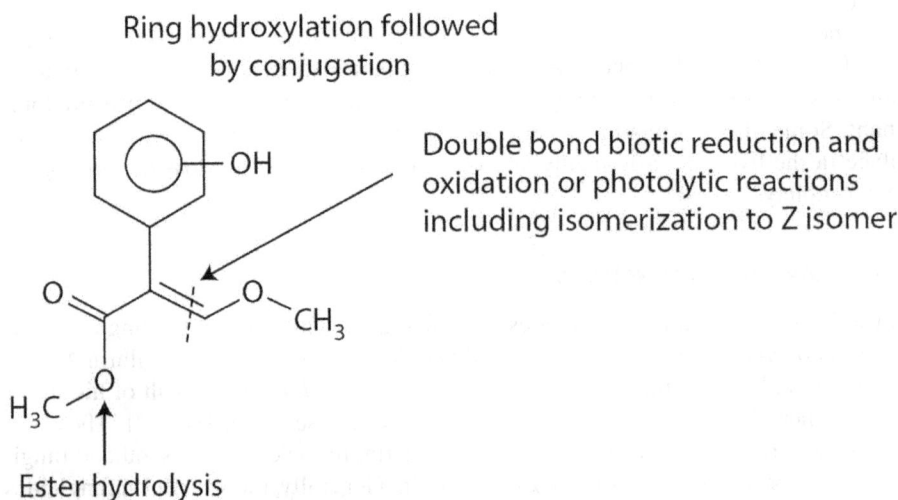

FIGURE 3.2 Basic degradation pathways of strobilurins.

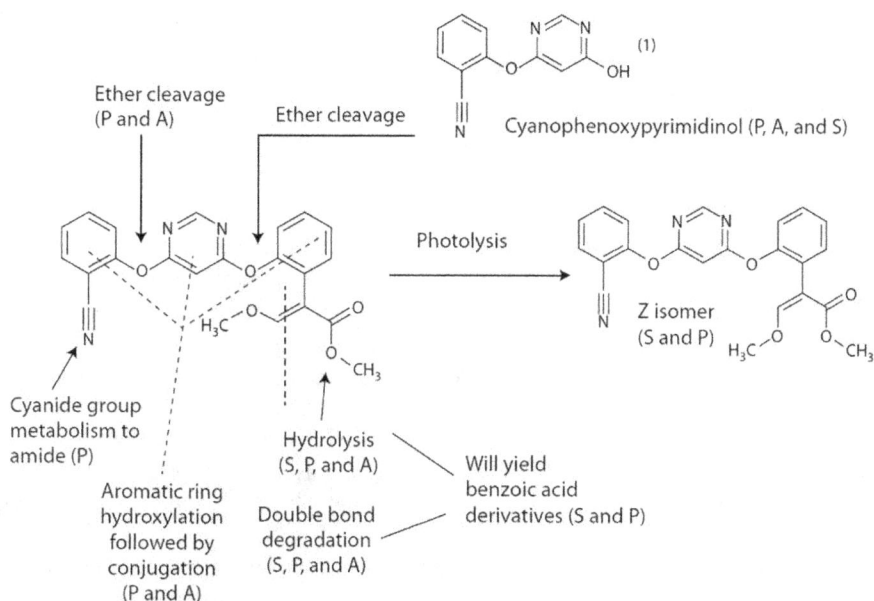

FIGURE 3.3   Metabolism of azoxystrobin.

3.4.2.2.1 Metabolism in Animals and Soil

In animals, the metabolism of azoxystrobin was rapid, and so was the excretion. The metabolic pathway was almost similar to that in plants. The major difference was in the conjugation of hydroxylated aromatic rings. In plants, it was done with glucose while in animals, it was done with mercapturic acid. The major metabolite in mammals was azoxystrobin carboxylic acid, which was later converted to glucuronide for excretion. Furthermore, in water, the metabolism of ^{14}C-azoxystrobin also gave the same metabolite, azoxystrobin acid [41].

In soil, the major metabolite of azoxystrobin was azoxystrobin carboxylic acid. The metabolic pathways of kresoxim methyl and trifloxystrobin were similar to that of azoxystrobin in soil, plants, and mammals.

By 2002, there were six strobilurin active ingredients commercially available for agricultural use [42]. The review of Bartlett et al. [42] describes in detail the properties of these active ingredients—their synthesis, biochemical mode of action, biokinetics, fungicidal activity, yield and quality benefits, resistance risk, and human and environmental safety. It also describes the clear technical differences that exist between these active ingredients, particularly in the areas of fungicidal activity and biokinetics [42].

Strobilurins are natural products isolated and identified from specific fungi. Natural strobilurins were named in the order of their discovery as strobilurin-A followed by strobilurin-B, C, D, etc. Their discovery opened the door for new chemistry of synthetic fungicides. Applying quantitative structural activity relationship (QSAR) on the structures of the natural strobilurins, many pesticide companies were able to discover many synthetic analogues that are more efficacious and more stable fungicides. At present, there are about eight synthetic strobilurins in the worldwide fungicides market [43]. Some of these products are registered for worldwide use as agrochemicals, and some are in the process of registration.

3.4.3 MACROCYCLIC LACTONES

The macrocyclic lactones are a group of compounds that are derived from soil bacteria *Streptomyces*. The two main classes are avermectins and milbemycins. The main examples of avermectins include abamectin, ivermectin, doramectin, and selamectin. As for milbemycins, the main ones are moxidectin and milbemycin oxime.

In invertebrates, avermectins increase the release of gamma aminobutyric acid (GABA), which increases the permeability of the cell membrane to chloride ions. This results in the disturbance of the nervous system of the invertebrate, which becomes paralyzed and can no longer feed [44]. It is also important to note that although the mechanism of action of milbemycins is poorly understood, it is suggested to be similar to that of avermectins.

The major metabolic pathways for macrocyclic lactones have been observed to take place by ether hydroxylation, epoxidation, N-demethylation, O-demethylation, and deglycosylation [45].

3.4.3.1 Avermectins

Avermectins are produced by the fermentation of *Streoptomyces avermitilis*. They can act as insecticides, acaricides, or nematicides. They are chosen for these purposes because of their specific physicochemical properties, such as low water solubility and low leaching potential. The avermectin complex comprises four major related components—A1a, B1a, A2a, and B2a—and four minor components: A1b, A2b, B1b, and B2b. Other known avermectins include milbermectin and emamectin.

3.4.3.1.1 Abamectin

The metabolism of abamectin varies with the animal under study. In sheep, lower metabolism occurred in the tissues; as a result, the parent compound was found in most of them. In lactating goats, similar results were obtained, since up to 99% of the residues in the tissues and excreta were unmetabolized abamectin. However minute quantities of the metabolites 2,4 hydroxymethyl and 3″-o-desmethyl B_{1A} were found in the tissues [46]. The fate of abamectin was determined in rats with a mixture of [^{3}H] and [^{14}C] abamectin. It was shown that regardless of the radioactive isotope used in the metabolism study, the result was the same. Furthermore, the metabolic pathway in rats and the metabolites formed were the same as those in other animals [47, 48].

The metabolism of abamectin in plants was determined in citrus fruits such as oranges, lemons, and grapefruit. The rate of metabolism was very high with a significant decrease to approximately 3–11% of the original applied. Moreover, the high metabolic rates of abamectin in plants can be considered the reason for its low toxicity to non-target organisms [44]. Moreover, oranges show the highest persistence. The study concluded that because of the low persistence of abamectin in citrus fruits, risks associated with exposure to residues in humans were minimal [49]. The same research group performed a similar study on citrus fruits to determine the differences in the rate of metabolism between field fruits and picked ones. It was observed that the metabolism of abamectin in both picked and field fruits was the same. However, in picked fruits, the residue levels might be higher. In conclusion, the findings of the study suggest that if a similar study is carried out on citrus fruits with another pesticide, the results will be the same, provided the degradation mechanisms of the pesticides are the same.

3.4.3.1.2 Emamectin Benzoate

Emamectin is also an avermectin consisting of greater than 90% of the B1a component.

3.4.3.1.2.1 Metabolism in Plants
Metabolism in plants takes place by N-demethylation, N-formylation, and conjugation. The rate of metabolism was observed to be very fast in plants such as lettuce and cabbage.

3.4.3.1.2.2 Metabolism in Animals
In animals, the rate of metabolism is very low because of the rapid excretion rates [45]. This was confirmed by a study by Kim-Kang et al. to determine the metabolism of emamectin B_{1A} in animals; Atlantic salmon was used. The findings of the study were similar to those obtained earlier when the study was carried out in rats. After oral administration of radioactive-labeled emamectin B_{1A}, it was observed that the major metabolite, desmethyl emamectin B_{1A}, and the compound itself were abundant in the tissues and the feces [50]. The metabolism takes place primarily by N-demethylation [45].

In studies carried out in hens, the results also tallied with those obtained in fish. The laying hens excreted approximately 70% of the administered dose. The major metabolites were 2,4 dihydroxymethyl derivative and the N demethylated product, amino avermectin B_{1A}. Furthermore, conjugation of the two major metabolites produced eight fatty acid products, which were found in 8–75% of the eggs and tissues [51].

See further references [52, 53].

3.4.3.1.3 Ivermectin

Ivermectin is also a member of the avermectins. Like abamectin, it is also used as an antiparasitic agent and an acaricide. Several studies on its metabolism were carried out in cattle, sheep, rats, and swine [54–56]. The findings of all the studies in the species confirmed the presence of undissociated ivermectin. However, the notable differences were in the metabolites. The major metabolite in cattle, sheep and rats was 24-(hydroxymethyl-H_2B_{1A}), while those in swine were 3″-O-desmethyl-H_2B_{1B} and 3″-O-desmethyl-H_2B_{1B}.

3.4.3.2 Milbemycins

The milbemycins are also macrocyclic lactones that have the same mode of action as avermectins. Their structure consists of a cyclohexane-tetrahydrofuran ring, a cyclohexene ring, and a bicyclic 6,6-membered spiroskeleton joined to a 16-membered macrocyclic ring. Furthermore, it is this feature that they share with avermectins: hence, the name macrocyclic lactones. They are mainly used as antiparasitic agents to kill fleas, worms, and ticks.

3.4.3.2.1 Moxidectin

Moxidectin is a parasiticide synthesized from nemadectin. The difference between moxidectin and the avermectins is in the microorganism from which they are produced [57, 58].

Moxidectin is thought to combine with the GABA receptors and, as a result, cause the binding of this neurotransmitter substance. This has the effect of paralysis and even death on the parasites [59]. Several studies have been carried out on the metabolism of moxidectin in different animals. The major objective of these studies was to determine the potential toxicity of the residues of moxidectin and its metabolites. The studies on the metabolic fate of moxidectin have primarily been focused on cattle, but comparative studies have also been carried out on rats, sheep, and horses.

In the studies carried out on cattle, moxidectin was not readily metabolized in fat (90%), liver (36%), kidney (77%), or muscle (50%); hence, it was the main component there. The metabolites identified were the products of hydroxylation: namely, monohydroxy, dihydroxy, and O-desmethyl-dihydroxymethyl products [57]. **Figure 3.4** properly illustrates all the metabolites and the metabolic pathways involved in their formation. Similarly, in horses, moxidectin was the chief component in the tissue, with the metabolic pathway not changing, as well as the metabolites [59].

Moreover, in similar studies carried out in sheep, similar results were obtained, as the major metabolic pathway was hydroxylation, with some demethylation also

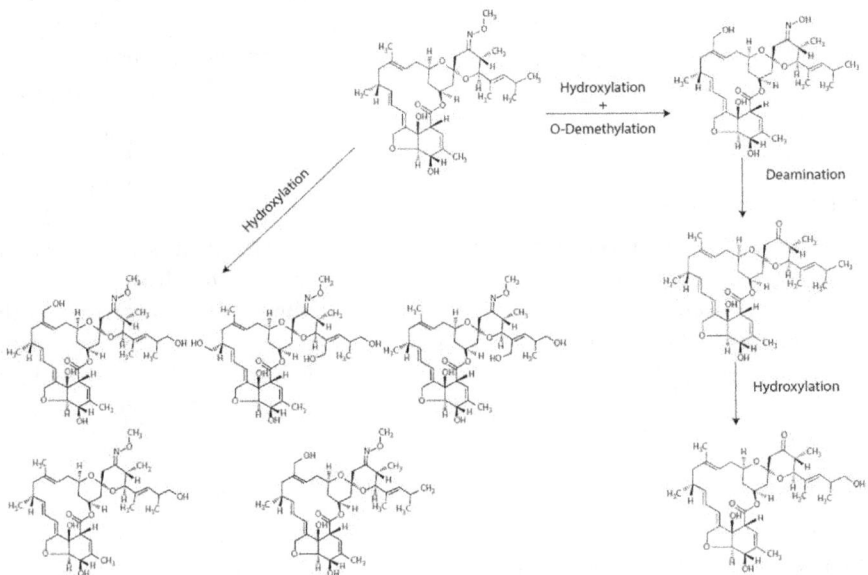

FIGURE 3.4   Metabolism of moxidectin in cattle.

occurring. As in the cattle study, the major metabolite was the C 29/30 monohydroxy methyl derivative of moxidectin. Furthermore, unmetabolized moxidectin was the major component in the tissue [60].

3.5 PLANT-DERIVED PESTICIDES

For this topic, the reader is also directed to Chapter 15 of this book.

Plant-derived pesticides are better known as botanical biopesticides. The major botanical biopesticides are pyrethrum, rotenone, neem, and other essential oils. Minor botanical pesticides include nicotine and sabadilla [1]. Botanicals are often used as insecticides (pyrethrum), repellants (citronella), fungicides (laminarin), and herbicides (pine oil) [61]. Despite their discovery long ago, most botanical insecticides have failed to complete with the synthetic insecticides due to their relatively slow action, lack of persistence, and inconsistency in their availability [62]. However, they may prove to be effective if used alongside other farming techniques such as crop rotation.

3.5.1 Pyrethrum

Pyrethrum is extracted from the dried flowers of the plant *Tanacetum cinerariae-folium*. The most common pyrethrins are pyrethrin I and pyrethrin II. However, there can be other pyrethrins such as cinerin I and II and jasmolin I and II. Pyrethrins are mainly used as insecticides, and as such, they produce a rapid knock-down effect on flying insects such as mosquitoes and flies. Not only are they used on flying insects, but they are also used as grain protectants and to control lice and fleas in poultry and dogs. They do so by blocking the sodium channels in the nerve axons [63].

3.5.1.1 Metabolism in Animals

As one of the most commonly used insecticides, pyrethrins have had research carried out on various aspects, such as their residue chemistry and metabolism. Several studies on their metabolism in animals have been carried out in mammals and in some insects. Generally, in some animals such as mammals, their metabolism has been found to be very rapid: hence, their low toxicity in these animals. In contrast, in insects, where they produce a rapid knockdown effect, the rate of toxicity is higher as a result of the slower metabolism [64].

To confirm the low toxicity of pyrethroids in mammals, a study was carried out on the metabolism of pyrethrin I, pyrethrin II, and allethrin. The study confirmed the findings of Yamamoto et al. 1969 that the metabolic pathways took place by oxidation of the alcohol and acid moieties of pyrethrin I and allethrin [65]. Furthermore, minimal hydrolysis of the methyl ester groups in mammals is another metabolic pathway. The study concluded that, as a result of the metabolic reactions on many pyrethrin sites, it was not likely that they would remain in the mammalian system: hence, the low toxicity. Moreover, a significant amount of the administered dose of pyrethrin II and I present in the urine was unmetabolized, also contributing to its low toxicity [66]. A similar study to confirm the metabolic pathway of pyrethrins was carried out in humans, and the results proved that metabolism of pyrethrins, just like in other mammals, occurs by hydrolysis and oxidation of the alcohol and acid moieties. The metabolic pathway is properly illustrated in **Figure 3.5**.

However, in a more recent study done to compare the metabolism of six pyrethrins in mouse and rat microsomes, it was discovered that selectivity for oxidation was greater at some sites than others in rat microsomes and in mouse microsomes for allethrin. Moreover, with cinerin I and jasmolin I, there was considerable hydroxylation of the methyl and methylene groups [64].

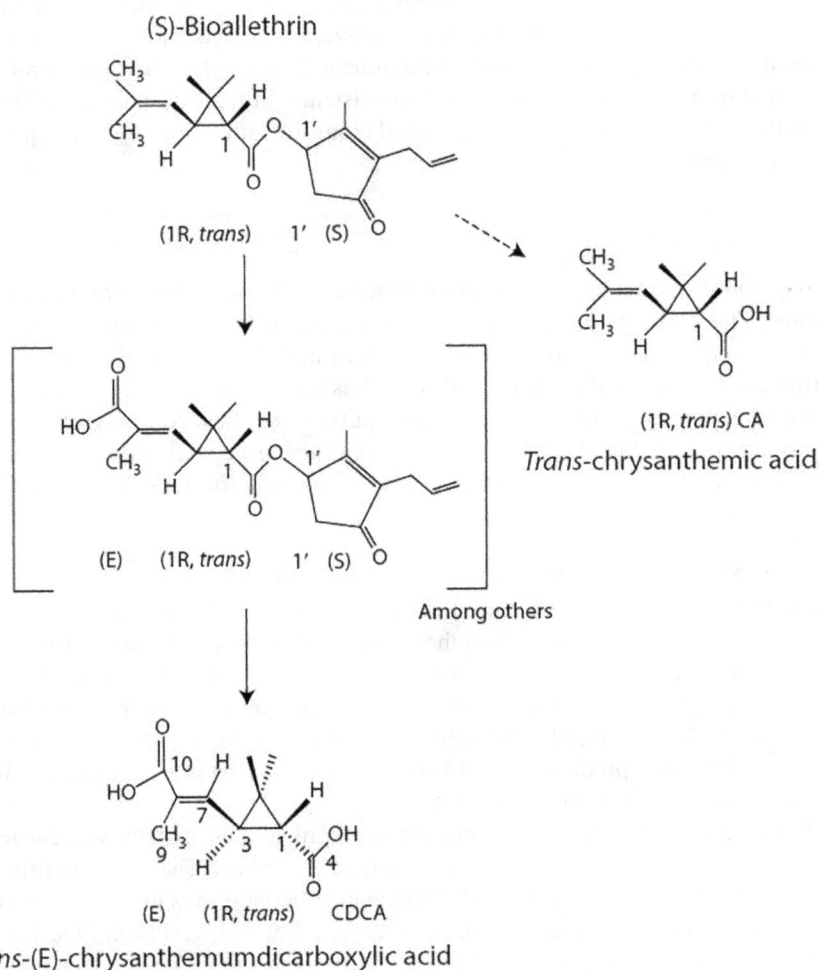

FIGURE 3.5 Bio-oxidation of the isobutenyl moiety in the tentative metabolic pathway for (S)-bioallethrin in humans.

Source: *J. Agric. Food Chem.*, vol. 61, K. Myung et al. Metabolism of strobilurins by wheat cell suspension cultures, pp. 47–52, copyright 2012, with permission from Elsevier

Metabolism of pyrethrins in plants was carried out in crops under storage. The objective of the research was to determine the factors affecting the degradation of natural pyrethrins in stored crops. It was concluded that temperature played a major role in the degradation of natural pyrethrins in stored crops. Contrary to previous studies, moisture and oxygen did not contribute to the degradation of pyrethrins [67].

3.5.2 ROTENONE

Rotenone is also one of the major natural pesticides that have insecticidal and acaricidal functions. It controls insects such as lice and ticks [1].

Rotenone stops respiration in mitochondria by blocking the NADH segment of the respiratory chain. The degree of toxicity of rotenone depends on the organism; that is, it is extremely toxic to insects but moderately toxic to mammals.

Apparently, the toxicity of rotenone is related to the rate of metabolism in living organisms. This was confirmed by a study carried out by Fukami et al. Metabolism studies were carried out in vitro and in vivo on rats, houseflies, and cockroaches. The major metabolic pathways were hydroxylation to give rotenolones and oxidation to give 8-hydroxy rotenone. The metabolic pathways were similar in all the organisms investigated. Moreover, the biological activity of the metabolites differed as some were more toxic to rats than others. They concluded that the detoxification of rotenone in organisms was facilitated by its high specificity when inhibiting enzymes such as NADH oxidase. Thus, in systems where this enzyme is present, there is bound to be greater metabolism and hence less toxicity than in systems where they are absent [68].

3.5.3 BIOCHEMICAL PESTICIDES

Biochemical pesticides are a special type of biopesticides that control pests without killing them. They include substances such as insect pheromones, plant growth regulators, and insect growth regulators.

3.5.3.1 Insect Growth Regulators

Insect growth is regulated by two main hormones, 20-hydroxyecdysone and the sesquiterpenoid juvenile hormone. Any slight disturbance to these hormones may result in stunted growth, abnormal growth, or changes in the reproduction pattern. Therefore, insect growth regulators mimic the action of these two hormones and thus result in a change in the growth pattern, possibly leading to death [69].

The bisacylhydrazines belong to a class of insecticides that mimic the action of the hormone 20-hydroxyecdysone. The main compounds of interest are methoxyfenozide, tebufenozide, chromafenozide, and fufenozide, as well as halofenozide. All of these are very active against lepidopteran larvae, while halofenozide is very active against coleopteran [70].

3.5.3.1.1 Methoxyfenozide

Methoxyfenozide finds very wide application in the control of insects in bulb vegetables, cereals, citrus crops, and leafy vegetables, among others. It is widely sold under the trade names Intrepid and Runner, and the formulations are usually wettable powder or, in some cases, suspension concentrates.

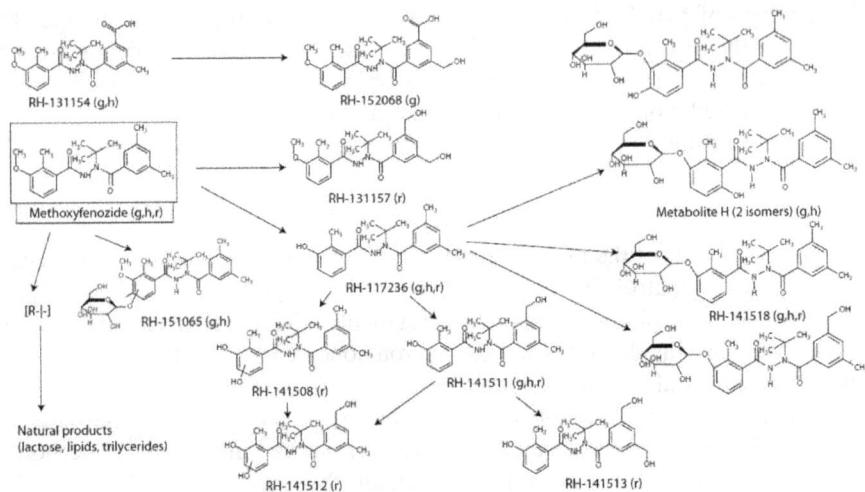

FIGURE 3.6   Metabolic pathway of methoxyfenozide in animals.

3.5.3.1.1.1 Metabolism of Methoxyfenozide

Numerous studies have been conducted on the metabolism of methoxyfenozide in plants, animals, and the environment. In animals, the metabolic rates were very high, with minimal residues occurring in the edible tissues of goats and hens. However, the major metabolites occurring in the liver was RH-141518 and minute quantities of other glucuronides. Furthermore, the metabolic pathways in rats proceeded by demethylation, hydroxylation, and glucuronation. The metabolic pathway is the same for hens and goats. **Figure 3.6** summarizes the metabolic pathways in animals and the respective metabolites [69].

In plants, studies were conducted on apples, cotton, grapes, and rice. The major metabolites observed were alcohols of methoxyfenozide on the methyl or methoxy groups of the biphenyls, which were identified as RH-117236 and RH-131157. It is important to note that the metabolic pathway was the same as in animals, but in a few cases, metabolism took place by conjugation.

3.5.3.2 Insect Pheromones

Insect pheromones are chemicals used by an insect to initiate communication with other insects in the same species. They are used in integrated pest management to disturb the mating process and to lead an insect to a trap containing a dangerous pesticide. Their main advantages include high specificity and low toxicity. However, their use is less efficient in the absence of other pest control measures [1].

3.5.3.3 Metabolism of Insect Pheromones

The housefly contains a sex pheromone made up of (Z)-9-Tricosene. The pheromone is produced by the female fly when the time for mating is right. The female housefly is known for its profuse breeding practices. Therefore, the use of the sex pheromone is to suppress the breeding process [71].

Metabolism of the compound was observed to take place in both male and female flies. Metabolism was found to occur by epoxidation of carbon 9 and 10 atoms. Furthermore, in male flies, the metabolism was faster than in females. In immobilized males, (Z)-9-tricosene was metabolized to the epoxides and the corresponding ketone by a Phase 1 microsomal cytochrome P450 polysubstrate mono-oxygenase enzyme. Moreover, the ketone and epoxides formed both have pheromone functions. However, the function of the conversion of the compound to its epoxides and ketone could not be ascertained [72]. In a related study, continuation was made, and it was observed that once the metabolites were internalized, they underwent conjugation by phase II enzymes. Moreover, the study showed that the metabolic rates were higher in males than in females during the mating period [73].

The juvenile hormone is secreted by the adult *Cecropia* moth in the corpora allata. The juvenile hormone's function is to prevent the developing of pupa and larvae into adult moths. Not only do they prevent the growth of the larva into adult moths, but they also disrupt the overall life cycle of the insect [74].

Metabolism of the *Cecropia* juvenile hormone was studied in eight types of insects. The studies were conducted in vivo and in vitro. The study showed that the rate of metabolism was very high in all the insect types. The metabolic pathway taken involved ester hydrolysis and epoxides hydration as the phase I reactions. The phase II reactions involved conjugation. The major metabolites were the juvenile hormone acid, juvenile hormone acid-diol, and conjugated metabolites such as glucuronides. Furthermore, in vitro studies showed the presence of two additional metabolites, the juvenile hormone tetrol metabolite and the juvenile hormone bisepoxide [75].

3.6 CONCLUSION

The research in the area of metabolism of biopesticides is still inadequate, and therefore a lot of gray areas exist as limited data is available. Research in this area has not grown as fast or been as wide ranging as the area of the utilization of the biopesticides. The metabolism of these biopesticides varies from pesticide to pesticide and species to species, and this determines their general toxicity and persistence in the environment. High metabolism of these pesticides in both plants and animals contributes to their low toxicity to non-target organisms. Whereas the microbial insecticide thuringiensin derived from the bacteria *Bacillus thuringiensis* did not show any significant metabolism in plants, it was quickly metabolized in animals, reducing its toxicity in animals and making it persistent in the plants; the fungicide azoxystrobin, on the other hand, was metabolized in both plants and animals, showing that its level of persistence is low, and so is its toxicity. Some insecticides, such as pyrethrum, are more rapidly metabolized in mammals than in insects; hence, plasma levels of the insecticide in mammals will be low and will not reach toxic levels, resulting in the low toxicity of the insecticide in mammals compared to insects (the target organisms). These biopesticides are generally metabolized via pathways for the general metabolism of foreign materials within the body, with conjugation to water soluble moieties enabling their excretion from animals and plants. An understanding of the metabolism of biopesticides is important in their effective use in plant protection and control of pests.

REFERENCES

[1] R. S. Tanwar, P. Dureja, H. S. Rathore, "Biopesticides," in *Pesticides: evaluation of environmental pollution*, L. M. Nollet and H. S. Rathore, Eds. Boca Raton: CRC Press, 2012, pp. 588–604.

[2] F. Dayan, C. Cantrell, S. Duke, "Natural products in crop protection," *Bioorganic Med Chem*, 17, 4022–4034, 2009.

[3] www.globenewswire.com/en/news-release/2022/09/22/2521188/0/en/Global-Biopesticides-Market-to-Hit-USD-12-2-Billion-by-2028-At-a-CAGR-of-15-3-Vantage-Market-Research.html#:~:text=22%2C%202022%20(GLOBE%20 NEWSWIRE),used%20substances%20in%20the%20world

[4] S. Manahan, *Toxicological chemistry and biochemistry*, 3rd ed. Boca Raton: Lewis Publishers, 2003.

[5] S. H. Chiu, R. Buhs, E. Sestokas, R. Taub, T. Jacob, "Determination of ivermectin residues in animal tissues by high performance liquid chromatography reverse isotope dilution assay," *J Agric Food Chem*, 33, 99–102, 1985.

[6] G. Leng, K. Kuhn, B. Wiesler, H. Idel, "Metabolism of (S)-bioallethrin and related compounds in humans," *Toxicol Lett*, 107, 109–121, 1999.

[7] K. Bailey, "Microbial weed control: an offbeat application of plant pathology," *Can J Plant Pathol*, 26, 239–244, 2004.

[8] A. Bravo, S. Gill, M. Soberon, "Mode of action of Bacillus thuringiensis Cry and Cyt toxins and their potential for insect control," *Toxicon*, 49, 423–435, 2007.

[9] A. Bravo, C. Rincon-Castro, J. Ibarra, M. Soberon, "Towards a healthy control of insects-potential use of microbial insecticides," in *Green trends in insect control*, vol. 11, O. Lopez and J. Fernandez- Bolanos, Eds. Royal Society of Chemistry, London: UK, 2011, pp. 266–299.

[10] A. Aronson, Y. Shai, "Why Bacillus thuringiensis insecticidal toxins are so effective: unique features of their mode of action," *FEMS Microbiol Lett*, 195, 1–8, 2001.

[11] Y. Bel, J. Ferré, P. Hernández-Martínez, "Bacillus thuringiensis toxins: functional characterization and mechanism of action," *Toxins*, 12(12), 785, 2020.

[12] D. G. Heckel, "How do toxins from Bacillus thuringiensis kill insects? An evolutionary perspective," *Arch Insect Biochem Physiol*, 104(2), e21673, 2020.

[13] A. Bravo, S. Likitvivatanavong, S. S. Gill, M. Soberón, "Bacillus thuringiensis: a story of a successful bioinsecticide," *Insect Biochem Mol Biol*, 41(7), 423–431, 2011.

[14] V. Vachon, R. Laprade, J. L. Schwartz, "Current models of the mode of action of Bacillus thuringiensis insecticidal crystal proteins: a critical review," *J Invert Pathol*, 111(1), 1–12, 2012.

[15] M. Singh, W. Mersie, "Absorption, translocation and metabolism of thuringiensin in potato," *Am. Potato J*, 66(1), 5–12, 1989.

[16] S. L. F. Wiest, H. L. P. Junior, L. M. Fiuza, "Thuringiensin: a toxin from Bacillus thuringiensis," *BT Res*, 6, 2015.

[17] W. Mersie, M. Singh, "Uptake, translocation, and metabolism of [14-C] thurigiensin in corn," *J Agric Food Chem*, 37(2), 481–483, 1989.

[18] W. Mersie, M. Singh, "Absorption, translocation and metabolism of 14-C thuringiensin (B-exotoxin) in snapbeans," *Florida Entomol*, 71(2), 105–111, 1988.

[19] D. Wolfenbarger, A. Guerra, H. Dulmage, R. Garcia, "Properties of the β-exotoxin of Bacillus thuringiensis IMC 10, 001 against the tobacco budworm," *J Econ Entomol*, 65, 1245–1248, 1972.

[20] C. Carsolio, N. Benhamou, S. Haran, C. Cortes, A. Gutierez, I. Chet, A. Herrera-Estrela, "Role of Trichoderma harzianumEndochitinase gene, ech42, in mycoparasitism," *Appl Environ Microbiol*, 65(3), 929–935, 1999.

[21] A. Litwin, M. Nowak, S. Różalska, "Entomopathogenic fungi: unconventional applications," *Rev Environ Sci Bio/Technol*, 19(1), 23–42, 2020.

[22] S. Mondal, S. Baksi, A. Koris, G. Vatai, "Journey of enzymes in entomopathogenic fungi," *Pac Sci Rev A*, 18(2), 85–99, 2016.

[23] T. M. Butt, C. J. Coates, I. M. Dubovskiy, N. A Ratcliffe, "Entomopathogenic fungi: new insights into host—pathogen interactions," *Adv Genet*, 94, 307–364, 2016.

[24] P. Binod, R. Sukumaran, S. Shirke, J. Rajput, A. Pandey, "Evaluation fungal culture filtrate containing chitinase as a biocontrol agent against Helicoverpa armigera," *J Appl Micribiol*, 103, 1845–1852, 2007.

[25] N. A. Zin, N. A. Badaluddin, "Biological functions of trichoderma spp. for agriculture applications," *Ann Agric Sci*, 65(2), 168–178, 2020.

[26] S. L. Woo, M. Ruocco, F. Vinale, M. Nigro, R. Marra, N. Lombardi, . . . M. Lorito, "Trichoderma-based products and their widespread use in agriculture," *Open Mycol J*, 8(1), 2014.

[27] E. Ghisalberti, M. Narbey, M. Dewan, K. Sivasithamparam, "Variability among strains of Trichoderma harzianum in their ability to reduce take-all and to produce pyrones," *Plant Soil*, 121, 287–291, 1990.

[28] F. Vinale, R. Marra, F. Scala, E. Ghisalberti, M. Lorito, K. Sivasithamparam, "Major Secondary metabolites produced by two commercial trichoderma strains active against different phytopathogens," *Lett Appl Microbiol*, 43, 143–148, 2006.

[29] K. Sivasithamparam, E. Ghisalberti, "Secondary metabolism in trichoderma and gliocladium," in *Trichoderma and gliocladium: basic biology, taxonomy and genetics*, vol. 1, C. Kubicek and G. Harman, Eds. London: Taylor and Francis, 1998, pp. 139–192.

[30] M. Hertlein, G. Thompson, B. Subrahmanyam, C. Athanassiou, "Spinosad: a new natural product for stored grain protection," *J Stored Prod Res*, 47, 131–146, 2011.

[31] H. Kirst, "The spinosyn family of insecticides: realizing the potential of natural products research," *J Antibiot (Tokyo)*, 63, 101–111, 2010.

[32] V. Salgado, "Studies on the mode of action of spinosad: insect symptoms and physiological correlates," *Pestic Biochem Physiol*, 60, 91–102, 1998.

[33] L. Bacci, D. Lupi, S. Savoldelli, B. Rossaro, "A review of Spinosyns, a derivative of biological acting substances as a class of insecticides with a broad range of action against many insect pests," *J Entomol Acarol Res*, 48(1), 40–52, 2016.

[34] J. E. Dripps, R. E. Boucher, A. Chloridis, C. B. Cleveland, C. V. DeAmicis, L. E. Gomez, . . . G. B. Watson, "The spinosyn insecticides," *Green Trends Insect Cont*, 11, 2011.

[35] R. Ramachanderan, B. Schaefer, "Spinosyn insecticides," *ChemTexts*, 6(3), 1–29, 2020.

[36] Y. Feng, Y. Huang, H. Zhan, P. Bhatt, S. Chen, "An overview of strobilurin fungicide degradation: current status and future perspective," *Front Microbiol*, 11, 389, 2020.

[37] J. Dripps, R. Boucher, A. Chloridis, C. Cleveland, C. Deamicis, L. Pavan, T. Sparks, G. Watson, "The spinosyn insecticides," in *Green trends in insect control*, vol. 11, O. Lopez and J. Fernandez-Bolanos, Eds. London: Royal Society of Chemistry, 2011.

[38] H. Balba, "Review of strobilurin fungicide chemicals," *J Environ Sci Heal Part B Pestic Food Contam Agric Wastes*, 42(4), 441–451, 2007.

[39] K. Myung, D. Williams, Q. Xiong, S. Thornburgh, "Metabolism of strobilurins by wheat cell suspension cultures," *J Agric Food Chem*, 61, 47–52, 2012.

[40] G. Von Jagow, W. Becker, "Novel inhibitors of cytochrome B as a valuable tool for a closer study of oxidative phosphorylation," *Bull Mol Biol Med*, 7(1–2), 1–16, 1982.

[41] N. Singh, S. Singh, I. Mukerjee, S. Gupta, V. Gajbhiye, P. Sharma, M. Goel, P. Dureja, "Metabolism of 14-C azoxystrobin in water at different pH," *J Environ Sci Heal Part B Pestic Food Contam Agric Wastes*, 45, 123–127, 2010.

[42] D. W. Bartlett, J. M. Clough, J. R. Godwin, A. A. Hall, M. Hamer, B. Parr-Dobrzanski, "The strobilurin fungicides," *Pest Manag Sci*, 58(7), 649–662, 2002.

[43] H. Balba, "Review of strobilurin fungicide chemicals," *J Environ Sci Heal Part B*, 42(4), 441–451, 2007.

[44] P. Reddy, "Avermectins," in *Recent advances in crop protection*, P. Reddy, Ed. New Delhi: Springer, 2013, pp. 13–24.

[45] T. Roberts, D. Hutson, *Metabolic pathways of agrochemicals- insecticies and fungi-cides*. London: Royal Society of Chemistry, 1999, pp. 79–99.

[46] M. Maynard, V. Gruber, W. Feely, R. Alvaro, P. Wislocki, "Fate of the 8,9-Z isomer of abamectin B1A in rats," *J Agric Food Chem*, 37(6), 1487–1491, 1989.

[47] M. Maynard, B. Halley, M. Green-Erwin, R. Alvaro, V. Gruber, S. Hwang, B. Bennett, P. Wislocki, "Fate of avermectin B1A in rats," *J Agric Food Chem*, 38, 864–870, 1990.

[48] M. S. Khalil, "Abamectin and azadirachtin as eco-friendly promising biorational tools in integrated nematodes management programs," *J Plant Pathol Microbiol*, 4(174), 2013.

[49] M. Maynard, Y. Iwata, P. Wislocki, C. Ku, T. Jacob, "Fate of Avermectin B1A on citrus fruits: 1 distribution and magnitude of the Avermectin B1A and 14C residues on citrus fruits from a field study," *J Agric Food Chem*, 37, 178–183, 1989.

[50] H. Kim-Kang, A. Bova, L. Crouch, P. Wislocki, R. Robinson, J. Wu, "Tissue distribu-tion, metabolism, and residue depletion study in Atlantic Salmon follwoing oral admin-stration of [3-H] emamectin benzoate," *J Agric Food Chem*, 52, 2108–2118, 2004.

[51] C. Wrzesinski, M. Mushtaq, T. Faidley, N. Johnson, B. Arison, L. Crouch, "Metabolism of 3H/14-c labelled 4″ deoxy epimethylaminoavermectin B1A benzoate in chickens," *Drug Metab Dispos*, 26(8), 786–794, 1998.

[52] X. Cheng, X. Liu, H. Wang, X. Ji, K. Wang, M. Wei, K. Qiao, "Effect of emamectin ben-zoate on root-knot nematodes and tomato yield," *PLoS One*, 10(10), e0141235, 2015.

[53] H. A. A. Khan, W. Akram, T. Khan, M. S. Haider, N. Iqbal, M. Zubair, "Risk assessment, cross-resistance potential, and biochemical mechanism of resistance to emamectin ben-zoate in a field strain of house fly (Musca domestica Linnaeus)," *Chemosphere*, 151, 133–137, 2016.

[54] S. H. Chiu, E. Sestokas, R. Taub, M. Green, F. Baylis, T. Jacob, A. H. Lu, "Metabolic disposition of ivermectin in swine," *J Agric Food Chem*, 38, 2079–2085, 1990.

[55] S. H. Chiu, E. Sestokas, R. Taub, R. Buhs, M. Green, R. Sestokas, W. J. Vandenheuvel, B. Arison, T. Jacob, "Metabolic disposition of ivermectin in tissues of steers, sheep and rats," *Drug Metab Dispos*, 14, 590–600, 1986.

[56] R. Laing, V. Gillan, E. Devaney, "Ivermectin—old drug, new tricks?," *Trends Parasitol*, 33(6), 463–472, 2017.

[57] J. Afzal, A. Burke, P. Batten, R. Delay, P. Miller, "Moxidectin: metabolic fate and blood pharmacokineticsof 14-C labelled moxidectin in horses," *J Agric Food Chem*, 45, 3627–3633, 1997.

[58] R. K. Prichard, T. G. Geary, "Perspectives on the utility of moxidectin for the control of parasitic nematodes in the face of developing anthelmintic resistance," *Int J Parasitol Drugs Drug Resist*, 10, 69–83, 2019.

[59] J. Zulalian, S. Stout, A. da Cunha, T. Garces, P. Miller, "Absorption, tissue distribution, metabolism and excretion of moxidectin in cattle," *J Agric Food Chem*, 42, 381–387, 1994.

[60] J. Afzal, S. Stout, A. da Cunha, P. Miller, "Moxidectin: absorption, tissue distribution, excretion and biotransformation of 14-C labelled moxidectin in sheep," *J Agric Food Chem*, 42, 1767–1773, 1994.

[61] I. Cavoski, P. Caboni, Y. Miano, "Natural pesticides and future perspectives," in *Pesticides in the modern world-pesticide use and Management*, M. Stoytcheva, Ed. Rijeka: Intech, 2011, pp. 163–190.

[62] M. Isman, "Botanical insecticides-for rich, for poorer," *Pestic Manag Sci*, 64, 8–11, 2008.

[63] M. Isman, "Botanical insecticides, deterrents, and repellents in modern agriculture and an increasingly regulated world," *Annu Rev Entomol*, 51, 45–66, 2006.

[64] T. Class, T. Ando, J. Casida, "Pyrethroid metabolism-microsomal oxidase metabolites of (S-) bioallethrin and the six natural pyrethrins," *J Agric Food Chem*, 38, 529–537, 1990.

[65] I. Yamamoto, E. Kimmel, J. Casida, "Oxidative metabolism of prethroids in houseflies," *J Agric Food Chem*, 17(6), 1227–1236, 1969.

[66] M. Elliot, N. Janes, E. Kimmel, J. Casida, "Metabolic fate of pyrethrin I, pyrethin II, and allethrin adminstered orally to rats," *J Agric Food Chem*, 20(2), 300–312, 1972.

[67] B. Atkinson, A. Blackman, H. Faber, "The degradation of natural pyrethrins in crop storage," *J Agric Food Chem*, 52, 280–287, 2004.

[68] J. Fukami, T. Shishido, K. Fukunaga, J. Casida, "Oxidative metabolism of rotenone in mammals, fish, insects and its relation to selective toxicity," *J Agric Food Chem*, 17(6), 1217–1226, 1969.

[69] G. Smagghe, L. Gomez, T. Dhadialla, "Bisacylhydrazine insecticides for selective pest control," in *Advances in insect physiology*, vol. 43, T. Dhadialla, Ed. Burlington: Elsevier, 2012, pp. 164–249.

[70] L. Gomez, K. Hastings, H. Yoshida, J. Dripps, J. Bailey, S. Rotondaro, S. Knowles, D. Paroonagian, T. Dhadialla, R. Boucher, "The bisacylhydrazine insecticides," in *Green trends in insect control*, O. Lopez and J. Fernandez-Bolanos, Eds. London: Royal Society of Chemistry, 2011, pp. 213–247.

[71] D. Carlson, M. Mayer, D. Silhacek, J. James, M. Beroza, B. Bierl, "Sex attractant pheromone of the housefly: isolation, identification and synthesis," *Science*, 174, 76–77, 1971.

[72] G. Blomquist, J. Dillwith, J. Pomonis, "Sex pheromone of the housefly: metabolism of (Z)-9-Tricosene to(Z)-9,10-epoxytricosane and (Z)-tricosen-10-one," *Insect Biochem*, 14(3), 279–284, 1984.

[73] S. Ahmad, M. Mackay, G. Blomquist, "Accumulation of the female sex pheromone and its transfer to and metabolism in the male housefly, Musca Domestica, L, during courtship and mating," *J Insect Physiol*, 35(10), 775–780, 1989.

[74] C. Williams, "The Juvenile hormone 1: endocrine activity of the Corpora Allata of the adult cecropia silkworm," *Biol Bull*, 116(2), 323–338, 1959.

[75] A. Ajami, L. Riddiford, "Comparative metabolism of the Cecropia Juvenile hormone," *J Insect Physiol*, 19, 635–645, 1973.

4 Biopesticides and Their Mode of Action
Communicating Sustainable Agricultural Practices amid Climate Change Threats

Shariq Ibrahim Sherwani and Haseeb Ahmad Khan

CONTENTS

DOI: 10.1201/9781003265139-5

4.1 INTRODUCTION

The use of biopesticides has been shown to be a sustainable pest control strategy in agricultural management and environmental protection (Fenibo et al., 2021). Climate change continues to cause unprecedented temperature increases, unpredictable rainfall patterns, and extreme human displacement (Ebi et al., 2021). It has also disrupted agricultural cycles leading to food insecurity, famine, and drought for a world population that just hit the eight billion mark. Invasive species of microorganisms, invertebrates, and vertebrates continue to spread among both flora and fauna, including humans, as is evidenced by the coronavirus pandemic (COVID-19). Synthetic pesticides still comprise a large market share despite their harmful effects on the plants, animals, and the environment (Romanazzi et al., 2022; Schleiffer & Speiser, 2022). Overcoming the environmental damage is unsustainable if the incessant application of synthetic pesticides continues at such a large scale due to their leaching into ground water. Their use can disrupt biodiversity and affect the consumers directly by causing health risks like cancers and compromising the quality of life (QOL) (Carvalho, 2017; Zúñiga-Venegas et al., 2022).

The health risks and compromised QOL due to chemically infested agricultural lands, crops, and products are borne by all consumers, especially the farmers who till their farms, grow the crops, and raise farm animals (livestock) (Carvalho, 2017; Zúñiga-Venegas et al., 2022). The resource-limited farmers from developing countries find themselves in dire situations in which they must decide between using the affordable and easily available synthetic pesticides or the costlier, hard-to-find, and challenging-to-calibrate biopesticides. The dilemma of whether to use the synthetic pesticides (which can increase agricultural output) and eradication of various plant disease vectors is not an easy one for farmers to overcome. But many studies have shown that synthetic pesticides are detrimental to plants, animals, humans, and the environment, thus compromising the food chain (Fenibo et al., 2021; Nicolopoulou-Stamati et al., 2016). Plant pests are also adept at developing resistance to synthetic pesticides and can make it difficult to implement pest control and management strategies (Souto et al., 2021; Stejskal et al., 2021).

Biopesticides provide a safer and environment-friendly alternative in promoting pest control, sustainable agriculture, and environmental management. They are derived from plants, animals, microorganisms, and natural products such as minerals, which can help in ecological agriculture (Šunjka & Mechora, 2022). Sustainable agriculture considers the topographical conditions, local farming practices, and waste minimization. Biopesticides offer major strengths, weaknesses, opportunities, and threats to the agricultural sector amid the risks that climate change entails (Jansson & Hofmockel, 2020; Seiber et al., 2014). The efficacy, efficiency, and performance of biopesticides is influenced by the ingredients, concentrations, and applications that comprise them (Siegwart et al., 2015). They can be used in both organic and general farming management practices. Upon uptake, biopesticides act by interfering with, compromising, or disrupting the physiology of plant pathogens, leading to their eventual death. The biological control of pests requires that the biopesticides be maintained in an optimum physical and chemical environment during storage; otherwise they can lose their potency and efficiency (Arthurs & Dara, 2019;

Siegwart et al., 2015). Products like salicylic acid, a naturally occurring phenolic compound and a precursor/metabolite of aspirin, is added to enhance the efficiency in controlling the fungus *Cryptococcus laurentii* in fruits like pear, apple, cherry, and cherry tomato (Hernández-Ruiz & Arnao, 2018; Lai et al., 2018; Qin et al., 2003). Sometimes, salicylic acid works better in combination with *Rhodotorula glutinis* and helps in controlling the natural spoilage of strawberries (Zhang et al., 2010). This updated chapter will focus on (1) biopesticides and climate change, (2) biopesticides and resistance, and (3) mode of action of biopesticides (retaining this section from the previous edition, with updated references).

4.2 BIOPESTICIDES AND CLIMATE CHANGE

Increases in carbon dioxide (CO_2) footprint, and temperatures cause adverse effects in global ecosystems, leading to negative outcomes for agriculture, environment, and food (in)security (Gomez-Zavaglia et al., 2020; Raza et al., 2019). These phenomena have also caused irregular and unpredictable precipitation patterns, which further exacerbate issues with crop diversity and productivity. The uptake and effectiveness of biopesticides has been shown to be influenced by climatic changes such as temperature, humidity, CO_2, and ionizing radiations (Kaka et al., 2021; Skendžic et al., 2021; Ziska et al., 2018). According to the United States Environmental Protection Agency (EPA, 2022), "climate changes refers to changes in global or regional climate patterns attributed largely to human-caused increased levels of atmospheric greenhouse gases" (n.p.). Crops and agricultural practices are regional in nature, depend upon the local geography/topography, and are highly sensitive to climate change.

The unprecedented industrialization by the industrialized (high-income) countries in the last century, especially the last 70 years, has contributed to the increase in greenhouse gases, leading the earth to trap more heat and causing global warming. The industrialization trend continues with low- and middle-income countries generating maximum greenhouse gases as they industrialize their economies and become food secure. Agriculture, deforestation, and farm animals generate maximum greenhouse gases and have global implications (Kristiansen et al., 2021). There is an urgent need for implementing global, regional, and national policies.

The effectiveness of biopesticides is highly influenced by precipitation and humidity as their efficacy depends upon host-pathogen interaction(s). Arid lands have received unprecedented rainfall while wetlands have experienced draught. Deforestation has led to soil erosion, landslides, flooding, and disruptions in flora, fauna, and indigenous livelihoods (Sacco et al., 2020). All these factors can influence agricultural practices, cropping systems, and animal farming. These disruptions, compounded by the incessant use of synthetic insecticides, have influenced pest behavior, making them resistant to chemicals and challenging to manage. Biopesticides are environment-friendly but highly sensitive to environmental conditions, causing them to be ineffective if the optimum conditions are not available or even with slight condition changes. The inconsistency in the optimum performance of biopesticides can be attributed to temperature, precipitation, humidity, formulation, and radiation sensitivity (Ebi et al., 2021; Gomez-Zavaglia et al., 2020; Raza et al., 2019).

4.2.1 Impact of Climate Change

Biopesticides are highly influenced by climate change, and a slight variation can make them ineffective or inefficacious (Ebi et al., 2021; Jansson & Hofmockel, 2020; Seiber et al., 2014). The percent of humidity that entomopathogens require for effective impact varies readily, as does the availability of the appropriate host. If all the right conditions are not met/available, the infection of the host pathogen by the biopesticide cannot occur. Climate change poses significant risks globally in terms of food (in)security, environmental degradation, and limited resources (Gomez-Zavaglia et al., 2020; Raza et al., 2019). Communities that live on the margins (marginalized, vulnerable, overburdened, and underserved) bear the brunt of climate change (Berberian et al., 2022; Ebi et al., 2021). Agricultural lands and crops cultivated in low-lying areas are often destroyed and communities disrupted due to seasonal flooding or unnatural disasters. Even though not a focus of this chapter, unending global wars and human-made disasters contribute immensely to global warming and climate change (Kemp et al., 2022).

4.2.2 Revolution in Sustainable Agriculture

Sustainable agricultural practices strive to meet the present needs of the society without compromising the needs of the future generations. It involves creating equity (social and economic), profitability, and a healthy environment. Individuals at all levels of the food chain/system can contribute toward sustainability (Toussaint et al., 2022). Farmers can use biopesticides, promote soil health, conserve water, and minimize pollution—all leading to sustainable agricultural practices and saving the environment (Stein & Santini, 2022). Reduced precipitation can compromise the water table, and other related factors can cause hypoxic water conditions. Consumers can consume values-based foods, support farmer well-being, strengthen the local economy, and follow the 3R motto (reduce, reuse, and recycle) for reducing waste (Stein & Santini, 2022; Toussaint et al., 2022).

4.2.3 Pest Management

Pest management is an environment-friendly methodology for effectively controlling pests. Integrated pest management practices include pest prevention and "as-needed" approaches. Inspecting pests, monitoring pest progress, and determining the best preventive measures can help in mitigating the impact of pests and avoiding the unnecessary use of synthetic pesticides. Environment-friendly biopesticides, on the other hand, have led to a revolution in pest management practices (Deguine et al., 2021). As climate change, population growth, and technological advancements continue to influence agricultural practices with impending food demand, biopesticides have provided a tool to the farmers to safeguard their soil health and grow sustainable crops. Scouting the field, tilling the land, rotating the crops, seasonal patterns specificity, and applying biopesticides are some of the practices common among farmers and also encouraged for pest management. Plant pathogens and weeds can cause insurmountable and irreparable damage to crops and farm animals (livestock) (Rizzo et al., 2021).

4.2.4 Current Challenges and Opportunities

Pest populations continue to infest crops, soils, and the environment at an unprecedented rate (Kaka et al., 2021; Skendžic et al., 2021; Ziska et al., 2018). They also compromise the health of farmers and livestock alike. If the pest populations are not regularly monitored, they can reach uncontrollable levels and quickly infest vast areas, making it difficult, if not impossible, to prevent and control pest infestation (Skendžic et al., 2021). Beyond the challenges associated with climate change, some of the current challenges accompanying pest control include (1) soil inspection, (2) crop monitoring, (3) reducing pest varieties, (4) economic considerations, and (5) application of biopesticides (Bueno et al., 2021; Kadoić Balaško et al., 2020). Biopesticides provide newer and better opportunities for further influencing human agricultural practices and the relationship of humans to their environment. As early human ancestors were gatherers and hunters, and agriculture provided them with predictable and stable food sources, modern-day humans can reap larger benefits by implementing biopesticides in their agricultural practices for healthier, environmentally safe, plentiful, and economically beneficial agricultural production. Applying agricultural technology and methods has the potential for sustainable agriculture—thus minimizing the gap between agrarian communities and urban lifestyles. There are opportunities for avoiding or minimizing environmental degradation while still maximizing agricultural and livestock output (Leinonen & Kyriazakis, 2016; Tanentzap et al., 2015).

4.2.5 Future Strategies

The research, technology, and availability associated with biopesticides make them a viable option to incorporate into regular agricultural practices (Fenibo et al., 2021; Nicolopoulou-Stamati et al., 2016). Environment-friendly biopesticides can positively impact crop production and outcome. The cost-benefit ratio can deter some farmers from using biopesticides as part of their agricultural practices and pest management strategies (Fenibo et al., 2021). But, as the price of biopesticides continues to go down, more competition comes into the market, and consumers demand environment-friendly and healthier agricultural and livestock products, future strategies will continue to focus on natural pest control options such as biopesticides (Nicolopoulou-Stamati et al., 2016; Šunjka & Mechora, 2022). Using natural ingredients while developing biopesticides can help in mitigating the detrimental effects on the crops, livestock, farmers, and the environment.

4.3 BIOPESTICIDES AND RESISTANCE

Insects are sturdy and highly capable of developing resistance over time—survival of the fittest (Brevik et al., 2018). Resistance to synthetic pesticides is common and detrimental to the environment, food chain, and sustainable agriculture. Despite the use of biopesticides as a tool in sustainable agricultural practices, some insects can develop resistance, just like they would to synthetic pesticides (Siegwart et al., 2015). For example, *Bacillus thuringiensis* (*Bt*) toxins have been shown to be ineffective in least 27 species of insects, which have already developed resistance to it (Kaze et al., 2021). However, certain properties of biopesticides can be utilized to make them still effective against many species and make them more sustainable.

Biopesticides usually comprise plant extracts, microorganisms, minerals, or a combination. In the 1990s, *Bt* was prevalent in most of the biopesticides (about 95%), and hence, the resistance to it by at least 27 species of insects (Kaze et al., 2021). Over the last several decades, scientists have focused on other microorganisms (entomopathogenic fungi, bacteria, and viruses) and other biological agents such as insects. Resistance to biopesticides is only one cause for concern, the others being their efficacy, efficiency, and robustness (Siegwart et al., 2015). Biopesticides are very sensitive, and their potency can be easily influenced by the presence/absence of optimum temperature, humidity, radiation, and storage conditions—all elements highly influenced by climate change (Zhao et al., 2022). So it is a challenge for researchers and scientists to develop more potent formulations that are sustainable under various conditions, easy to transport, and have a longer shelf life.

If a specific biopesticide is used constantly and consistently, chances are that the pests it is designed to destroy may, instead, develop resistance to it due to overuse and familiarity. Insects and other pests develop resistance to various chemicals (originating from synthetic and natural sources) by modifying their genetic, metabolic, and behavioral makeup (Hawkins et al., 2018). Genetically, pests may evolve over time (sometimes within a short span of time, through mutation) due to the presence of a specific chemical or moiety. Metabolically, pests may develop increased excretion rates, excreting the biopesticides before they can destroy the gut/abdomen of the pests. Similarly, pests may develop new signaling pathways or alter them altogether to avoid the binding of biopesticides to their target sites while compromising the optimum penetration of toxins, making them less effective. Despite the potential for developing resistance even to biopesticides, because the biopesticides operate via complex modes of action, the chances of developing a quick and sustained resistance are acutely minimized (Kaze et al., 2021; Siegwart et al., 2015).

4.3.1 Implications for Crop Protection

Biopesticides provide the farmers with tools to implement pest management programs and increase healthier and sustainable crop yields (Šunjka & Mechora, 2022). The higher costs associated with the procurement and use of biopesticides remain an important consideration in decision-making vis a vis their use. As the debate over the need for sustainable agriculture (resources), climate change, and saving the environment heats up and newer developments/technologies continue to help in reducing the cost of biopesticides, the future of biopesticides looks greener (more promising/better). As regulators such as the EPA continue to tighten the rules and regulations regarding the environment and consumers demand sustainable, healthier, safer, and environment-friendly products, biopesticides will continue to find a prominent place in decision-making regarding crop protection and sustainable agriculture.

Despite being "natural," all biopesticides are not safe for crops, soil, farmers, and livestock. So regulating agencies such as the EPA and the United States Food and Drug Administration (FDA) implement stringent codes to ensure safety across all levels. It takes several years, millions of dollars, and rigorous regulatory approvals as part of the research and development (R&D) process for a new synthetic insecticide to be launched into the market. The relative safety and sustainability of biopesticides makes them relatively easy and quicker to launch, with complex modes of action leading to a better control of pests, if not total elimination. Often, farmers use biopesticides in combination with synthetic pesticides for better control of pests, instead of using either as a standalone application (Zhao et al., 2022). Unlike synthetic pesticides, biopesticides can be used from the beginning (tilling/sowing) to the end (harvesting and storage) of the crop (Arthurs & Dara, 2019). Biopesticides do not leave residue and are exempt from residue tolerance levels as per the internationally agreed Codex Maximum Residue Levels (CXLs), which cover food standards related to pesticide residues on human food and animal feed (Wahab et al., 2022). However, the long-term safety of biopesticides is difficult to predict until humans, animals, and the environment experience them over decades and maybe even centuries.

4.3.2 FOOD (IN)SECURITY

Biopesticides can play a critical role in food security for communities, nations, and globally (Fenibo et al., 2021). The seriousness of food (in)security can be appreciated from the fact that the United Nations Organization (UNO) in its 2030 Agenda for Sustainable Development Goals (SDGs) has listed the top two SDGs as "No Poverty" (SDG1) and "Zero Hunger" (SDG2) (Tickner et al., 2021). Global poverty has continued to decline from 10.1% (2015) to 8.6% in 2018. However, during the COVID-19 pandemic, the poverty rate increased significantly from 8.3% in 2019 to 9.2% in 2020, which drastically compromised the steady progress of poverty reduction (United Nations, 2022). Similarly, in 2020, 720–811 million people globally experienced hunger, which was 161 million people more than in 2019, which coincided with the outbreak of the COVID-19 pandemic (United Nations, 2022).

Biopesticides can play a significant role in ensuring higher crop production, safer agricultural products, sustainable agricultural practices, the health of farmers and livestock, and a safer environment (Cox & Zeiss, 2022). Increased, healthier, and affordable agricultural produce can positively influence the lives of people globally, especially the vulnerable populations. Reduced poverty and hunger rates have been shown to be associated with citizen well-being, opportunities, and increased productivity (United Nations, 2022). Poverty, hunger, and national and international issues affect all communities, especially the vulnerable and marginalized. Sustainable farming can address the goals of both no poverty and zero hunger, among many other SDGs (United Nations, 2022). Along with biopesticides, there is an increased focus on developing and using biodegradable products (livestock feed, construction materials, and soil treatments) for environmental protections, effective use of resources, and minimal carbon footprints (Souto et al., 2021).

4.3.3 ENVIRONMENTAL POLLUTION

Biopesticides are designed/manufactured from natural resources and pose far less environmental threat than synthetic pesticides (Leinonen & Kyriazakis, 2016;

Šunjka & Mechora, 2022; Tanentzap et al., 2015). The mode of action of biopesticides is specific, and they target a narrow range of pests, unlike the broad-spectrum synthetic pesticides (Deshayes et al., 2017) (Table 4.1). Biopesticides break down easily and become a part of the natural environment and habitat. The financial and environmental benefits of using biopesticides are enormous. On synthetic pesticides, for every $1 investment, the returns are $4; however, on biopesticides, for every $1 investment, the returns are $30—$100 (Pimentel, 1989; Seiber et al., 2014; Zhao et al., 2022). There are about 60,000 known species of pests globally, but only about 0.2% of pests are treated with biopesticides, so there is an enormous potential for experimenting with and developing newer varieties/products (Pimentel, 1989; Seiber et al., 2014; Zhao et al., 2022).

There are enormous costs (financial and health) associated with using and monitoring synthetic pesticides; the EPA estimates that such costs amount to $1.2 billion annually for monitoring the water table. Most countries continue to take the risks of

TABLE 4.1
Summary of Biopesticides and Their Mode of Action

Biopesticide	Mode of Action	Examples	Control Agent
Bacteria	Produce toxins that are detrimental to certain insect pests when ingested.	*Bacillus thuringiensis* *Bacillus popilliae* *Agrobacterium radiobacter*	Lepidopterans Japanese beetles Crown gall disease
Fungi	Control insects by growing on them secreting enzymes that weaken the insect's outer coat and then getting inside the insect and continuing to grow, eventually killing the infected pest.	*Entomophaga praxibulli* *Zoophthora radicans* *Neozygites floridana*	Grasshoppers Aphids Cassava green mites
Viruses	Kill insects when ingested. Insect's feeding behavior is disrupted; thus, it starves and dies.	*Baculoviruses: Nuclear polyhedrosis virus (NPV)* *Baculoviruses: Granulosis virus (GV)* *Baculoviruses: Group C Entomopox*	Lepidopteran and Hymenopteran Lepidopteran Arthropods
Protozoa	Kill insects when ingested. Insect's feeding behavior is disrupted; thus it starves and dies.	*Nosema* *Vairimorpha* *Malamoeba*	Grasshoppers Lepidoptera Locusts
Nematodes	Kill their target organisms by entering natural body openings or penetrating the insect cuticle directly.	*Heterorhabditis bacteriophora* *Phasmarhabditis hermaphrodit* *Steinernema carpocapsae*	Black vine weevils Japanese beetles Various slugs and snails Black vine weevils Strawberry root weevils Cranberry girdlers

using synthetic pesticides very seriously due to their detrimental impact on sustaina-ble agriculture, the environment, and human and livestock health. Their seriousness and urgency are reflected in the passing/implementation of legislation pertaining to reducing/eliminating the use of synthetic pesticides by a certain time period (EPA, 2022). This opens the door further for the increased use of biopesticides in crop man-agement application and sustainable agricultural practices (Seiber et al., 2014; Zhao et al., 2022). There is no dearth of microorganisms and insects that can be used in developing biopesticides as scientists have barely scratched the surface (Arthurs & Dara, 2019). Native pests can be controlled better by applying specific biopesticides that have been found to be more effective (Souto et al., 2021). At least 27 species of insects are reported to have developed resistance to *Bt* toxins but beyond *Bt*, *Bacillus poppillat* has been used as an effective biocontrol agent (Kaze et al., 2021).

Some of the biopesticides and their modes of action can be a cause for concern (Glare et al., 2016). For example, a biopesticide that may work effectively against an insect like a caterpillar may destroy all caterpillar when all caterpillars are not bad for the crop. How do scientists and farmers navigate this dilemma? A case in point is the biopesticide containing *Bt*, which has been found to be detrimental to all such insects (caterpillars, moths, butterflies) and, if applied broadly, can kill them, with several of them being on the endangered species list. Genetically, gene transfer is a major issue between such closely related populations (microorganism to microor-ganism), which can be an environmental risk factor. Mutation is another concern that could be devastating to beneficial insects such as beetles, which help in controlling pest populations themselves. Finally, the effect(s) that genetically modified micro-organisms will have many years, decades, or centuries after their release into the environment is not always predictable, and they will not be 100% safe in the long run (Teem et al., 2020).

4.3.4 Environmental Protection

The behavior of microorganisms, especially after they have come in contact with other microorganisms in the environment, is unpredictable, despite years of R&D (Li et al., 2022). This detrimental phenomenon has been observed with the introduction of commercial crops from one region of the world to another (even within the same country) that suddenly/over time have mutated or evolved into weeds. There is a similar danger of biopesticides becoming pests themselves (Siegwart et al., 2015). With an ever-increasing need for food and shelter, humans continue to explore newer areas and are in contact with newer species of microorganisms, animals, and plants, increasing the chances of zoonotic diseases like COVID-19, Ebola, and West Nile virus, among many others (Murray et al., 2021; Rohr et al., 2019).

Despite the use of biopesticides, most naturally occurring pests cannot be elim-inated completely from the environment. They can be controlled in a specific area/region but not eliminated, and, when the conditions are right, they can make a triumphant return. It is a cat-and-mouse game—when the threat is greater (applica-tion of biopesticides) they disappear, become dormant, mutate, or move to greener pastures, and as soon as the threat is reduced (no application of biopesticides), they return with a vengeance. Only two species of pathogens have been eliminated thus far: namely, citrus canker and the Mediterranean fruit fly (Ference et al., 2018; Leftwich et al., 2014).

Synthetic pesticides do not reproduce while biopesticides can and do reproduce, so there is a danger of their mutating and their behavioral patterns shifting. We may not realize it at this juncture, but what the future holds is unknown. Genetic engineers and biotechnologists have a good idea about the design of biopesticides and their modes of action, but their safety, in the next 100 years or more, cannot be predicted with 100% accuracy. Biopesticides have implications in biological warfare, where rogue nations or individuals can, intentionally or unintentionally, compromise the environment and the food chain (Jansen et al., 2014).

4.3.5 POLICIES AND REGULATIONS

There are global, national, regional, and local concerns regarding the design, use, and sequestration of biopesticides. With minimal threat to agro-ecosystems and the environment, biopesticides have seen an uptick in their demand, production, and use. They appear to be a panacea for all agricultural and environmental ills. Policies, regulations, and laws allowing their use are as varied as the countries themselves. From the EPA to FDA to the International Organization for Biological Control (IOBC), the regulations have similarities and differences. It is imperative that regulating bodies work together and establish standards that neither favor high-income countries nor compromise the agricultural development of low-income countries.

In the US, Title 40 of the Code of Federal Regulations (CFR 40) deals with the mission of the EPA in protecting human health and the environment. The "EPA works to inform and educate the public about its policies and activities" (EPA, 2022). The Biopesticides and Pollution Prevention Division (BPPD) of the EPA is responsible for regulatory activities related to biopesticides, which are considered "reduced risk pesticides" (EPA, 2022). It is imperative for all stakeholders and shareholders to be transparent, honest, and collaborative in designing biopesticides that are healthy, cost effective, and sustainable in terms of agriculture, livestock, and the environment (Cox & Zeiss, 2022).

4.4 MODE OF ACTION OF BIOPESTICIDES

Biopesticides comprise diverse and naturally occurring oils and extracts derived from plant sources and can be used as both bioinsecticides and bioherbicides. Their mode of action varies greatly from product to product. As opposed to insect sex pheromones, which themselves do not kill a target pest but rather interrupt their reproductive cycles, plant oils and extracts act indirectly and nonspecifically. Plant extracts like floral essences lure and attract insects to traps. Sometimes, they are used as repellents and deterrents (cayenne) by generating foul odors, which keep the pests away. Lemongrass oils cause dehydration by removing the waxy coating from the leaves of plants while other extracts cause suffocation. On the other hand, some natural products help the crops enhance their immune systems by developing systemic acquired resistance. The insect mating cycle is disrupted due to the presence of higher concentrations of sex pheromones in the surroundings, which may throw the males off the scent of females, thus reducing/limiting their ability to mate (Ferracini et al., 2021).

4.4.1 BACTERIAL BIOPESTICIDES

These are the most common form of microbial pesticides. Typically used as insecticides, they can be used to control undesirable bacteria, fungi, or viruses. When being used as insecticides, they explicitly act on specific species of moths, butterflies, beetles, flies, mosquitoes, etc. In order to be effective, it is essential that these biopesticides come in contact with the target pest and be ingested. The subspecies and strains of *Bacillus thuringiensis* (*Bt*) account for nearly 90% of the biopesticides market and are the most commonly used microbial pesticides (Jouzani et al., 2017). A specific strain of *Bt* produces a specific blend of proteins that target a specific species of insect larvae. Upon being ingested by insect larvae, *Bt* releases endotoxins (proteins) that attach to the intestinal lining of the midgut, creating pores, thus paralyzing the digestive system and, ultimately, causing insect death (Jurat-Fuentes, n.d.). *Bt* is mostly used to control lepidopteran pests, which cause the most damage to the crops; these include moths and butterflies. *Bt* 27, a specific strain of *Bt*, is used to control a wide range of different pests, such as some species of mosquitoes, beetles, and flies. So far, about 500–600 strains of *Bt* have been identified, and about 525 different kinds of insects have been found to be infected by *Bt* toxins (Palma et al., 2014). *Bt* endotoxins are commonly used in DNA-recombinant technology for genetically modifying and creating pest-resistant varieties of crops. *Pasteuria* spp. is being used to control nematodes that cause major damage to agricultural crops by feeding on plant roots. These bacteria are obligate parasites that need a specific host to complete their life cycle. These are gram-positive, endospore-forming bacteria that parasitize the nematodes and reproduce inside them, leading to their death.

Naturally occurring in the soil and plants, *Bt* produces a crystal-like protein that is toxic to specific insects. Many strains of *Bt* are used primarily to control insect larvae belonging to the order Lepidoptera (moths and butterflies) and increasingly are being used to control mosquitos as well. *Bt* is mostly applied as a spray to the underside of the foliage as the larvae feed essentially in that region. Care should be taken to apply *Bt* away from direct sunlight; it will lose its effectiveness as it tends to break down easily in the presence of sunlight. Commercially, *Bt* was first made available as a biopesticide about 100 years ago; however, it has gained widespread usage since the 1980s when the genetically modified crops were developed using the *Bt* toxins via DNA-recombinant technology. Unfortunately, the overuse of the *Bt* toxin has led to recent reports of some insects developing resistance to it and its specific mode of action. It is imperative that researchers look at developing newer strains of *Bt* with a different mode of action, which may be less susceptible to evolutionary resistance.

Bacteria belonging to the genus *Bacillus* produce a wide array of biologically active molecules that inhibit the growth of plant pathogens. In the case of *B. subtilis*, 4–5% of its genome is responsible for synthesizing antibiotics and may produce about 25 different kinds of molecules. These spore-forming bacteria make excellent contenders for efficient biopesticide development programs. The spores formed by the representatives of the genus *Bacillus* show an elevated level of resistance to the lack of moisture that is important for developing dry and stable products. Some of the examples of Bacillus species commonly used in developing bacterial biopesticides are *Bt*, *Bacillus papilliae*, *Bacillus sphaericus*, etc. Biopesticides are commonly developed by using not only spore-forming bacteria such as those belonging to the Bacillaceae family but also those belonging to the non-spore-forming families like Pseudomonadaceae (e.g., *Pseudomonas rhodesiae*, *Pseudomonas fluorescens*, *Pseudomonas aeruginosa*) and Enterobacteriaceae (e.g., *Enterobacter cloacae*, *Enterobacter aerogenes*). Pseudomonads help in the control of soil-borne plant pathogens as they reside commonly in the rhizosphere. *P. fluorescens*, which colonizes wheat roots, is known to have antifungal activities (Berry et al., 2010).

The mode of action of bacterial biopesticides varies greatly depending on the target pest. In insects, the bacteria disrupt the digestive system by producing an endotoxin that is specific to the target pest. As a control for pathogenic bacteria or fungus, the bacterial biopesticides colonize plants and crowd out the pathogenic species (competition for space and nutrition). The mode of action of *Bt* is divided into three stages: (1) attachment of active crystal protein to the *Bt* spore, ingestion of this combination by a target pest, and its entry into the digestive tract and midgut; (2) detachment of the crystal surrounding the protein and its dissolution and toxin activation at a specific pH; and (3) binding of the released toxin to the receptors in the midgut and initiation of the boring of holes in the host membrane, leading to complete perforation.

After the *Bt* spores are ingested by caterpillars, the bacteria reproduce and produce crystalline toxins. These parasporal bodies enclose crystal-containing toxins, which are activated by the conversion of monomeric protoxins to active delta endotoxins that bind to the receptors in the midgut, creating pores that interfere with the ion transport system, causing complete paralysis and eventual insect death. The toxins rupture the inner cell wall lining of the midgut, leading to the spilling out of the contents into the circulatory system, causing tissue damage, starvation, and eventual death. It may take 12 hours to five days, depending on the amount and type of Bt consumed and the size and species of the specific insect. Besides the delta endotoxins, some of the other toxins involved are hemolysins, enterotoxins, and beta-exotoxins.

In certain instances, insect larvae damage crops like corn, which allows fungal spores to enter and lodge themselves inside the corn tissue and produce mycotoxins that, upon consumption, may cause detrimental effects in both animals and humans. *Bt*-modified hybrid corn crops are able to resist the insects and diseases that they may cause in naturally occurring bacteria such as *Bacillus popilliae*, which cause the milky disease in Japanese beetle larvae and are commonly used to control their propagation.

4.4.2 Fungal Biopesticides

Fungal biopesticides, or myco-biopesticides, comprise the largest group of microorganisms that are pathogenic to insects and act rather quickly, decimating their prey. They provoke fungal infections primarily in Lepidoptera, Homoptera, Coleoptera, Diptera, and Hymenoptera species upon being disseminated into the environment. Fungal biopesticides act as parasites on insects, bacteria, nematodes, weeds, and other fungi by producing biologically active molecules, including enzymes that digest and dissolve plant cell walls, causing death. The fungi used commonly in the development of fungal biopesticides include *Beauveria metarhizium, Paecilomyces nomuraea*, and *Entomophaga zoophthora*.

Unlike in bacterial and viral biopesticides, the efficacy of fungal biopesticides does not depend on their consumption, even though they require a specific range of favorable conditions like moisture and temperature to propagate. As a replacement for methyl bromide, *Muscodor albus* is used in postharvest engineering and technology as a treatment for food, feed, ornamentals, and the flower industry. *M. albus* strain QST 20799 occurs naturally and produces volatile compounds like alcohols, esters, and acids, which help in inhibiting and destroying various microorganisms that cause postharvest and soil-borne diseases.

Entomopathogenic fungi regulate diverse insect populations by penetrating their hosts through the cuticle, gaining access to the hemolymph, producing toxins, and propagating by utilizing the host's nutrients present in the haemocoel. These fungicides are applied as conidia or mycelia, which sporulate upon application. They can also be used in combination with other biopesticides for better results against crop pests. Some examples include Beauveria bassiana, *Lecanicillium lecanii, Nomuraea rileyi, Metarhizium anisopliae*, and *Paecilomyces spp.*

Trichoderma spp., present in most soils, are used as biopesticides and growth-promoting agents for many agricultural crops. They are easily established in different soil types and become an integral part of the soil ecosystem for several months. They act as a jack of all trades, including antagonist, rhizosphere colonizer, plant-growth promoter, and neutralizer of plant pathogen-induced infection.

The mode of action of fungal biopesticides is complex, which makes it highly unlikely that resistance to it could be developed. They use different modes of action like mycoparasitism, competing for nutrients with soil-borne plant pathogens, producing soluble metabolites, hyphal interactions, and producing both volatile and non-volatile compounds. The mode of action depends on the pesticidal fungus as well as the target pest. *B. bassiana* spores germinate and proliferate in the gut of the insect, producing toxins, draining nutrients, and causing eventual death. *Trichoderma* spp. are fungal antagonists that lodge themselves in the main tissues of pathogenic fungi, consume their nutrients, and generate their own spores while disintegrating the cell walls of the host fungi by releasing hydrolytic enzymes. These enzymes require the release of a diffusible factor by the host fungi, which facilitates the physical contact between them. The lectins present in the cell wall of the host prompt the coiling of the fungi around the host hyphae while forming appressoria that penetrate and destroy the pathogen (Ryder & Talbot, 2015).

4.4.3 VIRAL BIOPESTICIDES

Viral biopesticides, or baculoviruses, are pathogens that attack and kill insects and other arthropods. Unlike other microorganisms that are used in the development of biopesticides, these are not classified as living organisms but rather as parasitically replicating microscopic particles. Baculoviruses are composed of double-stranded DNA, the genetic material that is required for establishment and reproduction. Because this genetic material can be easily destroyed by exposure to sunlight or unfavorable conditions in the host's gut, the baculoviruses (virion) are protected by a protein coat referred to as the polyhedron (D'Amico, 2017). Baculoviruses are classified into two main families: granulosis virus (GV) and nucleopolyhedrosis virus (NPV). Their differences emanate from the complex structure and number of the protective protein coats. Baculoviruses are highly host specific, particularly against insects, and have not been reported to have any detrimental effects on other living organisms or surrounding ecosystems and hence are great candidates for the development of biopesticides (D'Amico, 2017). Baculoviruses can be used in place of antibiotics and also help in controlling the host population by inducing severe and sudden outbreaks in their hosts, leading to complete control (D'Amico, 2017). In order to be effective, baculoviruses need to be ingested by the host, which leads to lower efficacy. Baculoviruses are mass produced in vivo: hence, the associated high cost of production. Upon death, the target insect's body is available for other larvae for consumption as part of the ongoing cycle of life and death (Rolff et al., 2019).

Cydia pomonellagranulo virus (CpGV), belonging to the family Baculoviridae, is used for controlling codling moth, a pest that damages fruit trees such as pear and apple, whose larvae, upon coming in contact with the fruit, feed on the virus during the initial stage of their development. CpGV is highly target specific and does not harm other organisms. As little as one CpGV particle can be effective, but they tend to be highly UV sensitive, requiring multiple reapplications with low uptake. In order to be effective, these viruses need to be encapsulated so as not to disintegrate in the presence of moisture (rain) and are sprayed on the eggs before they hatch and get infected, causing eventual death. CpGV are used in both organic and conventional crop management practices and can replace organophosphates and pyrethroids. Bacteriophages are another category of viruses that infect bacteria that cause plant diseases and hence are used as pesticides. The pathogenic bacterium *Xanthamonas* spp. is effectively eradicated by bacteriophages, which can be used effectively as a replacement for antibiotics such as streptomycin; however, the pathogenic bacteria have a tendency to develop resistance to antibiotics.

Baculoviruses are target specific when acting on insects. Nucleopolyhedrosis viruses (NPV) are effective against Lepidoptera (butterflies and moths), Hymenoptera (ants, bees, and wasps), and Diptera (flies) while granulosis viruses (GV) target only Lepidoptera. Baculoviruses develop in the nuclei of the host insect cells. As part of their mode of action, baculoviruses invade the pest's body via the gut, causing an infection that interferes with its physiology, reproduction, and mobility, leading to eventual pest death. Upon being ingested, the viral protein coat disintegrates inside the body of the host, and the DNA particles become physiologically active. Soon, the host struggles with nutrient uptake and succumbs to eventual death (Palma et al., 2014). In the case of cytoplasmic polyhedrosis virus (CPV) or cypoviruses, on the other hand, upon being ingested by the insect pest, the polyhedral are dissolved, releasing the viral particles, which penetrate the insect midgut. Once in the host's cytoplasm, transcription and replication processes are initiated in the columnar epithelial cells. After a full assembly, the progeny viruses are excreted on the foliage, fatally infecting other insects, and the whole cycle repeats itself.

4.4.4 BIOPESTICIDES DERIVED FROM PROTOZOA AND NEMATODES

Microorganisms like protozoa and nematodes are also used as biopesticides in integrated pest management practices. Protozoa are microscopic single-celled organisms that are motile with the help of pseudopodia. They are excellent sources of biopesticides, particularly against many species of grasshopper. However, in the last decade, only one insecticidal protozoan, Nosema spp. has been registered with EPA. Nematodes, on the other hand, are microscopic parasitic organisms and about three dozen are used as insecticides. These biopesticides are not suited for short-term and quick-acting results and are mostly effective against mosquitoes and grasshoppers. As part of their mode of action, upon being ingested by a pest, the microorganisms spread to all the tissues and organs via the midgut and multiply quickly, causing septicemia and eventual insect death. They may interfere with insect reproductive or feeding cycles (competition for space and nutrients) rather than killing the pest right away. For effective results, biopesticides need to be applied before the outbreak of the disease (Migunova & Sasanelli, 2021).

Some of the protozoa commonly used as biopesticides include *Nosema* spp. and *Thelohania vairimorpha*. Similarly, nematodes belonging to the families Steinernematidae and Heterorhabdititae are used as biopesticides, but in association with bacteria belonging to the genus *Xenorhabdus*. The mode of action is such that the nematodes harbor the bacteria in their intestines while the larvae enter the host through natural openings and penetrate the haemocoel (Migunova & Sasanelli, 2021). Nematodes feed on the bacteria, liquefying insects while maturing into adults. The bacteria are then released inside the insect gut, causing septicemia and quick insect death. *Radopholus similis* is an endoparasitic root nematode that infects and damages the cytoplasm while interfering with water and nutrient uptake (Chaves et al., 2009).

4.4.5 BIOPESTICIDES DERIVED FROM NATURAL PRODUCTS

These biopesticides comprise diverse and naturally occurring oils and extracts derived from plant sources and can be used as both bioinsecticides and bioherbicides. Their mode of action varies greatly from product to product. As opposed to insect sex pheromones, which themselves do not kill a target pest but rather interrupt their reproductive cycles, plant oils and extracts act indirectly and nonspecifically. Plant extracts like floral essences lure and attract insects to traps. Sometimes, they are used as repellents and deterrents (e.g., cayenne) by generating foul odors that keep the pests away. Lemongrass oils cause dehydration by removing the waxy coating from the leaves of plants while others extracts cause suffocation. On the other hand, some natural products help crops enhance their immune systems by developing systemic acquired resistance. The insect mating cycle is disrupted due to the presence of higher concentrations of sex pheromones in the surroundings, which may throw the males off the scent of females, thus reducing/limiting their ability to mate (Ferracini et al., 2021).

Insect growth regulators from natural products may be regulated as biopesticides or conventional pesticides. Neem (*Azadirachta indica*), native to India and other South Asian countries, and its constituent azadirachtin, are considered biopesticides.

The Neem products offer broad-spectrum activity as they infect insects, mites, nematodes, fungi, bacteria, and viruses. Azadirachtin (limonoids) is a highly effective insect growth regulator, and it indirectly kills by altering the life cycle of the insect, which can no longer feed, breed, or undergo metamorphosis (Nicoletti, 2020). Azadirachtin disrupts molting by inhibiting biosynthesis and metabolism of the juvenile molting hormone, ecdysone (Ferracini et al., 2021).

Rotenone is another natural product–based biopesticide: a colorless, odorless, and crystalline ketone harvested from the stem and seeds of the jicama vine (*Pachyrhizus erosus*) and the roots of Fabaceae and has been used for decades as an insecticide (Zhang et al., 2022). Rotenone behaves in a pyrethrin-like manner but is stronger and more persistent (Richardson et al., 2019). It is an effective neurotoxin against insects like aphids, suckers, and thrips that infest fruits and vegetables and acts by interfering with their reproductive cycle (Zhao et al., 2022). Upon being ingested by the Lepidopteran insect larvae, rotenone interferes with nutrient uptake, causing eventual death (Sun et al., 2021). Rotenone is also used in fisheries management, including the treatment of rivers and river systems to exterminate parasites like *Gyrodactylus salaris*, a common pest in North Atlantic salmon and trout (Davidsen et al., 2013).

4.5 CONCLUDING REMARKS

Climate change continues to be a challenge for humanity. Developing sustainable agricultural practices that are safer for humans, livestock, and the environment has never been more important and urgent. Biopesticides present an excellent tool for farmers in keeping their crops and animals safe while maintaining a healthy environment for present and future generations. Biopesticides are typically developed from bacteria, fungi, viruses, protozoa, nematodes, and plant products and are used to control different kinds of pests. They are target specific and pose no or negligible risk to humans, animals, and the environment. Even though they are the natural manifestation of a substance, they do not necessarily have completely non-toxic modes of action and/or are completely safe for the environment, so it is advisable not to use them indiscriminately. The mode of action of various biopesticides is based on infection of the pest, physiological starvation, and death (Table 4.1). The use of biopesticides instead of synthetic pesticides has led to healthier and safer agricultural practices.

REFERENCES

Arthurs, S., & Dara, S. K. (2019). Microbial biopesticides for invertebrate pests and their markets in the United States. *Journal of Invertebrate Pathology*, *165*, 13–21. https://doi.org/10.1016/j.jip.2018.01.008

Berberian, A. G., Gonzalez, D. J. X., & Cushing, L. J. (2022). Racial disparities in climate change-related health effects in the United States. *Current Environmental Health Reports*, *9*(3), 451–464. https://doi.org/10.1007/s40572-022-00360-w

Berry, C., Fernando, W. G. D., Loewen, P. C., & de Kievit, T. R. (2010). Lipopeptides are essential for *Pseudomonas* sp. DF41 biocontrol of *Sclerotinia sclerotiorum*. *Biological Control*, *55*, 211–218.

Brevik, K., Schoville, S. D., Mota-Sanchez, D., & Chen, Y. H. (2018). Pesticide durability and the evolution of resistance: A novel application of survival analysis. *Pest Management Science*, 74(8), 1953–1963. https://doi.org/10.1002/ps.4899

Bueno, A. F., Panizzi, A. R., Hunt, T. E., Dourado, P. M., Pitta, R. M., & Gonçalves, J. (2021). Challenges for adoption of integrated pest management (IPM): The soybean example. *Neotropical Entomology*, 50(1), 5–20. https://doi.org/10.1007/s13744-020-00792-9

Carvalho, F. P. (2017). Pesticides, environment, and food safety. *Food and Energy Security*, 6(2), 48–60. https://doi.org/10.1002/fes3.108

Chaves, N. P., Pocasangre, L. E., Elango, F., Rosales, F. E., & Sikora, R. (2009). Combining endophytic fungi and bacteria for the biocontrol of *Radopholus similis* (Cobb) Thorne and for effects on plant growth. *Scientia Horticulturae*, 122, 472–478.

Cox, C., & Zeiss, M. (2022). Health, pesticide adjuvants, and inert ingredients: California case study illustrates need for data access. *Environmental Health Perspectives*, 130(8), 85001. https://doi.org/10.1289/EHP10634

D'Amico, V. (2017). *Baculoviruses in biological control: A guide to natural enemies in North America*. Ithaca, NY: Cornell University Press.

Davidsen, J. G., Thorstad, E. B., Baktoft, H., Aune, S., Økland, F., & Rikardsen, A. H. (2013). Can sea trout *Salmo trutta* compromise successful eradication of *Gyrodactylus salaris* by hiding from CFT legumin (rotenone) treatments? *Journal of Fish Biology*, 82, 1411–1418.

Deguine, J. P., Aubertot, J. N., Flor, R. J., & Lescourret, F. (2021). Integrated pest management: Good intentions, hard realities. A review. *Agronomy for Sustainable Development*, 41(38). https://doi.org/10.1007/s13593-021-00689-w

Deshayes, C., Siegwart, M., Pauron, D., Froger, J. A., Lapied, B., & Apaire-Marchais, V. (2017). Microbial pest control agents: Are they a specific and safe tool for insect pest management? *Current Medicinal Chemistry*, 24(27), 2959–2973. https://doi.org/10.2174/0929867324666170314144311

Ebi, K. L., Vanos, J., Baldwin, J. W., Bell, J. E., Hondula, D. M., Errett, N. A., Hayes, K., Reid, C. E., Saha, S., Spector, J., & Berry, P. (2021). Extreme weather and climate change: Population health and health system implications. *Annual Review of Public Health*, 42, 293–315. https://doi.org/10.1146/annurev-publhealth-012420-105026

Environmental Protection Agency (EPA) (2022). *Federal insecticide, fungicide, and rodenticide act and federal facilities*. Retrieved on December 17, 2022 from www.epa.gov/enforcement/federal-insecticide-fungicide-and-rodenticide-act-fifra-and-federal-facilities

Fenibo, E. O., Ijoma, G. N., & Matambo, T. (2021). Biopesticides in sustainable agriculture: A critical sustainable development driver governed by green chemistry principles. *Frontiers in Sustainable Food Systems*, 5, 619058. https://doi:10.3389/fsufs.2021.619058

Ference, C. M., Gochez, A. M., Behlau, F., Wang, N., Graham, J. H., & Jones, J. B. (2018). Recent advances in the understanding of Xanthomonas citri ssp. citri pathogenesis and citrus canker disease management. *Molecular Plant Pathology*, 19(6), 1302–1318. https://doi.org/10.1111/mpp.12638

Ferracini, C., Pogolotti, C., Rama, F., Lentini, G., Saitta, V., Mereghetti, P., Mancardi, P., & Alma, A. (2021). Pheromone-mediated mating disruption as management option for *Cydia* spp. in chestnut orchard. *Insects*, 12(10), 905. https://doi.org/10.3390/insects12100905

Glare, T. R., Gwynn, R. L., & Moran-Diez, M. E. (2016). Development of biopesticides and future opportunities. *Methods in Molecular Biology (Clifton, N.J.)*, 1477, 211–221. https://doi.org/10.1007/978-1-4939-6367-6_16

Gomez-Zavaglia, A., Mejuto, J. C., & Simal-Gandara, J. (2020). Mitigation of emerging implications of climate change on food production systems. *Food Research International (Ottawa, Ontario)*, 134, 109256. https://doi.org/10.1016/j.foodres.2020.109256

Hawkins, N. J., Bass, C., Dixon, A., & Neve, P. (2018). The evolutionary origins of pesticide resistance. *Biological Reviews of the Cambridge Philosophical Society, 94*(1), 135–155. https://doi.org/10.1111/brv.12440

Hernández-Ruiz, J., & Arnao, M. (2018). Relationship of melatonin and salicylic acid in biotic/abiotic plant stress responses. *Agronomy, 8*(4), 33. http://dx.doi.org/10.3390/agronomy8040033

Jansen, H. J., Breeveld, F. J., Stijnis, C., & Grobusch, M. P. (2014). Biological warfare, bioterrorism, and biocrime. *Clinical Microbiology and Infection: The official publication of the European Society of Clinical Microbiology and Infectious Diseases, 20*(6), 488–496. https://doi.org/10.1111/1469-0691.12699

Jansson, J. K., & Hofmockel, K. S. (2020). Soil microbiomes and climate change. *Nature Reviews Microbiology, 18*, 35–46. https://doi.org/10.1038/s41579-019-0265-7

Jouzani, G. S., Valijanian, E., & Sharafi, R. (2017). *Bacillus thuringiensis*: A successful insecticide with new environmental features and tidings. *Applied Microbiology and Biotechnology, 101*(7), 2691–2711. https://doi.org/10.1007/s00253-017-8175-y

Jurat-Fuentes, J. L. (n.d.). *Characterization of cry insecticidal protein mode of action.* Retrieved on December 17, 2022 from http://juratfuenteslab.utk.edu/Btresearchtable.html

Kadoić Balaško, M., Bažok, R., Mikac, K. M., Lemic, D., & Pajač Živković, I. (2020). Pest management challenges and control practices in codling moth: A review. *Insects, 11*(1), 38. https://doi.org/10.3390/insects11010038

Kaka, H., Opute, P. A., & Maboeta, M. S. (2021). Potential impacts of climate change on the toxicity of pesticides towards earthworms. *Journal of Toxicology, 2021*, 8527991. https://doi.org/10.1155/2021/8527991

Kaze, M., Brooks, L., & Sistrom, M. (2021). Antimicrobial resistance in *Bacillus*-based biopesticide products. *Microbiology (Reading, England), 167*(8). https://doi.org/10.1099/mic.0.001074

Kemp, L., Xu, C., Depledge, J., Ebi, K. L., Gibbins, G., Kohler, T. A., Rockström, J., Scheffer, M., Schellnhuber, H. J., Steffen, W., & Lenton, T. M. (2022). Climate endgame: Exploring catastrophic climate change scenarios. *Proceedings of the National Academy of Sciences of the United States of America, 119*(34), e2108146119. https://doi.org/10.1073/pnas.2108146119

Kristiansen, S., Painter, J., & Shea, M. (2021). Animal agriculture and climate change in the US and UK elite media: Volume, responsibilities, causes and solutions. *Environmental Communication, 15*(2), 153–172. https://doi.org/10.1080/17524032.2020.1805344

Lai, J., Cao, X., Yu, T., Wang, Q., Zhang, Y., Zheng, X., & Lu, H. (2018). Effect of *Cryptococcus laurentii* on inducing disease resistance in cherry tomato fruit with focus on the expression of defense-related genes. *Food Chemistry, 254*, 208–216. https://doi.org/10.1016/j.foodchem.2018.01.100

Leftwich, P. T., Koukidou, M., Rempoulakis, P., Gong, H. F., Zacharopoulou, A., Fu, G., Chapman, T., Economopoulos, A., Vontas, J., & Alphey, L. (2014). Genetic elimination of field-cage populations of Mediterranean fruit flies. *Proceedings of the Royal Society of Biological Sciences, 281*(1792), 20141372. https://doi.org/10.1098/rspb.2014.1372

Leinonen, I., & Kyriazakis, I. (2016). How can we improve the environmental sustainability of poultry production? *The Proceedings of the Nutrition Society, 75*(3), 265–273. https://doi.org/10.1017/S0029665116000094

Li, Y., Wang, C., Ge, L., Hu, C., Wu, G., Sun, Y., Song, L., Wu, X., Pan, A., Xu, Q., Shi, J., Liang, J., & Li, P. (2022). Environmental behaviors of *Bacillus thuringiensis* (Bt) insecticidal proteins and their effects on microbial ecology. *Plants (Basel, Switzerland), 11*(9), 1212. https://doi.org/10.3390/plants11091212

Migunova, V. D., & Sasanelli, N. (2021). Bacteria as biocontrol tool against phytoparasitic nematodes. *Plants (Basel, Switzerland), 10*(2), 389. https://doi.org/10.3390/plants10020389

Murray, G. G. R., Balmer, A. J., Herbert, J., Hadjirin, N. F., Kemp, C. L., Matuszewska, M., Bruchmann, S., Hossain, A. S. M. M., Gottschalk, M., Tucker, A. W., Miller, E., & Weinert, L. A. (2021). Mutation rate dynamics reflect ecological change in an emerging zoonotic pathogen. *PLOS Genetics, 17*(11), e1009864. https://doi.org/10.1371/journal.pgen.1009864

Nicoletti, M. (2020). New solutions using natural products. *Insect-Borne Diseases in the 21st Century*, 263–351. https://doi.org/10.1016/B978-0-12-818706-7.00007-3

Nicolopoulou-Stamati, P., Maipas, S., Kotampasi, C., Stamatis, P., & Hens, L. (2016). Chemical pesticides and human health: The urgent need for a new concept in agriculture. *Frontiers in Public Health, 4*, 148. https://doi.org/10.3389/fpubh.2016.00148

Palma, L., Muñoz, D., Berry, C., Murillo, J., & Caballero, P. (2014). *Bacillus thuringiensis* toxins: An overview of their biocidal activity. *Toxins, 6*(12), 3296–3325. https://doi.org/10.3390/toxins6123296

Pimentel, D. (1989). *Biopesticides and the environment*. Retrieved on December 17, 2022 from https://ecommons.cornell.edu/handle/1813/49659

Qin, G. Z., Tian, S., Xu, Y., & Wan, Y. K. (2003). Enhancement of biocontrol efficacy of antagonistic yeasts by salicylic acid in sweet cherry fruit. *Physiological and Molecular Plant Pathology, 62*(3), 147–154. https://doi.org/10.1016/S0885-5765(03)00046-8

Raza, A., Razzaq, A., Mehmood, S. S., Zou, X., Zhang, X., Lv, Y., & Xu, J. (2019). Impact of climate change on crops adaptation and strategies to tackle its outcome: A review. *Plants (Basel, Switzerland), 8*(2), 34. https://doi.org/10.3390/plants8020034

Richardson, J. R., Fitsanakis, V., Westerink, R. H. S., & Kanthasamy, A. G. (2019). Neurotoxicity of pesticides. *Acta Neuropathologica, 138*(3), 343–362. https://doi.org/10.1007/s00401-019-02033-9

Rizzo, D. M., Lichtveld, M., Mazet, J. A. K., Togami, E., & Miller, S. A. (2021). Plant health and its effects on food safety and security in a One Health framework: Four case studies. *One Health Outlook, 3*(6). https://doi.org/10.1186/s42522-021-00038-7

Rohr, J. R., Barrett, C. B., Civitello, D. J., Craft, M. E., Delius, B., DeLeo, G. A., Hudson, P. J., Jouanard, N., Nguyen, K. H., Ostfeld, R. S., Remais, J. V., Riveau, G., Sokolow, S. H., & Tilman, D. (2019). Emerging human infectious diseases and the links to global food production. *Nature Sustainability, 2*(6), 445–456. https://doi.org/10.1038/s41893-019-0293-3

Rolff, J., Johnston, P. R., & Reynolds, S. (2019). Complete metamorphosis of insects. *Philosophical Transactions of the Royal Society of London. Series B, Biological Sciences, 374*(1783), 20190063. https://doi.org/10.1098/rstb.2019.0063

Romanazzi, G., Orçonneau, Y., Moumni, M., Davillerd, Y., & Marchand, P. A. (2022). Basic substances, a sustainable tool to complement and eventually replace synthetic pesticides in the management of pre and postharvest diseases: Reviewed instructions for users. *Molecules (Basel, Switzerland), 27*(11), 3484. https://doi.org/10.3390/molecules27113484

Ryder, L. S., & Talbot, N. J. (2015). Regulation of appressorium development in pathogenic fungi. *Current Opinion in Plant Biology, 26*, 8–13. https://doi.org/10.1016/j.pbi.2015.05.013

Sacco, A. D., Hardwick, K. A., Blakesley, D., Brancalion, P. H. S., Breman, E., Rebola, L. C., Chomba, S., Dixon, K., Elliott, S., Ruyonga, G., Shaw, K., Smith, P., Smith, R. J., & Antonelli, A. (2020). Ten golden rules for reforestation to optimize carbon sequestration, biodiversity recovery and livelihood benefits. *Global Change Biology, 27*, 1328–1348. https://doi.org/10.1111/gcb.15498

Schleiffer, M., & Speiser, B. (2022). Presence of pesticides in the environment, transition into organic food, and implications for quality assurance along the European organic food chain—A review. *Environmental Pollution (Barking, Essex: 1987), 313*, 120116. https://doi.org/10.1016/j.envpol.2022.120116

Seiber, J. N., Coats, J., Duke, S. O., & Gross, A. D. (2014). Biopesticides: State of the art and future opportunities. *Journal of Agricultural and Food Chemistry, 62*(48), 11613–11619. https://doi.org/10.1021/jf504252n

Siegwart, M., Graillot, B., Blachere Lopez, C., Besse, S., Bardin, M., Nicot, P. C., & Lopez-Ferber, M. (2015). Resistance to bio-insecticides or how to enhance their sustainability: A review. *Frontiers in Plant Science*, *6*, 381. https://doi.org/10.3389/fpls.2015.00381

Skendžic, S., Zovko, M., Živkovic, I. P., Lešic, V., & Lemic, D. (2021). The impact of climate change on agricultural insect pests. *Insects*, *12*, 440. https://doi.org/10.3390/insects12050440

Souto, A. L., Sylvestre, M., Tölke, E. D., Tavares, J. F., Barbosa-Filho, J. M., & Cebrián-Torrejón, G. (2021). Plant-derived pesticides as an alternative to pest management and sustainable agricultural production: Prospects, applications and challenges. *Molecules (Basel, Switzerland)*, *26*(16), 4835. https://doi.org/10.3390/molecules26164835

Stein, A. J., & Santini, F. (2022). The sustainability of "local" food: A review for policymakers. *Review of Agricultural, Food and Environmental Studies*, *103*, 77–89. https://doi.org/10.1007/s41130-021-00148-w

Stejskal, V., Vendl, T., Aulicky, R., & Athanassiou, C. (2021). Synthetic and natural insecticides: Gas, liquid, gel and solid formulations for stored-product and food-industry pest control. *Insects*, *12*(7), 590. https://doi.org/10.3390/insects12070590

Sun, Z., Xue, L., Li, Y., Cui, G., Sun, R., Hu, M., & Zhong, G. (2021). Rotenone-induced necrosis in insect cells via the cytoplasmic membrane damage and mitochondrial dysfunction. *Pesticide Biochemistry and Physiology*, *173*, 104801. https://doi.org/10.1016/j.pestbp.2021.104801

Šunjka, D., & Mechora, Š. (2022). An alternative source of biopesticides and improvement in their formulation-recent advances. *Plants (Basel, Switzerland)*, *11*(22), 3172. https://doi.org/10.3390/plants11223172

Tanentzap, A. J., Lamb, A., Walker, S., & Farmer, A. (2015). Resolving conflicts between agriculture and the natural environment. *PLOS Biology*, *13*(9), e1002242. https://doi.org/10.1371/journal.pbio.1002242

Teem, J. L., Alphey, L., Descamps, S., Edgington, M. P., Edwards, O., Gemmell, N., Harvey-Samuel, T., Melnick, R. L., Oh, K. P., Piaggio, A. J., Saah, J. R., Schill, D., Thomas, P., Smith, T., & Roberts, A. (2020). Genetic biocontrol for invasive species. *Frontiers in Bioengineering and Biotechnology*, *8*, 452. https://doi.org/10.3389/fbioe.2020.00452

Tickner, J. A., Simon, R. V., Jacobs, M., Pollard, L. D., and van Bergen, S. K. (2021). The nexus between alternatives assessment and green chemistry: Supporting the development and adoption of safer chemicals. *Green Chemistry Letters and Reviews*, *14*, 21–42. https://doi.org/10.1080/17518253.2020.1856427

Toussaint, M., Cabanelas, P., & Muñoz-Dueñas, P. (2022). Social sustainability in the food value chain: What is and how to adopt an integrative approach? *Quality & Quantity*, *56*, 2477–2500. https://doi.org/10.1007/s11135-021-01236-1

United Nations (UN). (2022). *The sustainable development goals report 2022*. Retrieved on December 17, 2022 from www.un.org/sustainabledevelopment/progress-report/

Wahab, S., Muzammil, K., Nasir, N., Khan, M. S., Ahmad, M. F., Khalid, M., Ahmad, W., Dawria, A., Reddy, L. K. V., & Busayli, A. M. (2022). Advancement and new trends in analysis of pesticide residues in food: A comprehensive review. *Plants (Basel, Switzerland)*, *11*(9), 1106. https://doi.org/10.3390/plants11091106

Zhang, H., Ma, L., Jiang, S., Lin, H., Zhang, X., Ge, L., & Xu, Z. (2010). Enhancement of biocontrol efficacy of *Rhodotorula glutinis* by salicylic acid against gray mold spoilage of strawberries. *International Journal of Food Microbiology*, *141*, 122–125.

Zhang, P., Zhang, M., Mellich, T. A., Pearson, B. J., Chen, J., & Zhang, Z. (2022). Variation in rotenone and deguelin contents among strains across four *Tephrosia* species and their activities against aphids and whiteflies. *Toxins*, *14*(5), 339. https://doi.org/10.3390/toxins14050339

Zhao, J., Liang, D., Li, W., Yan, X., Qiao, J., & Caiyin, Q. (2022). Research progress on the synthetic biology of botanical biopesticides. *Bioengineering (Basel, Switzerland)*, *9*(5), 207. https://doi.org/10.3390/bioengineering9050207

Ziska, L. H., Bradley, B. A., Wallace, R. D., Bargeron, C. T., LaForest, J. H., Choudhury, R. A., Garrett, K. A., & Vega, F. E. (2018). Climate change, carbon dioxide, and pest biology, managing the future: Coffee as a case study. *Agronomy, 8*(8), 152. http://dx.doi.org/10.3390/agronomy8080152

Zúñiga-Venegas, L. A., Hyland, C., Muñoz-Quezada, M. T., Quirós-Alcalá, L., Butinof, M., Buralli, R., Cardenas, A., Fernandez, R. A., Foerster, C., Gouveia, N., Gutiérrez Jara, J. P., Lucero, B. A., Muñoz, M. P., Ramírez-Santana, M., Smith, A. R., Tirado, N., van Wendel de Joode, B., Calaf, G. M., Handal, A. J., Soares da Silva, A., . . . Mora, A. M. (2022). Health effects of pesticide exposure in Latin American and the Caribbean populations: A scoping review. *Environmental Health Perspectives, 130*(9), 96002. https://doi.org/10.1289/EHP9934

5 Biopesticides Regulatory Schemes

Kumari Neha, Faraat Ali, Hasan Ali,
Neelam Singh, Manisha Trivedi, and
Arvind Kumar Sharma

CONTENTS

5.1 BIOPESTICIDE REGULATIONS, LEGISLATION, AND ORGANIZATIONS

Biopesticides (microbial pesticides, bio-derived compounds, and plant-incorporated protectants) have distinct and complex mechanisms of action that allow them to compete with chemical pesticides in plant productivity. Biopesticides include a diverse variety of living and non-living creatures that differ significantly in basic qualities such as composition, mechanism of action, fate and behavior in the environment, and so on. Governments categorize them together in order to regulate their authorization and use. These laws are in place to protect human and environmental safety, as well as to describe goods and ensure that manufacturers provide biopesticides of consistent and trustworthy quality. Biopesticides can control crop pests effectively with minimal environmental impact when used as part of an integrated pest management (IPM) programme. However, biopesticide regulation is managed by a system established for chemical pesticides, which might operate as a barrier to investment in biopesticide research and development (R&D).

There are many systemic problems arising in regulation of biopesticides, and one main problem is that it is based on the conventional chemical pesticide models

According to Chandler et al. [1], in the EU system, the regulatory failure arises from the application of an inappropriate synthetic pesticide model and lack of regulatory innovation. So, Chandler et al. [1] studied two novel biopesticide regulatory frameworks in the United Kingdom and the Netherlands that could assist in overcoming this hurdle.

Farmers are attempting to use less conventional pesticide in response to requests from merchants and customers for more ecologically friendly pesticide alternatives. To maintain yields and productivity, they require ongoing access to a wide variety of plant protection chemicals. In a recent UK study, a team of political and scientific specialists analyzed the obstacles to the use of biopesticides by speaking with regulators, biopesticide producers, farmers, retailers, EU officials, and environmental organizations.

This resulted in revisions to biopesticide regulation in the UK in order to boost research and development [2]. The Pesticides Safety Directorate (PSD) has selected a "biopesticides champion." The champion is a person who acts as a point of contact for biopesticide manufacturers during the product approval process. A pilot program offering discounted registration fees for new biopesticides was also established. This recent study compares UK advances to those of the Genoeg scheme in the Netherlands [3], which likewise aimed to enhance the use of biopesticides. The study suggests that an improved comprehension of biopesticides' modes of action and impacts, as well as the regulatory challenges that occur during their adoption, may help increase their profile among policymakers, allowing them to realize their potential contribution to sustainable crop production.

The regulator's needed biopesticide registration data portfolio is often a modified version of the one in existence for conventional chemical pesticides and is utilized to conduct a risk assessment. It contains details on the mechanism of action, toxicological and eco-toxicological evaluations, host range testing, and so on. This information is costly for businesses to produce, and it may discourage them from commercializing biopesticides, which are often niche market items. As a result, the regulator's job is to put in place an adequate system for biopesticides that ensures their safety and uniformity while not impeding commercialization [4].

Registration is one of the most expensive aspects of commercializing bioherbicides [5]. Pesticide registration is governed by various institutions in various nations; see the following paragraphs. In general, there has been a push to expedite biopesticide registration and harmonize regulations across nations [6, 7]. Threats to human health must be assessed, including toxicological issues, non-target impacts, and the dissemination of the proposed biopesticide [8–10]. Companies must consider all these costs when considering the commercialization of a bioherbicide.

The inaugural Australia and New Zealand Biocontrol Conference, held in Sydney in February 2008, inspired a special issue [11] of *Biological Control*. This special issue represented the perspectives of a varied group of biological control researchers, suppliers, and practitioners. The goal of this conference and this collection of papers is to help biological control personnel understand the vast range of factors that influence the availability of biological control agents and the adoption of IPM systems that use them. The biological knowledge of which species of agent may have an influence on a specific target is simply one component of the information set required for effective biological control to be implemented in a commercial agricultural set-ting. Sundh and Goettel [12] explore the ecological principles of biological con-trol, potential dangers and threats, and biocontrol agent safeness and legislation. The study concludes that the possible environmental dangers for microbial and microbial control agents are identical and are best assessed using biological and ecological concepts.

The fundamental distinction between invertebrate biocontrol agents (IBCAs) and microbial biocontrol agents (MBCAs) in terms of potential human dangers is that some bacteria can cause infectious disease. Although the absence of pathogenicity in new environmental isolates may never be 100% established, existing standards pro-vide acceptable assurance. Furthermore, attention to the safety of producers, makers, and applicators can presumably be ensured in part by labor laws for working with microbes and to consumers by food safety regulations. Both practical practice of biological control and scientific awareness of the nature of potential risks to humans or the environment associated with the use of MBCAs have existed for a long time but were given insufficient attention in the 1970s and 1980s when MBCAs (but not IBCAs) were incorporated into existing frameworks for chemical pesticides.

Regulation of MBCAs within chemical systems has stifled the development of new microbial pesticide alternatives, particularly in Europe. It is debatable if the shared resources used for MBCA approval target the most relevant dangers, which can be a societal issue. Furthermore, the long and expensive registration processes have been poorly matched with the small specialty enterprises creating and manufacturing MBCAs. Several initiatives have been launched to improve the regulatory environ-ment for MBCAs by simplifying their registration within existing frameworks. How-ever, as long as these microbes are treated (and controlled) as "pesticides," with many similarities to chemical pesticides, the risk of undue emphasis on chemical threats and thus difficult and confusing safety evaluations is maintained.

The new EU regulatory framework for plant protection goods requires IPM, which is projected to raise consumption for biocontrol products. Despite governments' gen-eral encouragement of risk-reduced pest management and biocontrol techniques, we suggest that there has not been enough effort to address this through more appro-priate regulatory frameworks. Concerning the current scenario, risk assessments of MBCAs could place less emphasis on the toxicology and ecotoxicology of specific groups of metabolites and instead be based on a characterization of biological/path-ogenic properties and ecology of the organism, as well as realistic circumstances for a potential advancement in human and environmental exposure.

The authors propose the following as future priorities:

1. They highly endorse the notion of developing new regulatory oversight sys-tems for MBCAs, possibly a single system for all live BCAs, based on the biological and ecological features of the organism and realistic exposure scenarios for the intended application. An idea for a harmonized European system for IBCA regulation [13] could serve as a starting point for a new framework for evaluating the environmental safety of MBCAs.

   Present laws for microorganisms used for other purposes, such as biore-mediation or plant growth encouragement, can also be a good starting point.

2. Latest data standards, test protocols, and recommendations for evaluating the possible impact of MBCAs on human health or the environment are urgently needed. These should be predicated more on actual risks posed by these species and should take the latest improvements in knowledge of ecological principles of biocontrol and microbial ecology in general into account.

Further research is required on:

1. The degree to which MBCA use contributes to increased total exposure of humans or other species in target habitats, including a comparison to "normal" background microbe exposure.
2. The ecology and biogeography of microbial groups with high biocontrol potential, giving a foundation for determining how to appropriately use the words *native/indigenous* versus *alien/exotic* to biocontrol microorganisms.
3. Creation of more efficient and dependable methodologies for determining the potential human toxicity and pathogenicity (or lack thereof) of MBCA candidates.

The Biopesticide Industry Alliance (BPIA) is focused on promoting the use of biopesticide technology by raising awareness of its effectiveness and the full spectrum of benefits to a proactive pest control program [14]. The reader can access several documents concerning the EPA on the webpage Regulatory Resources.

The primary goals of the International Biocontrol Manufacturers' Association (IBMA's) are [15]:

- To bring together all enterprises, organizations, and individuals involved in the development and application of biocontrol activities.
- To serve as the industry's spokesperson in interactions with national and international organizations, policymakers, the media, and the general public.
- To establish a platform for IBMA members to exchange ideas and discuss matters of mutual interest. Position papers are produced and brought to the attention of the parties involved as a result of this exchange of opinions.
- To establish and enforce ethical professional guidelines.
- To meet market-required product quality requirements.
- To helping expand and intensify the application of biological crop protection, animal health, and public hygiene.
- To provide training and information at the request of member firm staff in order to improve their abilities and, as a result, their business performance.
- To provide the possibility to perform joint initiatives that meet the needs of IBMA members.
- To ensure the exchange of information among all relevant parties.

Biopesticide organizations who collaborate with IBMA on a regular basis include:

- Aarhus Convention Working Group [16]
- Association for Promotion of Sustainable Agriculture (APAD) [17]

- EU DG SANCO [18]
- Food and Agriculture Organisation of the United Nations (FAO) [19]
- Forshungsinstitut für Biologische Landbau (FIBL) [20]
- GLOBALGAP [21]
- Internatioanl Federation of Organic Agriculture Movements (IFOAM) [22]
- Global International Organisation for Biological Control (IOBC) [23]
- Institut Technique de l'Agriculture Biologique (ITAB) [24]
- New Ag International [25]
- OECD Work on Biological Pesticides [26]

5.2 THE REGULATION SYSTEM OF THE UNITED KINGDOM (UK)

The Chemicals Regulation Directorate (CRD) of the Health and Safety Executive (HSE) manages European and national permissioning programmes for biocides and non-agricultural pesticides for the UK government [27]. These authorization processes are in place to ensure that possible risks from these compounds to people and the environment are effectively controlled. The UK Competent Authority (CA) for the implementation of the EU Biocides Regulation (EU-BPR) (528/2012) is the HSE. The regulations must be implemented into national legislation by each EU member state. In the United Kingdom, this is governed by the Biocidal Products Regulations 2001 (BPR) and the BPR (Northern Ireland) 2001. The BPR was created in accordance with the Health and Safety at Work Act of 1974. (HSWA). More data about CRD's biocides and non-agricultural pesticides activities may be accessed on the HSE's official site [27].

Any pesticide must be authorized for use before it can be used, sold, supplied, promoted, or stored. The term *biopesticides* refers to a wide range of possible plant protection chemicals; however, for the purposes of the HSE's biopesticide system, they are classified into four categories: products based on pheromones and other semi chemicals (for mass trapping or trap farming); products containing a microorganism (e.g., bacterium, fungus, protozoa, virus, viroid); products based on plant extracts; and other unique alternative products

The HSE grants approvals on behalf of ministers under a variety of pesticide-related legislation.

Before an approval can be granted, applicants supporting products must offer evidence that pesticides are safe and effective. Biopesticides are included. The reader is directed to ref. [13] for details.

To be placed on the market, a biocidal substance or family must be approved. The precise active chemicals in the product determine when the conditions for authorization (national authorization, union authorization, simplified authorization, or permission under EU Biocides Regulations (EU BPR) apply to the product. In general, the EU BPR authorization necessities will adapt to a product if the active substance(s) in the product are listed in the EU BPR Article 9 Approved List of Active Substances (union list) [28], and the admittance includes the product type [29] acceptable for that product.

The EU Biocides Regulation (Regulation 528/2012) [30] governs a wide range of products, comprising disinfectants, pesticides, and stabilizers. From September 1, 2013, it revoked and replaced the Biocidal Products Directive 98/8/EEC (the BPD) and the associated UK Biocidal Products Regulations (BPR) [31]. The Management of Pesticides Regulations (COPR) [32] is an earlier UK national scheme that regulates a wide range of products for pest control, including bioactive components that are not yet regulated by BPR. COPR-controlled products are gradually being absorbed into the scope of the Biocidal Products Regulations, with COPR finally becoming obsolete.

5.3 THE UNITED STATES (US) REGULATION SYSTEM

Pesticides are regulated by the Environmental Protection Agency (EPA) in the United States to ensure they do not harm individuals or the environment. The EPA's Biopesticides and Pollution Prevention Division (BPPD) oversees all regulatory actions involving biopesticides. Biopesticides are classified as a regulatory category, and different data are needed than for conventional chemicals. The EPA classifies biopesticides as "lower risk pesticides," meaning they pose a lower risk than conventional pesticides. The EPA normally requires fewer data for biopesticides than for conventional pesticides (most pheromones, for example, are exempt from testing), and biopesticides may be excluded from residual tolerance thresholds. This third argument, however, is debatable, since it has the potential to stymie global trade to regions with the highest residue levels in food (such as the EU), decreasing the potential economic potential of novel biopesticides and, hence, their inclusion in global crop protection programs. Despite their natural origin, biopesticides are not always harmless: for example, the development of plant-produced poisons triggered by biopesticides. According to ref. [33], before a pesticide can be commercialized and utilized in the United States, the Federal Insecticide, Fungicide, and Rodenticide Act (FIFRA) demands that the EPA analyze the suggested pesticide to ensure that its use will not pose unfair risks or harm to either the environment or human safety. This legislation necessitates a thorough examination of health and safety information. Biopesticides comprise naturally existing compounds that manage pests (biochemical pesticides), microorganisms that regulate pests (microbial pesticides), and pesticidal substances created by plants with incorporated genetic material (plant-incorporated protectants, or PIPs).

The *Pesticide Registration Manual* (Blue Book) and Biopesticide Registration Tools links provide access to resources to help applicants. Each of the biopesticide's active constituents is described in detail in the Fact Sheets sections. The Product Lists section contains numerous lists of specific biopesticide products to help the public identify the best one for their pest problem. Finally, the PIPs section has detailed information on the management of genetically altered plants.

- **Biopesticide Registration Tools**

The government pre-marketing authorization of pesticides, known as registration, is a difficult process. The basic authentication process related to the registration of biopesticides is addressed.

- **Biopesticide Active Ingredient Fact Sheets**

This summary report collection offers chemical-specific information regarding biopesticides' active components. A full technical report, bibliographies, legislative history, Federal Register announcements, and/or registrant and product lists are also included in some of the fact sheets.

- **Biopesticide Ingredient and Product Lists**

There are lists of EPA-certified (approved) biopesticide components and products, as well as listings of biopesticides' active components by year first accepted.

- **Plant-Incorporated Protectants (PIPs)**

Plant-incorporated protectants are pesticides generated by plants as well as the genetic materials needed for the plant to manufacture the chemical. For example, scientists can transfer the gene for a specific *Bt* pesticidal protein into the plant's genetic code. The plant subsequently produces pesticidal protein, which suppresses the pest when it feasts on the plant. The EPA regulates both the protein and its gene product ingredients; the plant itself is not regulated.

The reader is directed to ref. [33] for more details.

5.4 THE EUROPEAN REGULATION SYSTEM

Regulations is implemented in tandem with other EU regulations and directives (for example, the Regulation on Maximum Residue Levels (MRLs) in Food [Regulation 396/2005]) and the Directive on Sustainable Pesticide Use [Directive 2009/128/EC]). The EU Biocides Regulation 528/2012 (EU BPR) [34] replaces Directive 98/8/EC and applies to a wide range of products, comprising disinfectants, pesticides, and stabilizers. It also includes links to the EU BPR data protection, data sharing laws, and alternative active ingredient distributors.

Pesticide law in the EU can be found in several regulations:

- Data necessities for active substances are set by Regulation EU 283/2013 [35, 36].
- Data necessities for plant protection compounds are set by Regulation EU 284/2013 [37, 38].
- Marketing of plant protection substances is governed by Regulation EC 1107/2009 [39].
- Sustainable use of pesticides is covered by Directive 2009/128/EC [40].
- The evaluation, authorization, and approval of active substances at the EU level and national authorization of products is covered by Directive 91/414/EEC [41].
- Regulation EC 33/2008 [42] implements Directive 91/414/EEC regulations and expedites substance evaluation.
- Regulation EC 1095/2007 [43] implements Directive 91/414/EEC rules.

5.5 THE REGULATION SYSTEM OF CANADA

Each year, the public is informed about new registration initiatives by the Pest Management Regulatory Agency (PMRA), which is in charge of overseeing the registration of pest management products [44]. The patterns of usage of pesticides and biopesticides is likely related to safety issues, the phase-out of harmful pesticides, stricter regulation, environmental impacts, and the advancement of new techniques [45]. However, these justifications frequently rest on speculation and lack supporting data.

Bailey et al. [46] discuss the precise social and economic factors that gave rise to the Canadian microbial biopesticide sector. They outline a model system that may have an impact on how biopesticide research is developed globally in the years to come. The act also supported PMRA policy reforms that promoted the registration of lower-risk pest management products, with biopesticides being taken into consideration as part of the initiative for reduced-risk products [44, 47].

By enhancing the accessibility to biopesticides already reported in the US, synchronizing the biopesticide application process with the US and eventually with other OECD countries, starting pre-submission counselling with registrants of biopesticides, shortening the review timeframes for registration of biopesticides, and waiving service charges for scientific analysis of microbial biopesticide, this program depicted PMRA's intention to eliminate discrimination in biopesticide authorization. Despite the fact that the number of biological pest control solutions has grown over time, Canadian producers have been shown to have difficulty adopting lower-risk pest management strategies because of the comparatively modest number of biological pest control products available on the country's market.

It was acknowledged that support for the registration and use of biological control products was required in order for sustainable pest management options to be made available in Canada. In order to increase the accessibility of biological control ingredients for use by Canadian farmers, another federal initiative launched the Pest Management Centre (PMC) in 2003. The PMC is a facility to carry out the Pesticide Risk Reduction Program (jointly with PMRA) by assisting in the development, registration, and adoption of biological pest control products for use in conventional and organic agriculture. The regulatory assistance offered to biocontrol solutions complements this study.

5.6 THE INDIAN REGULATION SYSTEM

Government agencies have a crucial and significant role in the control of pesticide application because both producers and users are unlikely to set strict limits on the pesticides they sell and use [48]. In India, the Ministry of Agriculture, Department of Agriculture and Cooperation passed the Insecticide Act in 1968 and the Insecticides Rules in 1971; the Ministry of Environment and Forest passed the Environment Protection Act in 1986, and the Ministry of Health and Family Welfare passed the Prevention of Food Adulteration Act in 1954 for the regulation of biopesticides.

5.7 INTERNATIONAL REGULATION SYSTEMS

Pesticides have historically been covered by international law in a number of areas. The appropriate use, manufacture, export, import, and application technologies for pesticides are all covered by laws and regulations on human health, environmental protection, agricultural practices, international trade, and border control [49–51]. The Food and Agriculture Organization of the United Nations (FAO) adopted the international code of conduct and use of pesticides at its 23rd session in 1985. It was created to set universal standards of ethics for all parties, but particularly for national governments and pesticide-related industries. Since then, the code of conduct has been reviewed in 2002 and changed once, in 1989, to add the prior informed consent method.

Since 1985, a number of additional international conventions that directly or indirectly relate to the control of pesticides have been in force. The most important are the Rotterdam conference on the prior and informed consent process for certain potentially dangerous chemicals and pesticides in international trade (Rotterdam convention), the Stockholm convention on persistent organic contaminants (Stockholm convention), the Basel convention on the transboundary transportation of dangerous effluents and their disposal (Basel convention), the Montreal protocol on substances that deplete the ozone Layer (Montreal protocol), the International Labor Organization convention No. 184 on safety and well-being in farming (ILO convention 184), and various Codex Alimentarius Commission regulations on pesticide residues in food concerns. Additionally, a new, worldwide standardized system for classifying and labeling chemicals (GHS) has been developed to better safeguard the environment and public health while handling, transporting, and using chemicals.

5.8 THE REGULATION SYSTEM OF AUSTRALIA

The Australian Pesticides and Veterinary Medicines Authority (APVMA) is a statutory authority formed by the Australian government in 1993 to centralize the registration of all farming and veterinary chemical goods into the Australian market. Previously, each state and territory government had its own registration system [52]. The APVMA evaluates applications from chemical businesses and individuals seeking registration so that their product can be sold in the market. Applications are rigorously evaluated using the APVMA's scientific staff's experience as well as the technical expertise of other related scientific bodies, commonwealth government agencies, and ministry of agriculture departments. The APVMA will register a product if it functions as intended, and scientific data demonstrates that when used as instructed on the product label, it will have no negative or unforeseen impacts on people, animals, the environment, or international trade.

5.9 CONCLUSION

The indiscriminate use of chemical pesticides in agriculture and related industries has led to soil, water, and environmental pollution, resulting in not only both human and animal sickness but also habitat loss. Thus, the late 1980s provided fresh opportunities for alternate pest management approaches. Biopesticides generated from biological sources were tested, and they have since become a vital element of pest management strategies, reducing the need for chemical pesticides. The registration of biopesticides has been made mandatory in order to assure the safety of human health, valuable non-target creatures, and the environment. Countries that promote biopesticides set up various regulatory organizations, such as committees, boards, and special authorities, to manage biopesticide regulatory activities [53]. These regulatory authorities develop biopesticide dossier criteria and update them on a regular basis based on local and worldwide needs. Strong, strict, but user-friendly regulatory requirements will assure the availability of high-quality, safe biopesticide formulations.

The current dossier requirements are dynamic and vary by country. The majority of industries are unable to absorb the costs of data generation and registration of biopesticides due to their high costs and complexity. The development of an internationally accepted tier system document is required. The requirement for technical competence and a separate division or body for biopesticide registration in many countries is urgent. Diverse formulations and easy consumer availability are two critical aspects of encouraging biopesticide use. To reduce reliance on conventional chemical pesticides, regulators should endeavor to promote cost-effective, stable, and bio-effective biopesticide formulations with single or a consortia of technological ingredients. This will secure the global availability of high-quality, safe biopesticide formulations for the benefit of end users, resulting in a cleaner environment.

Several organizations, including FAO, the International Organization for Biological Control (IOBC), the European and Mediterranean Plant Protection Organization (EPPO), the Organization for Economic and Co-operative Development (OECD), the European Union, and the United States Environmental Protection Agency (EPA), are revising registration guidance for biopesticides at the international, national, and regional levels. Many (but not all) countries and organizations distinguish between biopesticide and conventional pesticide registration and advocate for a unified approach to biopesticide registration, either through biopesticide provisions in existing pesticide legislation or through separate legislation. Harmonization allows data to be exchanged across countries and regulatory agencies, and regulations encouraging the registration of low-risk compounds will aid in the commercialization and use of biopesticides in crop protection strategies.

REFERENCES

1. Chandler, D., Davidson, G., Grant, W.P., Greaves, J. and Tatchell, G.M., 2008. Microbial biopesticides for integrated crop management: an assessment of environmental and regulatory sustainability. Trends in Food Science & Technology, 19(5), pp. 275–283.
2. www.pesticides.gov.uk/environment.asp?id=1846
3. www.genoeg.net
4. Chandler, D., Bailey, A.S., Tatchell, G.M., Davidson, G., Greaves, J. and Grant, W.P., 2011. The development, regulation and use of biopesticides for integrated pest management. Philosophical Transactions of the Royal Society B: Biological Sciences, 366(1573), pp. 1987–1998.

5. Ash, G.J., 2010. The science, art and business of successful bioherbicides. Biological Control, 52(3), pp. 230–240.

6. Neale, M., 2000. The regulation of natural products as crop-protection agents. Pest Management Science: Formerly Pesticide Science, 56(8), pp. 677–680.

7. OECD, 2003. Guidance for Registration Requirements for Microbial Pesticides. OECD Series on Pesticides, vol. 18. OECD, p. 51.

8. Bourdôt, G.W., Baird, D., Hurrell, G.A. and De Jong, M.D., 2006. Safety zones for a Sclerotinia sclerotiorum-based mycoherbicide: Accounting for regional and yearly variation in climate. Biocontrol Science and Technology, 16(4), pp. 345–358.

9. de Jong, M.D., Bourdôt, G.W., Powell, J. and Goudriaan, J., 2002. A model of the escape of Sclerotinia sclerotiorum ascospores from pasture. Ecological Modelling, 150(1–2), pp. 83–105.

10. Hoagland, R.E., Boyette, C.D., Weaver, M.A. and Abbas, H.K., 2007. Bioherbicides: research and risks. Toxin Reviews, 26(4), pp. 313–342.

11. Gurr, G.M., Horne, P., Page, J., Pilkington, L.J. and Ash, G.J. 2010. (Special Issue Introduction) Australia and New Zealand biological control conference: emerging themes, future prospects. Biological Control, 52, pp. 195–197.

12. Sundh, I. and Goettel, M.S., 2013. Regulating biocontrol agents: a historical perspective and a critical examination comparing microbial and microbial agents. BioControl, 58, pp. 575–593.

13. Hunt, E.J., Kuhlmann, U., Sheppard, A., Qin, T.K., Barratt, B.I.P., Harrison, L., Mason, P.G., Parker, D., Flanders, R.V. and Goolsby, J., 2008. Review of invertebrate biological control agent regulation in Australia, New Zealand, Canada and the USA: recommendations for a harmonized European system. Journal of Applied Entomology, 132(2), pp. 89–123.

14. www.biopesticideindustryalliance.org/

15. www.ibma-global.org/index.html

16. www.unece.org/env/pp/introduction.html

17. www.apad.asso.fr/

18. http://ec.europa.eu/food/plant/index_en.htm

19. www.fao.org/home/en/

20. www.fibl.org/

21. http://www2.globalgap.org/about.html

22. www.ifoam.org/

23. www.iobc-global.org/

24. www.itab.asso.fr/

25. www.newaginternational.com/

26. www.oecd.org/chemicalsafety/pesticides-biocides/biologicalpesticideregistration.htm

27. www.hse.gov.uk/biocides/index.htm

28. http://ec.europa.eu/environment/chemicals/biocides/active-substances/approved-substances_en.htm

29. www.hse.gov.uk/biocides/eu-bpr/product-types.htm

30. www.hse.gov.uk/biocides/eu-bpr/index.htm

31. www.hse.gov.uk/biocides/bpd/index.htm

32. www.hse.gov.uk/biocides/copr/index.htm

33. www.epa.gov/pesticides/biopesticides/

34. http://eurlex.europa.eu/LexUriServ/LexUriServ.do?uri=OJ:L:2012:167:0001:01:EN:HTML

35. http://eurlex.europa.eu/LexUriServ/LexUriServ.do?uri=OJ:L:2013:093:0001:01:EN:HTML

36. http://eurlex.europa.eu/LexUriServ/LexUriServ.do?uri=OJ:C:2013:095:0001:01:EN:HTML

37. http://eurlex.europa.eu/LexUriServ/LexUriServ.do?uri=OJ:L:2013:093:0085:01:EN:H
TML

38. http://eurlex.europa.eu/LexUriServ/LexUriServ.do?uri=OJ:C:2013:095:0021:01:EN:H
TML

39. http://eurlex.europa.eu/LexUriServ/LexUriServ.do?uri=OJ:L:2009:309:0001:01:EN:H
TML

40. http://eurlex.europa.eu/LexUriServ/LexUriServ.do?uri=OJ:L:2009:309:0071:01:EN:H
TML

41. http://ec.europa.eu/food/plant/protection/evaluation/legal_en.htm

42. http://eurlex.europa.eu/LexUriServ/LexUriServ.do?uri=OJ:L:2008:015:0005:01:EN:H
TML

43. http://eurlex.europa.eu/LexUriServ/LexUriServ.do?uri=CELEX:32007R1095:EN:HTML

44. www.hc-sc.gc.ca/cps-spc/pest/index-eng.php

45. Thakore, Y., 2006. The biopesticide market for global agricultural use. Industrial
Biotechnology, 2(3), pp. 194–208.

46. Bailey, K.L., Boyetchko, S.M. and Längle, T.J.B.C., 2010. Social and economic drivers
shaping the future of biological control: a Canadian perspective on the factors affecting the
development and use of microbial biopesticides. Biological Control, 52(3), pp. 221–229.

47. PMRA, 2004. Pest Management Strategic Plan 2003–2008. Pest Management Regulatory
Agency, Catalogue Number H114–14/2003. www.pmra-arla.gc.ca/english/aboutpmra/
plansandreports-e.html

48. Abhilash, P.C. and Singh, N., 2009. Pesticide use and application: an Indian scenario.
Journal of Hazardous Materials, 165(1–3), pp. 1–12.

49. Lallas, P.L., 2001. The Stockholm Convention on persistent organic pollutants. American
Journal of International Law, 95(3), pp. 692–708.

50. Rotterdam Convention Secretariat, 2004. Guideline on the Development of National Laws
to Implement the Rotterdam Convention, Rome and Geneva (revised 2005). United Nation
Treaty Collections.

51. FAO, 2006. Strategic Program 2006–2011 for the Implementation by FAO of the Revised
Version of the International Code of Conduct on the Distribution and Use of Pesticides,
Rome. FAO.

52. www.apvma.gov.au/about/index.php

53. Desai, S., Kumar, G.P., Amalraj, E.L.D., Talluri, V.R. and Peter, A.J., 2016. Challenges
in regulation and registration of biopesticides: an overview. In Microbial Inoculants in
Sustainable Agricultural Productivity: Vol. 2: Functional Applications, pp. 301–308.
Springer.

Section 2

Residues of Biopesticides

6 Biopesticide Residues in Foodstuffs

Javed Ahamad, Adil Ahamad, Mohd Javed Naim,
Javed Ahmad, and Showkat Rasool Mir

CONTENTS

6.1 INTRODUCTION

Biopesticides consist of natural products such as microorganisms, plant-derived phytochemicals, plant-incorporated protectants, and semiochemicals as active ingredients. Biopesticides provide a safer alternative to synthetic chemical pesticides, which are known for their carcinogenic and mutagenic properties, for being toxic to reproductive and endocrine systems, and for accumulating in living organisms (Czaja et al., 2015; Dayan et al., 2009). A major challenge of agriculture is to increase food production to meet the needs of the growing world population without damaging the environment (Kumar and Singh, 2015). In current agricultural practices, the control of pests is often accomplished through excessive use of synthetic chemical pesticides such as organochlorides, organophosphates, and carbamates, which cause severe health problems in human and animals, including acute and chronic poisoning in farmers and consumers; destruction of fish, birds, and wildlife; disruption of natural biological control and pollination; extensive ground water contamination; and the development of pesticide-resistant pests (Kumar, 2012; Isman, 2006). Due to these

DOI: 10.1201/9781003265139-8

major issues, there is an urgent need to find alternative methods of controlling pests. In this context, biopesticides can offer a better alternative to synthetic pesticides, enabling safer control of pests (Isman, 2006; Chandler et al., 2011).

Due to environmental and health issues with synthetic pesticides, several countries responded to ban the use of synthetic pesticides and promote legalization of pesticides derived from natural products. Plant-derived compounds or botanical insecticides have a long history of use, mostly in India, China, Egypt, and Greece (Dev and Koul, 1997). After application to food crops, biopesticides are not entirely consumed by insects/target animals and remain as residue in food products (Munjanja et al., 2015). These food products are the main source of biopesticides in human and animals and are known to cause toxicity. Hence, these chemicals should be analyzed, and maximum permissible residue limits (MRL) should be established. In this chapter, we comprehensively review biopesticide residues in foodstuffs, its regulation, sources of biopesticides, and methods of analysis.

6.2 REGULATION OF BIOPESTICIDES RESIDUES IN FOODSTUFFS

For safer use of biopesticides, maximum residue limits (MRL) in foodstuffs were adapted worldwide. The United States Environmental Protection Agency (EPA) and the European Union (EU) developed a framework for safer use of biopesticides and set a residue limit for each registered biopesticide in foods (Kumar and Singh, 2014). The maximum residue limit is defined as "the maximum amount of biopesticide legally permitted in food." The MRL of some important biopesticides present in foodstuffs are mentioned in Table 6.1 (Munjanja et al., 2015).

6.3 CLASSES OF BIOPESTICIDES

Biopesticides are derived from natural sources, mainly microbes and plants, and are classified into different categories: (1) plant-derived biopesticides, (2) microbial biopesticides, and (3) semiochemicals.

TABLE 6.1
Maximum Allowed Limit of Biopesticide Residues in Foodstuffs

	Maximum Biopesticide Residue Limit (ppm)										
	Apricot		Apple		Avocado		Banana		Nectarine		Honey	
Biopesticides	US	EU	US	EU	US	EU	US	EU	US	EU	US	EU
Azoxystrobin	1.5	2.0	–	–	2.0	0.05	2.0	2.0	1.5	2.0	–	–
Abamectin	0.09	0.02	0.02	0.01	0.02	0.01	0.01	0.01	0.09	0.02	0.01	0.01
Pyrethrins	1.0	1.0	1.0	1.0	1.0	1.0	1.0	1.0	1.0	1.0	1.0	0.05
Spinetoram	0.2	0.2	0.2	0.2	0.3	0.05	0.25	0.05	0.2	0.2	–	–
Spinosad	0.2	1.0	0.2	1.0	0.3	0.02	0.25	2.0	0.2	1.0	0.02	0.05
Trifloxystrobin	2.0	1.0	0.5	0.5	–	–	–	–	2.0	1.0	–	–
Pyraclostrobin	2.5	1.0	1.5	0.5	0.6	0.02	0.04	0.02	2.5	0.3	–	–

6.3.1 Plant-Derived Biopesticides and Plant-Incorporated Protectants

Plant secondary metabolites are major sources of drugs, foods, spices, flavours, and fragrances. They are also known for their antibacterial, antiseptic, insecticidal, and nematocidal properties and are currently used as biopesticides for preserving and protecting food crops. For example, neem, garlic, rapeseed, citronella, eucalyptus, and clove oils are currently available on the market for preserving foodstuffs, and these act as nematocidal and insecticidal substances (Ahmad and Ahamad, 2020). Plant-incorporated protectants (PIP) are substances produced in plants after genetic manipulation and are used as biopesticides in controlling pests. In this technique, the *Bacillus thuringiensis* protein and gene are incorporated into the plant genome, and then the plant produces specific phytochemicals responsible for pesticidal activity (Czaja et al., 2015). Some natural biopesticides are discussed later in detail and are also summarized in Table 6.2.

6.3.1.1 *Azadirachta indica* (Neem)

Azadirachta indica (family: Meliaceae), commonly known as neem, is an important source of remedy (bark, seeds, and leaves) for the treatment of several human ailments, including malaria, fever, infections, diabetes, and digestive and liver complications (Kilani-Morakchi et al., 2021; Trease and Evans, 1996; Britto and Sheeba, 2011). Among the currently marketed plant-derived insecticides, neem oil is one of the least toxic to humans and shows very low toxic potential to beneficial organisms. The leaves are rich in polyphenols and flavonoids, and the seeds are rich source of oil. Neem mainly contains azadirachtin (Figure 6.1) as a major chemical compound. It also contains limonoids (gedunin), which is reported as antimalarial compound (Lin et al., 2016; Prieto et al., 1999). Nimbolide B and nimbolic acid also possess phytotoxic activity (Kato-Noguchi et al., 2014). Nematodes such as *Meloidogyne incognita* are considered major plant parasites damaging food crops. Neem has been reported to exhibit significant nematocidal activity (in vitro and in vivo) against *M. incognita* (Chaudhary et al., 2017; Rabiu and Subhasish, 2009; Okumu et al., 2007; Nile et al., 2018). The larvicidal activity of a neem oil–based formulation was evaluated against *Anopheles, Culex,* and *Aedes* mosquitoes. The LC_{50} values of neem oil were 1.6, 1.8, and 1.7 ppm against *Anopheles, Culex,* and *Aedes*, respectively, 48 hours after treatment (Dua et al., 2009). Azadirachtin has been reported as a potent biopesticide, and it is effective at a very low concentration. It controls pests through ecdysone disruption and controls larval/nymphal instars, including the pupal stage (Immaraju, 1998). The deterrence effect of azadirachtin and neem seed extract was assessed in fifth instar *Spodoptera litura* larvae in cabbage (*Brassica oleracea*) leaf disc assay. Azadirachtin demonstrated significant deterrence in this study (Bomford and Isman, 1996).

6.3.1.2 *Allium sativum* (Garlic)

Garlic consists of bulbs of *Allium sativum* Linn (family: Liliaceae). It is cultivated in Central Asia, Southern Europe, the US, and India. In India, it is found in almost all states and cultivated as a spice crop. Garlic is a well-known spice in the Indian sub-continent. Garlic is used in the treatment of fever, pulmonary

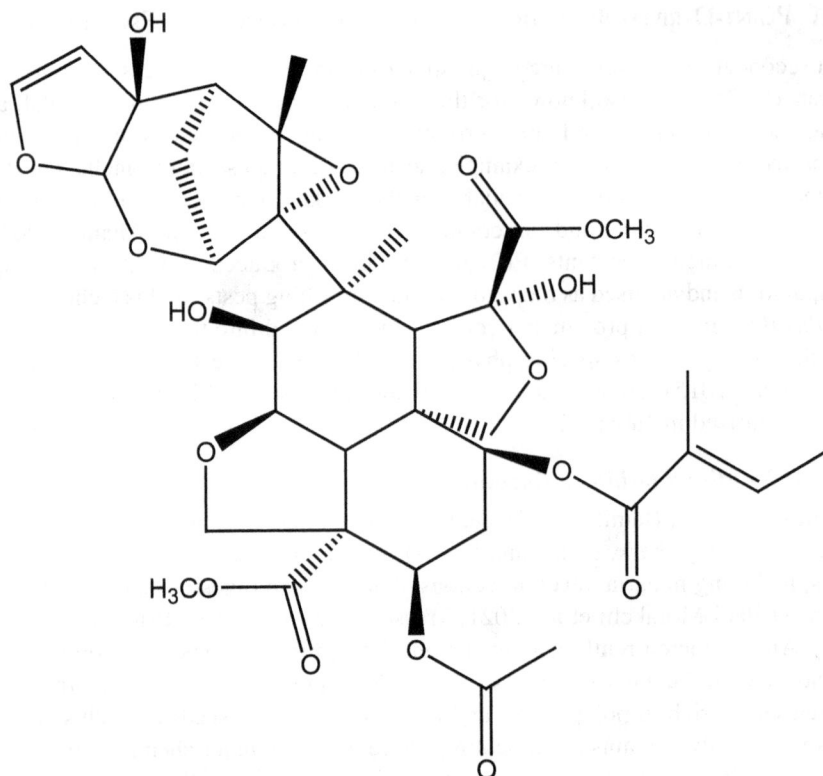

FIGURE 6.1 Chemical structure of azadirachtin.

infections, diabetes, etc. It is also used as carminative, aphrodisiac, expectorant, stimulant, and antiseptic substance. Garlic is a rich source of vitamins, minerals, and proteins. It mainly contains allin, which is converted to allicin by the alliinase enzyme. It also contains diallyl disulphide and diallyl trisulphide (Figure 6.2)

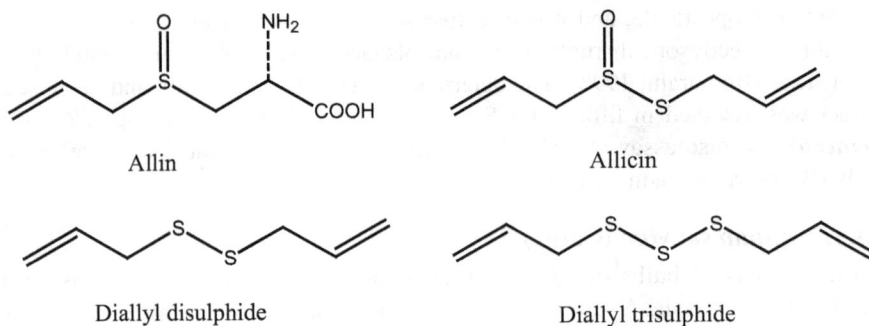

Allin

Allicin

Diallyl disulphide

Diallyl trisulphide

FIGURE 6.2 Structure of the major chemical constituents of garlic.

(Trease and Evans, 1996). Garlic oil is used as anthelmintic and rubefacient substance. It is also beneficial in hypertension and atherosclerosis (Ahamad, 2021). Garlic extract may be used as an insecticide, nematicide, fungicide, molluscicide, and as a bird or mammal repellent (European Food Safety Authority, 2012). Mangang et al. (2020) studied the effect of an aqueous extract of garlic, which showed 86.67% repellency against *Tribolium castaneum* (red flour beetle). The antiparasitic activity of garlic oil was assessed by Anthony et al. (2005) against *Trypanosoma*, *Plasmodium*, *Giardia*, and *Leishmania* in vitro, and it caused significant inhibition of these parasites. Prowse et al. (2006) reported a significant antifeedant and repellant effect in garlic. Garlic juice was also found to inhibit molting and respiration and disrupt the cuticle.

6.3.1.3 Essential Oils from Plants

Essential oils are obtained from different plant sources known for their aroma. Plants containing essential oils are used as spices, condiments, medicines, flavouring and fragrance (Trease and Evans, 1996). Due to their insecticidal, nematocidal, and microbial properties, they are also used as biopesticides. These plants are citronella, eucalyptus, clove, mentha, thymus, and lemongrass.

Kumar et al. (2012), studied the insecticidal activity of *Eucalyptus globulus* essential oil against housefly (*Musca domestica*) larvae and pupae in contact toxicity and fumigation assays. The contact toxicity assay with larvae showed LC_{50} of 2.73 and 0.60 µL/cm^2 for different observation days, and LT_{50} varied from 6.0 to 1.7 days. In a fumigation assay for housefly larvae, LC_{50} values of 66.1 and 50.17 µL/L were obtained in 24 hours and 48 hours, respectively. The results showed *E. globulus* efficacy in controlling houseflies. The insecticidal effects of the nanoemulsified oils of lemongrass and eucalyptus were assessed against *M. domestica* and *Lucilia cuprina* in a topical application method and impregnated paper exposure. The essential oils of both plants demonstrated significant insecticidal activity against *M. domestica* and *L. cuprina* (Velho et al., 2021). The larvicidal activity of *Mentha piperita*, *Cymbopogon citratus*, *Eucalyptus globulus*, and *Citrus sinensis* essential oils and their combinations was evaluated against housefly (*M. domestica*) and mosquito (*Anopheles stephensi*). *M. piperita* was found to be most effective larvicidal against housefly and mosquito with LC_{50} values of 0.66 µL/cm^2 and 44.66 ppm, respectively (Chauhan et al., 2016).

6.3.1.4 *Tanacetum parthenium* (Pyrethrum)

Pyrethrum is an oleoresin extracted from the dried flowers of *Tanacetum parthenium* (family: Asteraceae) and is commonly known as feverfew. Pyrethrum is the most used biopesticide and constitutes about 80% of the botanical pesticide market. It is an oleoresin consisting of three esters of chrysanthemic acid and three esters of pyrethric acid. Among the six esters, those incorporating the alcohol pyrethrolone are mainly pyrethrin I and pyrethrin II (Figure 6.3). Pyrethrins produce insecticidal activity through rapid knockdown, hyperactivity, and convulsions, especially in flying insects. It causes neurotoxic action through blocking voltage-gated sodium channels in nerve axons (Isman, 2006).

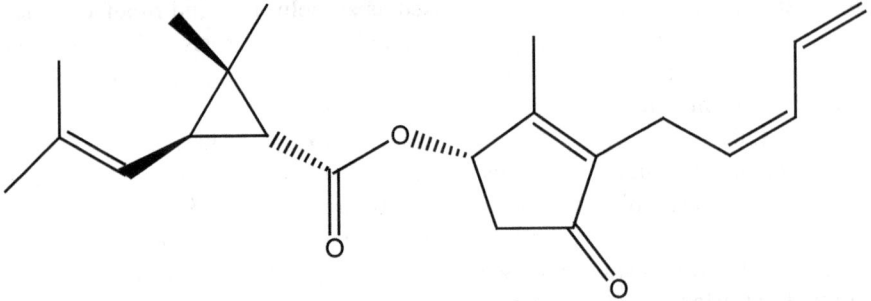

FIGURE 6.3 Chemical structure of pyrethrin.

6.3.1.5 Rotenone

Rotenone (Figure 6.4) is an isoflavone obtained from *Pachyrhizus erosus* (family: Leguminosae; jicama) and other topical legumes such as *Derris*, *Lonchocarpus*, and *Tephrosia*. It was previously used as fish poison. Its insecticidal property was reported in 1933 (Nature, 1933). Rotenone blocks the electron transport chain and prevents energy production and hence works as a mitochondrial poison (Hollingworth et al., 1994). Cabras et al. (2002) determined the rotenone residue in olives and found that the rotenone residue decreased with a half-life of four days after a pre-harvest time of ten days.

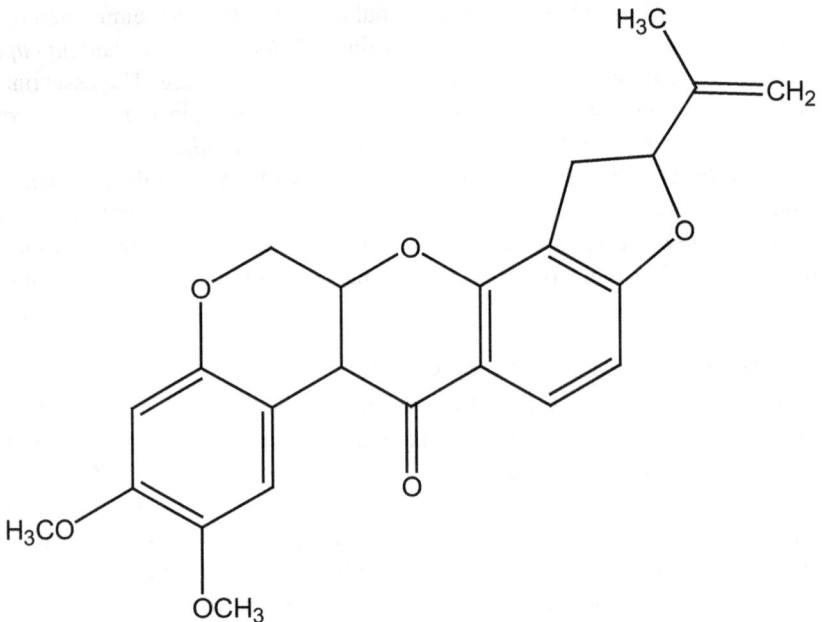

FIGURE 6.4 Chemical structure of rotenone.

TABLE 6.2
Plant-Derived Biopesticides and Their Mechanisms of Action

S. No.	Plant	Active Phytochemicals	Effective Against	Mechanism of Action	Ref.
1	*Azadirachta indica* (neem)	Azadirachtin	*Lepidoptera*	Antifeeding effect and increased larvae mortality.	Immaraju, 1998
			Hemiptera	Early death of nymphs due to inhibition of development and ecdysis defects.	Nile et al., 2018
			Hymenoptera	Food intake decreases, reduced larval and pupal development, larvae death during the molting process.	Dua et al., 2009
			Neuroptera	Severe damage in the midgut cells of larvae, injury and cell death during the replacement of midgut epithelium, and changes in cocoons, with increased porosity and decreased wall thickness affecting pupation.	Rabiu and Subhasish, 2009
			Meloidogyne incognita	Nematocidal effects.	Okumu et al., 2007
			Anopheles, culex, and *Aedes*		
			Spodoptera litura	Larvicidal activity, deterrence effect.	Bomford and Isman, 1996
2	*Allium sativum* (Garlic)	Allin and allicin	*Tribolium castaneum*	Repellent activity.	Mangang et al., 2020
			Trypanosoma, Plasmodium, Giardia and *Leishmania*	Antiparasitic activity.	Anthony et al., 2005
			P. falciparum and *L. amazonensis*	Antifeedant, repellant effect.	Prowse et al., 2006

(Continued)

TABLE 6.2 (Continued)

S. No.	Plant	Active Phytochemicals	Effective Against	Mechanism of Action	Ref.
3	Essential oils				
	Eucalyptus globulus	Eucalyptol	*Musca domestica* *M. domestica* and *Lucilia cuprina*	Contact toxicity and fumigation assay. Insecticidal activity.	Kumar et al., 2012 Velho et al., 2021
	Mentha piperita, Cymbopogon citratus, E. globulus, Citrus sinensis, clove, and thymus	Menthol, citronellal, eucalyptol, eugenol, thymol, and carvacrol	*M. domestica* and *Anopheles stephensi*	Larvicidal activity. fumigant, and contact insecticidal property; interferes with the neuromodulator octopamine and GABA-gates chloride channels.	Chauhan et al., 2016; Koul et al. 2008
4	*Tanacetum parthenium* (feverfew)	Pyrethrum (oleoresin): pyrethrin	Flying insects	Neurotoxic action through blocking voltage-gated sodium channels.	Isman, 2006
5	*Pachyrhizus erosus* (jicama)	Rotenone	Invertebrates, fish	Blocks the electron transport chain and prevents energy production.	Dudgeon, 1990
6	*Curcuma longa* (turmeric)	Curcuminoids (*ar*-turmerone)	Beetles (*Coleoptera*), *Nilaparvata lugens,* and *Plutella xylostella*	Antifeedant.	Tripathi et al., 2002; Lee et al., 2001
7	*Lawsonia inermis* (henna)	Lawsone	Fungi	Antifungal.	Jeyaseelan et al., 2012
8	*Ferula asafoetida* (asafoetida)	Oleo-gum-resins	*Ectomyelois ceratoniae*	Insect repellant.	Kavianpour et al. 2014
9	*Momordica charantia* (bitter melon)	Charantin	*Plutella xylostella*	Antifeedant.	Ling et al., 2008

6.3.2 Microbial Biopesticides

Microbial biopesticides consist of microorganisms (bacteria, fungi, virus, and proto-zoans) as active ingredient and have the potential to inhibit or kill the growth of other microorganisms such as bacteria and fungi (Czaja et al., 2015; Mann and Kaufman, 2012). Microbial biopesticides are used against several kind of pests, but these pesticides are specific against particular microbes (pests). Table 6.3 incorporates the list of microbial biopesticides.

For further information, view the list of registered biopesticides' active ingredients at the EPA's web page Biopesticide Active Ingredients (www.epa.gov/ingredients-used-pesticide-products/biopesticide-active-ingredients).

6.3.3 Semiochemicals

Semiochemicals are naturally occurring substances that control pests by non-toxic mechanisms, and they include substances that interfere with mating, such as phero-mones (e.g., lepidopteran) that attract pests to traps. These biopesticides either work intra-specifically (e.g., pheromones, population-regulating auto-inhibitors, autotox-ins, and necromones) or inter-specifically (e.g., allomones, kairomones, depressors, and synomones) (Duke et al., 2010; Czaja et al., 2015).

Semiochemicals have become an important part of integrated pest management (IPM) programmes and are used in various ways *viz.* as mass trapping, monitoring, attract-and-kill, push-pull, and disruption strategies (El-Ghany, 2019). Lepidopteran pheromone is a straight-chain biopesticide commonly used to trap insects in agricultural practice. Semiochemicals have an advantage as biopesticides as they decompose quickly after application (Czaja et al., 2015). In Table 6.4, pest control strategies are presented with insect targets.

6.4 TECHNIQUES FOR ANALYSIS OF BIOPESTICIDE RESIDUES IN FOODSTUFFS

The recent development in pest control through use of biopesticides derived from natural sources has had a very big impact on the safety of humans and animals because biopesticides provide a safer alternative for control of pests than synthetic

TABLE 6.3
Microbial Biopesticides

S. No.	Microbes	Example	Protection Against
1	Bacteria	*Bacillus thuringiensis subsp. tenebrionis* (strain NB-176)	Kills larvae of *Leptinotarsa decemlineata* (larva feeds on potatoes)
2	Fungus	*Beauveria bassiana* (strain GHA)	Controls sucking insects and insects feeding on greenhouse vegetables
3	Virus	*Cydia pomonella* (granulovirus)	Protects fruit trees from codling moth larvae

TABLE 6.4

Pest Control Strategies through the Use of Semiochemicals

Pest Control Strategy (Semiochemicals)	Target Insects	Ref.
Monitoring	Male spruce budworm (*Choristoneura fumiferana* Clemens; *Lepidoptera*: Tortricidae)	Rhainds et al., 2016
	Date moth (*Batrachedra amydraula* Meyrick; *Lepidoptera*: Batrachedridae)	Levi-Zada et al., 2018
	Meyrick (*Tuta absoluta*; *Lepidoptera*: Gelechiidae)	Abd El-Ghany et al., 2016
Mass trapping	Japanese beetle (*Popillia japonica* Newman; *Coleoptera*: Scarabaeidae)	Piñero and Dudenhoeffer, 2018
Attract-and-kill (Ammonium carbonate with volatile components of apple)	Apple maggot fly	Morrison et al. 2016
Mating disruption	Bark beetles (*Coleoptera*: Scolytidae)	Borden, 1997
Push-pull strategy (aphid alarm pheromone)	Bark beetles	Borden et al., 2008
Push-pull strategy (1-Octen-3-ol)	*Drosophila suzukii* Matsumura (Diptera: Drosophilidae)	Wallingford et al., 2017

pesticides. However, biopesticide residues in foodstuffs have been reported to cause toxic effects in humans and animals (Nollet and Rathore, 2015). Hence, methods for the detection and quantification of biopesticides in foodstuffs should be developed, and maximum residue limits (MRL) for each biopesticide must be established.

Biopesticides are complex organic compounds present in the matrix of natural materials; hence, detection and quantification require multiple steps, starting with extraction and purification from the complex matrix, and then quantification is performed by liquid chromatography or gas chromatography (Ahamad et al., 2022). Liquid chromatography emerged as the method of choice for detection and quantification of biopesticides. High-performance liquid chromatography (HPLC) and liquid chromatography-mass chromatography (LC-MS) are the main liquid chromatography methods employed in analysis of biopesticides (Ahamad et al., 2020). Gas chromatography (GC) and gas chromatography-mass spectroscopy (GC-MS) also emerged as methods for analysis of low molecular weight compounds present as biopesticides in food products (Munjanja et al., 2015). In Table 6.5, analytical methods for detection and quantification of biopesticides in foodstuffs are highlighted.

TABLE 6.5

Analytical Methods for Detection and Quantification of Biopesticides in Foodstuffs

Biopesticides	Foodstuffs	Analytical Condition	Ref.
HPLC			
Spinosad and its metabolites	Citrus fruits and foods containing citrus	Column: ODS-AM; Mobile phase: methanol: acetonitrile: ammonium acetate; Scanning at: 250 nm	West and Turner, 2000
	Leafy vegetables, peppers, and tomatoes	Column: ODS-AM; Mobile phase: methanol: acetonitrile: ammonium acetate (2%); Scanning at: 250 nm	Yeh et al., 1997
	Meat, milk, cream, and eggs	Column: ODS-AM; Mobile phase: methanol (44%): acetonitrile (12%): aq. ammonium acetate (2%)/acetonitrile (67:33); Scanning at: 250 nm	West and Turner, 1998
LC-MS			
Spinosad and its derivatives	Olive oils	Detection: LC-ESI-MS/MS, MRM Mode, positive ion mode; Column: Polaris C18; Mobile phase: gradient elution with aq. formic acid: acetonitrile (90:10)	Benincasa et al., 2011
	Food and environmental matrix	Detection: LC-MS ion trap, SIM Mode, positive ion mode; Column: ODS AM; Mobile phase: acetonitrile: methanol: 2% aq. ammonium acetate (42:42:16)	Schwedler et al., 2000
Azadirachtin and related compounds	Fruits and vegetables	Detection: LC-ESI-MS/MS, MRM Mode; Column: Polaris C18; Mobile phase: gradient elution with acetonitrile: formic acid, 0.01% sodium acetate	Sarais et al., 2008
	Oranges	Detection: LC-ESI-MS/MS; Column: Nucleosil C18; Mobile phase: gradient elution with methanol: water	Pozo et al., 2003
Pyrethrins	Lemons and apricots	Detection: LC-MS/MS, MRM Mode, ESI, PI; Column: Zorbax C18; Mobile phase: ammonium formate: acetonitrile	Ruiz et al., 2011
	Tea	Detection: LC-ESI/MS/MS, MRM Mode, ESI, PI; Column: acquity UPLC BEH C18; Mobile phase: ammonium formate: acetonitrile	Lu et al., 2010

(Continued)

TABLE 6.5 (Continued)

Biopesticides	Foodstuffs	Analytical Condition	Ref.
Rotenone	Fruits and vegetables	Detection: LC-MS/MS, MRM Mode, ESI, PI; Column: C18; Mobile phase: gradient elution with 0.1% formic acid: acetonitrile	Dun-Ming et al., 2010
	Vegetables	Detection: LC-MS, APCI, PI mode; Column: Supelco-Rx-C18; Mobile phase: gradient elution with methanol: ammonium acetate	Zang et al., 1998
Azoxystrobin	Processed fruits and vegetables	Detection: LC-ESI-MS/MS, MRM Mode, PI; Column: Synergy RP; Mobile phase: gradient elution with aqueous formic acid: acetonitrile	Samino et al., 2004
Veratridine	Lettuce and cucumbers	Detection: LC-MS, APCI, SIM, PI; Column: Supelco-Rx-C18; Mobile phase: gradient elution with methanol: acetonitrile: ammonium acetate	Zang et al., 1997
Abamectin	Milk and liver	Detection: LC-MS/MS, MRM, ESI, PI; Column: Prodigy C18; Mobile phase: gradient elution with ammonium acetate in acetonitrile: methanol (50:50)	Kinsella et al., 2009
GC & GC-MS			
Azoxystrobin	Fruits and vegetables	GC-MS, SIM Mode; splitless injection; Column: DB-35, EI	Stajnbaher and Zupancic-Kralj, 2003
	Wheat, apples, and grapes	GC-ECD & GC-NPD, SIM Mode; splitless injection; Column: DB-5	Christensen and Granby, 2001
	Vegetables	GC-QqQ-MS, SRM mode; splitless injection; Column: VF-5MS	Garrido Frenich et al., 2005
	Dietary supplements	GC-MS/MS, Column: Agilent capillary; PTV injection	Hayward et al., 2013

6.5 CONCLUSION

Biopesticides have emerged as an important class of compounds for management of various kinds of pests. Due to their safer nature, they can be used as an alternative or can be incorporated into integrated pest management (IPM) strategies. Biopesticides do not pose a big problem regarding residue, which is the major concern with synthetic pesticides. The biopesticide concept is still in the development phase; hence, new and potent biopesticides should be discovered, and analytical methods for their detection and analysis should be developed. Finally, the permissible maximum residue limits for each biopesticide should be established and regulated by concerned countries.

REFERENCES

Abd El-Ghany NM, Abdel-Wahab ES, Ibrahim SS. 2016. Population fluctuation and evaluation the efficacy of pheromone-based traps with different color on tomato leaf miner moth, *Tuta absoluta* (Lepidoptera: Gelechiidae) in Egypt. Research Journal of Pharmaceutical, Biological and Chemical Sciences, 7(4): 1533–1539.

Ahamad J. 2021. Text Book of Advanced Pharmacognosy and Natural Products, 1st ed. Mahi Publication, Ahmedabad.

Ahamad J, Mir SR, Kaskoos RA, Ahmad J. 2020. Chapter 11: Proteomics in Authentication of Wine. In Proteomics for Food Authentication (Edited by Leo M.L. Nollet, Semih Ötleş). CRS Press, Boca Raton.

Ahamad J, Uthirapathy S, Ahmad J. 2022. Mass Spectrometry in Food Authentication. In Mass Spectrometry in Food Analysis, 1st ed. CRS Press, Boca Raton.

Ahmad J, Ahamad J, editors. 2020. Bioactive Phytochemicals: Drug Discovery to Product Development. Bentham Science Publishers, Potomac, MD.

Anthony JP, Fyfe L, Smith H. 2005. Plant active components—a resource for antiparasitic agents?. Trends in Parasitology, 21(10): 462–468.

Benincasa C, Perri E, Iannotta N, and Scalercio S. 2011. LC/ESI-MS/MS method for the identification and quantification of spinosad residues in olive oils. Food Chemistry, 125: 1116–1120.

Bomford MK, Isman MB. 1996. Desensitization of fifth instar *Spodoptera litura* to azadirachtin and neem. Entomologia Experimentalis et Applicata, 81(3): 307–313.

Borden JH. 1997. Disruption of semiochemical mediated aggregation in bark beetles (Coleoptera: Scolytidae). In Insect Pheromone Research: New Directions (Edited by R.T. Cardé, A.K. Minks). Springer, Boston, pp. 421–438.

Borden JH, Lafontaine JP, Pureswaran DS. 2008. Synergistic blends of monoterpenes for aggregation pheromones of the mountain pine beetle (Coleoptera: Curculionidae). Journal of Economic Entomology, 101(4): 1266–1275.

Britto AJ, Sheeba DH. 2011. *Azadirachta indica* Juss—a potential antimicrobial agent. International Journal of Applied Biology and Pharmaceutical Technology, 2(3): 4550–4557.

Cabras P, Caboni P, Cabras M, Angioni A, Russo M. 2002. Rotenone residues on olives and in olive oil. Journal of Agricultural and Food Chemistry, 50: 2576–2580.

Chandler D, Bailey AS, Tatchell GM, Davidson G, Greaves J, Grant WP. 2011. The development, regulation and use of biopesticides for integrated pest management. Philosophical Transactions of the Royal Society B, 366: 1987–1998.

Chaudhary S, Kanwar RK, Sehgal A, Cahill DM, Barrow CJ, Sehgal R, Kanwar JR. 2017. Progress on *Azadirachta indica* based biopesticides in replacing synthetic toxic pesticides. Frontiers in Plant Science, 8: 610. [DOI: 10.3389/fpls.2017.00610]

Chauhan N, Malik A, Sharma S, Dhiman RC. 2016. Larvicidal potential of essential oils against *Musca domestica* and *Anopheles stephensi*. Parasitology Research, 115(6): 2223–2231.

Christensen H, Granby K. 2001. Method validation for strobilurin fungicides in cereals and fruits. Food Additives & Contaminants, 18(10): 866–874.

Czaja K, Góralczyk K, Struciński P, Hernik A, Korcz W, Minorczyk M, Łyczewska M, Ludwicki JK. 2015. Biopesticides–Towards increased consumer safety in the European Union. Pest Management Science, 71(1): 3–6.

Dayan FE, Cantrell CL, Duke SO. 2009. Natural product in crop protection. Bioorganic & Medicinal Chemistry, 17: 4022–4034.

Dev S, Koul O. 1997. Insecticides of Natural Origin. Harwood Academic Publishers, Amsterdam, p. 365.

Dua VK, Pandey AC, Raghavendra K, Gupta A, Sharma T, Dash AP. 2009. Larvicidal activity of neem oil (*Azadirachta indica*) formulation against mosquitoes. Malaria Journal, 8: 124. [DOI: 10.1186/1475-2875-8-124]

Dudgeon, D. 1990. Benthic community structure and the effect of rotenone piscicide on invertebrate drift and standing stocks in two Papua New Guinea streams. Archiv fur Hydrobiologie, 119: 35–53.

Duke SO, Cantrell CL, Meepagala KM, Wedge DE, Tabanca N, Schrader KK. 2010. Natural toxins for use in pest management. Toxins, 2: 1943–1962.

Dun-Ming X, Yu Z, Li-Yi L, Zhi-Gang Z, Jin Z, Sheng-yu L, Fang Y, Peng-Ying H. 2010. Determination of rotenone residues in foodstuffs by solid phase extraction (SPE) and liquid chromatography/tandem mass spectrometry (LC-MS/MS). Agricultural Sciences in China, 9(9): 1299–1308.

El-Ghany NM. 2019. Semiochemicals for controlling insect pests. Journal of Plant Protection Research, 59(1): 1–11.

European Food Safety Authority. 2012. Conclusion on the peer review of the pesticide risk assessment of the active substance garlic extract. EFSA Journal, 10(2): 2520.

Garrido Frenich A, Gonzalez-Rodriguez MJ, Arrebola F, Martinez Vidal J. 2005. Potentiality of gas chromatography-triple quadrupole mass spectrometry in vanguard and rearguard methods of pesticide residues in vegetables. Analytical Chemistry, 77: 4640–4648.

Hayward D, Wong J, Shi F, Zhang K, Lee N, DiBenedetto A, Hengel M. 2013. Multiresidue pesticide analysis of botanical dietary supplements using salt-out acetonitrile extraction, solid phase extraction and gas chromatography triple quadrupole mass spectrometry. Analytical Chemistry, 85: 4686–4693.

Hollingworth R, Ahmmadsahib K, Gedelhak G, McLaughlin J. 1994. New inhibitors of complex I of the mitochondrial electron transport chain with activity as pesticides. Biochemical Society Transactions 22: 230–233.

Immaraju JA. 1998. The commercial use of azadirachtin and its integration into viable pest control programmes. Journal of Pesticide Science, 54: 285–289.

Isman MB. 2006. Botanical insecticides, deterrents, and repellents in modern agriculture and an increasingly regulated world. Annual Review of Entomology, 51: 45–66.

Jeyaseelan EC, Vinuja T, Pathmanathan K, Jeyadevan JP. 2012. Control of plant pathogenic fungi using organic solvent extracts of leaf, flower and fruit of *Lawsonia inermis* L. International Journal of Pharmaceutical and Biological Science Archive, 3: 783–788.

Kato-Noguchi H, Salam MA, Ohno O, Suenaga K. 2014. Nimbolide B and Nimbic Acid B, phytotoxic substances in neem leaves with allelopathic activity. Molecules, 19: 6929–6940.

Kavianpour M, Dabbagh G, Taki M, Shirdeli M, Mohammadi S. 2014. Effect of fresh gum of asafoetida on the damage reduction of pomegranate fruit moth, *Ectomyelois ceratoniae* (Lep., Pyralidae) in Shahreza City. International Journal of Biosciences, 5: 86–91.

Kilani-Morakchi S, Morakchi-Goudjil H, Sifi K. 2021. Azadirachtin-based insecticide: Overview, risk assessments, and future directions. Frontiers in Agronomy, 3: 676208. [DOI: 10.3389/fagro.2021.676208].

Kinsella B, Lehotay SJ, Mastovska K, Lightfield A, Furey A, Danaher M. 2009. New method for the analysis of flukicide and other anthelmintic residues in bovine milk and liver using liquid chromatography tandem mass spectrometry. Analytica Chimica Acta, 637: 196–207.

Koul O, Walia S, Dhaliwal G. 2008. Essential oils as green pesticides: Potential and constraints. Biopesticides International, 4: 63–84.

Kumar P, Mishra S, Malik A, Satya S. 2012. Compositional analysis and insecticidal activity of *Eucalyptus globulus* (family: Myrtaceae) essential oil against housefly (*Musca domestica*). Acta Tropica, 122(2): 212–218.

Kumar S. 2012. Biopesticides: A need for food and environmental safety. Journal of Biofertilizers & Biopesticides 3: 4. [DOI: 10.4172/2155–6202.1000e107].

Kumar S, Singh A. 2014. Biopesticides for integrated crop management: Environmental and regulatory aspects. Journal of Fertilizers & Pesticides 5: e121. [DOI: 10.4172/2155–6202.1000e121].

Kumar S, Singh A. 2015. Biopesticides: Present status and the future prospects. Journal of Fertilizers & Pesticides, 6: 2. [DOI: 10.4172/jbfbp.1000e129].

Lee H-S, Shin W-K, Song C, Cho K-Y, Ahn Y-J. 2001. Insecticidal activities of ar-turmerone identified in *Curcuma longa* rhizome against *Nilaparvata lugens* (Homoptera: Delphacidae) and *Plutella xylostella* (Lepidoptera: Yponomeutidae). Journal of Asia-Pacific Entomology, 4: 181–185.

Levi-Zada A, Sadowsky A, Dobrinin S, Ticuchinski T, David M, Fefer D, Dunkelblum E, Byers JA. 2018. Monitoring and mass-trapping methodologies using pheromones: The lesser date moth *Batrachedra amydraula*. Bulletin of Entomological Research, 108(1): 58–68.

Lin T, Liu Q, Chen J. 2016. Identification of differentially expressed genes in *Monochamus alternatus* digested with azadirachtin. Scientific Reports, 6: 33484. [DOI: 10.1038/srep33484].

Ling B, Wang G-C, Ya J, Zhang, M-X, Liang G-W. 2008. Studies on antifeeding activity and active ingredients against *Plutella xylostella* from *Momordica charantia* leaves]. Scientia Agricultura Sinica, 10: 32.

Lu C, Liu X, Dong F, Xu J, Song W, Zhang C, Li Y, Zheng Y. 2010. Simultaneous determination of pyrethrin residues in teas by ultra-performance liquid chromatography tandem mass spectrometry. Analytica Chimica Acta, 678: 56–62.

Mangang IB, Tiwari A, Rajamani M, Manickam L. 2020. Comparative laboratory efficacy of novel botanical extracts against *Tribolium castaneum*. Journal of the Science of Food and Agriculture, 100(4): 1541–1546.

Mann RS, Kaufman PE. 2012. Natural product pesticides: Their development, delivery and use against insect vectors. Mini-Reviews in Organic Chemistry, 9: 185–202.

Morrison WR, Lee DH, Reissig WH, Combs D, Leahy K, Tuttle A, Cooley D, Leskey TC. 2016. Inclusion of specialist and generalist stimuli in attract-and-kill programs: Their relative efficacy in apple maggot fly (Diptera: Tephritidae) pest management. Environmental Entomology, 45(4): 974–982.

Munjanja B, Chaparadza A, Majoni S. 2015. Biopesticide residues in foodstuffs. In Handbook of Biopesticide, 1st ed. CRC Press, Boca Raton, pp. 71–92.

Nile AS, Nile SH, Keum YS, Kim DH, Venkidasamy B, Ramalingam S. 2018. Nematicidal potential and specific enzyme activity enhancement potential of neem (*Azadirachta indica* A. Juss.) aerial parts. Environmental Science and Pollution Research, 25(5): 4204–4213.

Nollet LML, Rathore HS. 2015. Biopesticides Handbook, 1st ed. CRC Press, Boca Raton.

Okumu FO, Knols BG, Fillinger U. 2007. Larvicidal effects of a neem (*Azadirachta indica*) oil formulation on the malaria vector Anopheles gambiae. Malaria Journal, 6: 63.

Piñero JC, Dudenhoeffer AP. 2018. Mass trapping designs for organic control of the Japanese beetle, *Popillia japonica* (Coleoptera: Scarabaeidae). Pest Management Science, 74(7): 1687–1693.

Pozo O, Marin J, Sancho J, Hernandez F. 2003. Determination of abamectin and azadirachtin residues in orange samples by liquid chromatography electrospray tandem mass spectrometry. Journal of Chromatography A, 992: 133–140.

Prieto P, Pineda M, Aguilar M. 1999. Activity of a standardized neem (*Azadirachta Indica*) seed extract on the rodent malaria parasite *Plasmodium berghei*. Analytical Biochemistry, 269: 337–341.

Prowse GM, Galloway TS, Foggo A. 2006. Insecticidal activity of garlic juice in two dipteran pests. Agricultural and Forest Entomology, 8: 1–6.

Rabiu H, Subhasish M. 2009. Investigation of *in-vitro* anthelmintic activity of *Azadirachta indica* leaves. International Journal of Drug Development and Research, 31: 94–99.

Rhainds M, Therrien P, Morneau L. 2016. Pheromone-based monitoring of spruce budworm (Lepidoptera: Tortricidae) larvae in relation to trap position. Journal of Economic Entomology, 109(2): 717–723.

Rotenone as an Insecticide. Nature, 1933; 132:167. [DOI: 10.1038/132167b0]

Ruiz I, Morales A, Oliva J, Barba A. 2011. Validation of an analytical method for the quantification of pyrethrins on lemons and apricots using high-performance liquid chromatography/mass spectrometry. Journal of Environmental Science and Health, Part B, 46: 530–534.

Sannino A, Bolzoni L, Bandini M. 2004. Application of liquid chromatography with electrospray tandem mass spectrometry to the determination of a new generation of pesticides in processed fruits and vegetables. Journal of Chromatography A, 1036: 161–169.

Sarais G, Caboni P, Sarritzu E, Russo M, Cabras P. 2008. A simple and selective method for the measurement of azadirachtin and related azadirachtoid levels in fruits and vegetables using liquid chromatography electrospray ionization tandem mass spectrometry. Journal of Agricultural and Food Chemistry, 56: 2939–2943.

Schwedler D, Thomas A, Yeh L. 2000. Determination of spinosad and its metabolites in food and environmental matrices. 2. Liquid chromatography mass spectrometry methods. Journal of Agricultural and Food Chemistry, 48: 5138–5145.

Stajnbaher D, Zupancic-Kralj L. 2003. Multiresidue method for determination of 90 pesticides in fresh fruits and vegetables using solid-phase extraction and gas chromatography-mass spectrometry. Journal of Chromatography A, 1015: 185–198.

Trease and Evans. 1996. Pharmacognosy, 14th ed. Elsevier, Edinburg, London, New York, Philadelphia, St Lois, Sydney, Toronto.

Tripathi A, Prajapati V, Verma N, Bahl J, Bansal R, Khanuja S, Evans, W, 2002. Bioactivities of the leaf essential oil of *Curcuma longa* (var. ch-66) on three species of stored-product beetles (Coleoptera). Journal of Economic Entomology, 95: 183–189.

Velho MC, Cossetin LF, Godoi SN, Santos RC, Gündel A, Monteiro SG, Ourique AF. 2021. Nanobiopesticides: Development and insecticidal activity of nanoemulsions containing lemongrass or eucalyptus oils. Natural Product Research, 35(24): 6210–6215.

Wallingford AK, Cha DH, Loeb GM. 2017. Evaluating a push-pull strategy for management of *Drosophila suzukii* Matsumura in red raspberry. Pest Management Science, 74(1): 120–125.

West D, Turner L. 1998. Determination of spinosad and its metabolites in meat, milk, cream and eggs by high performance liquid chromatography with ultraviolet detection. Journal of Agricultural and Food Chemistry, 46: 4620–4627.

West S, Turner L. 2000. Determination of spinosad and its metabolites in citrus crops and orange processed commodities by HPLC with UV detection. Journal of Agricultural and Food Chemistry, 48: 366–372.

Yeh L, Schwedler D, Schelle G, Balcer J. 1997. Application of empore disk extraction for trace analysis of spinosad and metabolites in leafy vegetables, peppers, and tomatoes by high performance liquid chromatography with ultraviolet detection. Journal of Agricultural and Food Chemistry, 45: 1746–1751.

Zang X, Fukuda E, Rosen J. 1997. Method for the determination of veratridine and vevadine, major components of the natural insecticide sabadilla, in lettuce and cucumbers. Journal of Agricultural and Food Chemistry, 45: 1758–1761.

Zang X, Fukuda E, Rosen J. 1998. Multiresidue analytical method for insecticides used by organic farmers. Journal of Agricultural and Food Chemistry, 46: 2206–2210.

7 Biopesticide Residues in Water

Debarati Rakshit and Awanish Mishra

CONTENTS

7.1 INTRODUCTION

Biopesticides are defined as pesticides that are derived from natural sources and are used in agriculture for pest management. These can be obtained from plants, animals, minerals, or natural predators and genetically modified plants producing pesticide chemicals, microbes, or microbial products [1]. Agriculture is a major food source and contributes to the socio-economic development of developing and developed countries [2]. To fulfil the growing demand for food, farmers use fertilizers for plant growth. Management of harmful pests is another essential aspect of this. Chemical pesticides have been used to prevent crop loss due to harmful insects, rodents, fungi, and nematodes [3]. But several adverse effects on the ecosystem and human health are associated with the use of chemical pesticides [4]. These conventional pesticides work less specifically as they harm non-targets also. These are considered organic pollutants as they produce more toxicity and persist longer in the environment. These limitations of chemical pesticides led to the development of alternate pesticides

DOI: 10.1201/9781003265139-9

termed biopesticides. These have target-specific modes of action and are less toxic to humans and the environment [5].

In India, some of the biopesticides which are registered under the Insecticides Act, 1968 are neem-based pesticides, Cymbopogon, *Bacillus thuringensis*, *Trichoderma viride*, and nuclear polyhedrosis virus [6]. The global market size of biopesticides was US$4.3 billion in 2019. It is expected to touch US$8.3 billion by 2025, showing a compound annual growth rate (CAGR) of 14.7% during the forecast period [7]. Based on the sources, biopesticides can be clustered into five main groups. These are botanical pesticides (neem products, rotenone, etc.), microbial pesticides (*Bacillus thuringensis*, *Beauveria bassiana*, etc.), biochemicals (pheromones, allelochemicals) [8], plant-incorporated protectants (PIP) (pesticidal agents obtained from genetically modified plants) [9], and biotics (natural predators).

Biopesticides are available in dry formulations, such as powders, dust, micro granules, granules, water-dispersible granules, and wettable powder, and liquid formulations, such as emulsions, suspo-emulsions, oil dispersion, suspension concentration, and capsule suspension [10]. Various additives are used to formulate the biopesticides to increase their stability and bioactivity. Residues of some biopesticides have been traced in ground and surface water as they are drained from agricultural field to nearby water bodies [11]. The quantity of biopesticide residues in waterways is increasing as people have gradually started producing and using biopesticides as a substitute for chemical pesticides. Accumulation of the residues of biopesticides and their additives can pose a risk to the ecosystem by affecting aquatic animals as well as human health [4]. So there is a need to study biopesticide residue in water.

In this chapter, we will discuss the sources depositing biopesticide residues into the water; the physicochemical properties related to the accumulation, occurrence, and degradation of biopesticides in the aquatic environment; the toxic effects of biopesticides on marine lives; and techniques for analyzing biopesticide residues in water.

7.2 SOURCES DEPOSITING BIOPESTICIDES INTO WATER

Biopesticides are used in agriculture and horticulture as well as in animal husbandry to control various pests. These anthropogenic activities contribute to the accumulation of biopesticides across multiple water bodies [12]. Several biopesticides have been detected in different waterways in the present day [11, 13]. After biopesticides are applied, they undergo metabolism by target organisms, and the surplus amount gets swept away from the surface of plants or soil. Then they are eluted to streams, rivers, lakes, etc., primarily by the surface runoff process. In this process, biopesticides move from the application site across the soil surface, are dissolved or suspended in runoff waters, and continue on to surface water bodies, such as ponds, lakes, rivers, etc.) Leaching is another process by which biopesticide residues can dissipate into water bodies. In this process, biopesticides, which are water soluble, flow downward through the soil into the groundwater [14].

These two processes can also pass to surface and groundwater [15]. Another way biopesticides enter the water is erosion, in which soil particles adsorb biopesticides and are driven to bodies of water by heavy rain or excess irrigation. Runoff or biopesticide leaching can also occur when it is used in excess amounts or spilled on

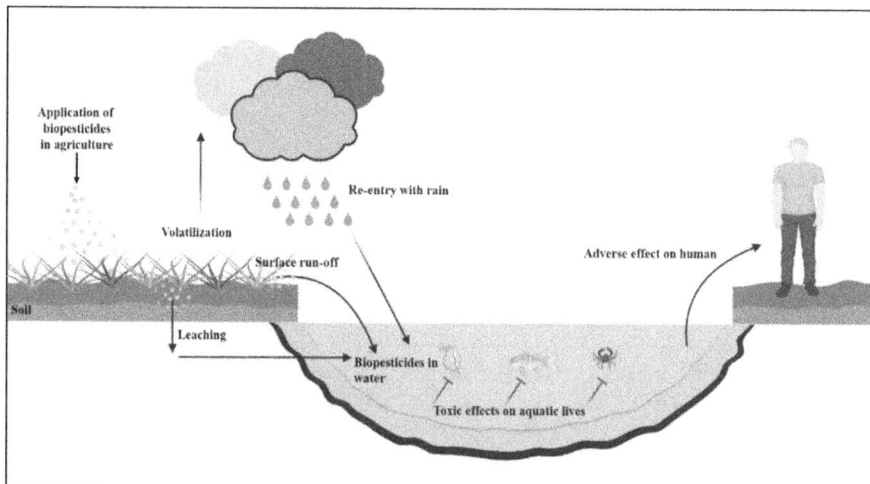

FIGURE 7.1 Sources of biopesticides in water.

the agricultural field or by massive rain and irrigation. Volatile pesticide compounds quickly evaporate into the atmosphere but are deposited into the water during rainfall [16]. In the case of PIP, after reaping, the leftovers decay and release biopesticides, which can enter aquifers and affect aquatic life [4] (Figure 7.1).

7.3 PHYSICOCHEMICAL PROPERTIES OF BIOPESTICIDES

After application in agricultural works, the residue of these contaminants enters the ground and surface water. Biopesticides can freely solubilize in water, or they can remain as suspended particles or accumulate in non-target living organisms. Generally, the entry and resident time of biopesticides depend on their own physicochemical properties. These properties include aqueous solubility, polarity, volatility, partition co-efficient, and half-life [14].

Water solubility and polarity are closely related factors that affect the distribution of biopesticides in water. Biopesticides with higher polarity possess higher water solubility, whereas biopesticides with lesser polarity get adsorbed on suspended matter or precipitate. Another factor that affects entry into the environment is volatilization. Volatile biopesticides are readily dispersed into the atmosphere. The physicochemical properties of biopesticides also guide the selection of techniques for analyzing biopesticide residues in water [4]. The physicochemical properties of various biopesticides are include in **Table 7.1**.

7.4 DEGRADATION OF BIOPESTICIDES IN WATER

After entering the aqueous environment, biopesticides undergo various chemical, physical, and microbial reactions [4, 16]. Numerous factors such as sunlight, pH, temperature, and water turbidity control the degradation of these organic pollutants

TABLE 7.1

Physicochemical Properties and Applications of Some Commonly Used Biopesticides

Biopesticide	Water Solubility at 20 °C (mg/L)	Partition co-efficient (pH 7, 20 °C)	Log P	Half-life in Soil (days)	Application
Abamectin	0.020	2.51×10^{04}	4.4	30	Insecticide, Acaricide, Nematicide
Azadirachtin A	2900	9.77×10^{00}	0.99	5	Insecticide, Acaricide, Fungicide
Bacillus thuringiensis subsp. *israelensis*	10	–	–	2.7	Insecticide, Bactericide, Nematicide
Capsaicin	10.3	3.02×10^{00}	0.48	5	Repellent, Insecticide, Miticide, Rodenticide
Garlic extract	1000000	3.24×10^{-02}	-1.49	–	Insecticide, Repellent
Nicotine	1000000	1.48×10^{01}	1.17	2	Insecticide
Pyrethrin I	0.96	7.94×10^{05}	5.9	2	Insecticide, Acaricide, Veterinary treatment
Rotenone	15.0	1.45×10^{04}	4.16	2	Insecticide, Acaricide, Veterinary treatment
Spinosad	7.6	1.26×10^{04}	4.1	14	Insecticide

Source: Bio-Pesticide DataBase, University of Hertfordshire, 2021 [17]

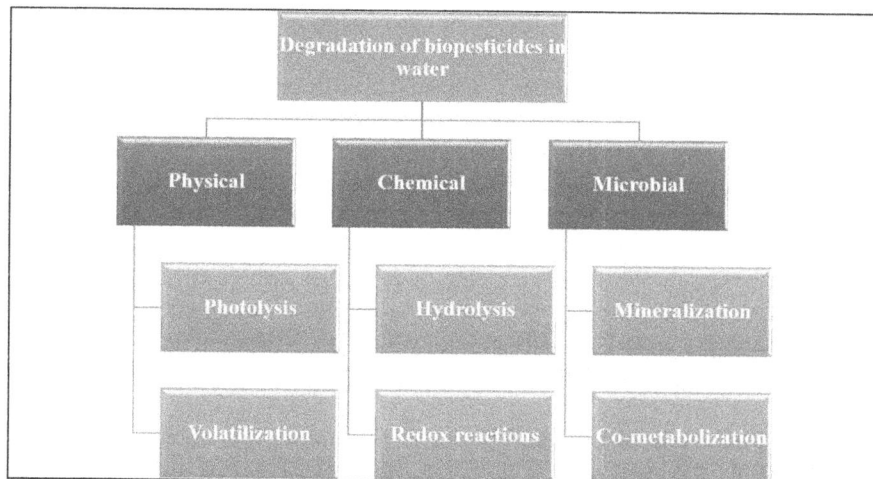

FIGURE 7.2 Degradation processes of biopesticides in the aquatic environment.

in water [18]. Compounds having a low biodegradation rate have a longer persistence in the environment and may contaminate water [16, 19]. Chemical reactions include hydrolysis and redox reactions. Hydrolysis is the degradation of a compound in the presence of water. Mainly compounds having ester bonds undergo hydrolysis in the presence of microbial esterase enzyme [20].

Photolysis is the physical process of decomposition, which occurs in the presence of light. Natural pyrethrins and rotenone are examples of biopesticides that undergo photodegradation and oxidation [21, 22]. Apart from oxidation and hydrolysis, biopesticides may experience microbe-induced decomposition reactions, mainly mineralization and co-metabolization processes [23]. The microbial species involved in the breakdown of biopesticides are *Bacillus, Serratia, Pseudomonas, Spingobium, Aerobacter*, and others [20]. In the case of mineralization, microbes break down the biopesticides into carbon dioxide, while co-metabolization, another microbial reaction, converts the biopesticides into other chemicals [16]. The microbial degradation of biopesticides depends on the complexity of the compound and the potential of microbial enzymes [24] (Figure 7.2).

7.5 POSSIBLE TOXIC EFFECTS OF BIOPESTICIDE RESIDUES IN WATER

Chemical pesticides are well known for causing toxicity to non-target species and triggering various human diseases. Biopesticides are comparatively safer in this aspect. But researchers have detected the presence of biopesticide residues in different ground and surface water sources and their toxic effects on aquatic life. Examples of such biopesticides include abamectin, azadirachtin, and spinosad [4]. Chemicals extracted from biopesticides are also harmful to humans [25].

Biopesticides Handbook

Rotenone is a biopesticide that has been reported to be highly toxic to fish [22]. Pyrethroids have a poisonous effect on axons and are 1,000 times more harmful to fish but less harmful to birds and mammals [26]. A potent nematicide derived from Spanish populations of *Artemisia absinthium* (var. Candial) has shown its ecotoxicity on non-target aquatic lives. The aqueous extract of this plant resulted in higher toxicity in its 0.2% dilution on *Daphnia magna*, *Vibrio fischeri*, and *Chlamydomonas reinhardtii*. The organic extract was highly toxic to *Daphnia magna* [27]. Saponin-rich botanical biopesticides also have aquatoxicity. Quillaja saponins, tea saponins, quinoa saponins, and Quil-A are harmful to the embryo of *Dania rerio* fish, followed by the embryo of *Lymnaea stagnalis*, and *Lymnaea stagnalis* itself [28]. Further data on the toxicity of different biopesticides are provided in Tables 7.2 and 7.3.

TABLE 7.2
Toxicity of Various Biopesticides on Aquatic Lives

Biopesticide	LD_{50} in Fish (mg/L)	Source	EC_{50} in Aquatic Invertebrates (mg/L)	Source
Abamectin	0.0087 (96 h)	*Oncorhynchus mykiss*	0.00056 (48 h)	*Daphnia pulex*
Azadirachtin A	>2.22 (96 h)	*Oncorhynchus mykiss*	3.54 (48 h)	*Daphnia magna*
Bacillus thuringiensis subsp. **Israelensis**	>370 (96 h)	*Oncorhynchus mykiss*	>50 (48 h)	*Daphnia magna*
Beauveria bassiana Vuillemin	7300 (96 h)	*Oncorhynchus mykiss*	4100 (48 h)	*Daphnia magna*
Citronella oil	17.3 (96 h)	*Oncorhynchus mykiss*	26.4 (48 h)	*Daphnia magna*
Garlic extract	>19.64 (96 h)	*Cyprinus Carpio*	> 9.3 (48 h)	*Daphnia magna*
Linalool	> 28.8 (96 h)	*Oncorhynchus mykiss*	36.7 (48 h)	Unknown species
Nicotine	4.0 (96 h)	*Oncorhynchus mykiss*	0.242 (48 h)	*Daphnia pulex*
Pyrethrin I	0.005 (96 h)	*Oncorhynchus mykiss*	0.012 (48 h)	*Daphnia magna*
Rotenone	0.0019 (96 h)	*Oncorhynchus mykiss*	0.004 (48 h)	*Daphnia magna*
Spinosad	27 (96 h)	*Oncorhynchus mykiss*	> 1.0 (48 h)	*Daphnia magna*

Source: Bio-Pesticide DataBase, University of Hertfordshire, 2021 [17]

TABLE 7.3
Terrestrial Ecotoxicology of Some Common Biopesticides

Biopesticide	Acute Oral LD_{50} (mg/kg) in Mammals	Source	Acute LD_{50} (mg/kg) in Birds	Source
Abamectin	8.7	*Rattus novergicus*	26	*Anas platyrhynchos*
Azadirachtin A	> 5000	*Rattus novergicus*	> 1000	*Colinus virginianus*
Bacillus thuringiensis subsp. Israelensis	> 5000	*Rattus novergicus*	> 5000	*Anas platyrhynchos*
Beauveria bassiana Vuillemin	> 1.8 x 10^{08}	*Rattus novergicus*	> 2667	*Coturnix japonica*
Citronella oil	> 4380	*Rattus novergicus*	> 2250	*Colinus virginianus*
Garlic extract	850	*Rattus novergicus*	–	–
Linalool	> 1700	*Rattus novergicus*	> 5620	*Colinus virginianus*
Nicotine	50	*Rattus novergicus*	> 17.8	Unknown species
Pyrethrin I	700	*Rattus novergicus*	> 51151	*Anas platyrhynchos*
Rotenone	> 132	*Rattus novergicus*	> 2600	*Anas platyrhynchos*
Spinosad	> 2000	*Rattus novergicus*	> 2000	*Anas platyrhynchos*

Source: Bio-Pesticide DataBase, University of Hertfordshire, 2021. [17]

7.6 ANALYTICAL TECHNIQUES FOR DETECTION AND DETERMINATION OF BIOPESTICIDE RESIDUES IN WATER

The selection of procedures for analyzing biopesticide residues in water samples depends on the biopesticides' chemical structure and physicochemical properties. The whole analytical process is divided into the following steps: collection of water samples, sample preparation, detection, and quantification of the biopesticide residue in that sample (see Figure 7.3).

7.6.1 SAMPLE COLLECTION

Water is collected from different water bodies in clean, dry 1L amber-colour glass containers and kept in ice and dark conditions until they are delivered to laboratories for further analysis [29].

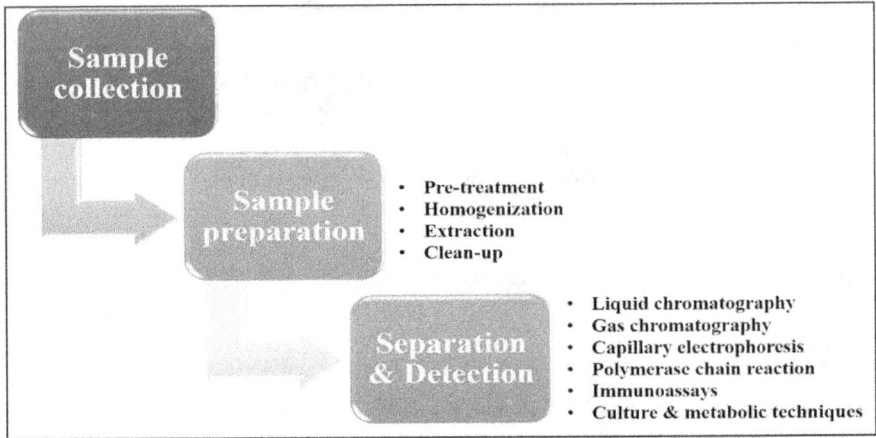

FIGURE 7.3 The flow of analysis of biopesticide residues in water samples.

7.6.2 SAMPLE PREPARATION

Sample preparation aims for the efficient extraction of the desired biopesticides and their residues from the collected water modules. The presence of other substances along with the biopesticides in the water samples (termed the matrix effect) creates problem during quantification of that target analyte. Therefore, proper removal of those interfering agents by proper sample preparation, using isotope-tagged internal standards and optimizing the extraction procedure, is necessary prior to analysis [30].

However, it is difficult as biopesticide residues are present in very low amounts in the aquatic environment. Sample preparation includes sample pre-treatment, homogenization, extraction, and clean-up [4].

7.6.2.1 Sample Pre-Treatment

This is the initial step in sample preparation. Pre-treatment makes the subsamples homogenous and representative of the collected samples. In the case of biopesticide residues, this is commonly done by filtration, which removes suspended particles that interfere with analysis.

7.6.2.2 Homogenization

Homogenization of the pre-treated subsamples is done or enhanced using sonication or mechanical shaking.

7.6.2.3 Extraction

After sample pre-treatment and homogenization, biopesticide residues are extracted from the water samples. Extraction may be done by liquid-liquid extraction (LLE), ultrasonic-assisted extraction (UAE), solid phase extraction (SPE), solid phase microextraction (SPME), or molecular-imprinted polymers (MIP).

LLE, also known as solvent extraction, is a classic technique for extracting biopesticide residues from water [31]. The water sample to be evaluated is mixed with an organic solvent in a separation funnel. The target biopesticide distributes between the immiscible liquids based on partition coefficient [32]. Factors contributing to the efficiency of LLE are the solubility of the biopesticide in that solvent and the number of repetitions of the extraction process. Cyclohexane, acetone, n-hexane, and dichloromethane extract non-polar compounds, and methanol, ethyl acetate, and acetonitrile extract polar compounds. But, in the case of polar biopesticides, recovery is poor, and it can be bettered by using multiple solvents and pH adjustment [33].

Hollow fiber liquid phase microextraction (HF-LPME) is a technique for analyzing biopesticides in water. It involves one donor phase and one acceptor. The acceptor phase is placed inside a porous hollow fiber, usually prepared of polypropylene. Biopesticide residue in an aqueous sample is determined using its partition coefficient. HF-LPME can be both two phased and three phased [34].

UAE is an ultrasound technique for separating biopesticide residue from the water. In this process, a solvent is added to the water sample to be tested, and extraction is done by placing that liquid mixture in a sonic bath [35]. This technique is also used to support any other extraction technique and in the process of extract clean-up [36].

SPE is the most common extraction technique used for analyzing biopesticide residues. In this technique, the analyte is adsorbed on a sorbent and then isolated by desorption using an eluent. Three steps involved in SPE are cartridge conditioning with solvents, sample loading, and elution of the analyte. Sometimes a washing step is required after loading to remove any interferents [37]. In general, no clean-up is needed after SPE. Commonly used sorbents include alumina, activated carbon, C18, Florisil, and hydrophilic-lipophilic balance copolymer [38].

7.6.2.4 Clean-Up

Cleaning up of extracts is done when the sample is not clean enough. Cleaning can be done by using gel permeation chromatography (GPC), SPE, micro-silica plus Florisil, micro-silica plus HPLC, etc. GPC removes high molecular weight interference such as organic compounds from the extract. Micro-silica and Florisil can be used to clean non-polar or polar interferences from the extracts [39].

7.6.3 Detection

This is the last step in the analytical procedure. Separation and detection of biopesticides are based on the type of biopesticides to be detected. Mainly chromatographic techniques such as liquid chromatography (LC) and gas chromatography (GC) are used for botanical and biochemical biopesticides, and immunoassays and polymerase chain reactions (PCR) are used to detect microbial biopesticides and PIPs. The choice of technique is influenced by boiling point, thermal stability, the solubility of the biopesticide in particular organic or aqueous solvents, molecular structure, electron ionization, and fragmentation. Biopesticides with high boiling temperatures and thermolabile compounds cannot be analyzed using GC. For LC analysis of a compound, it has to be soluble in the solvent. To serve the purpose, the technique should have a proper limit of detection (LOD) and quantitation (LOQ). The instrument used for analysis should have high selectivity and sensitivity. Different detectors are used to detect the presence of biopesticide residues in water samples based on specific functional groups in their structure [4] (Table 7.4).

The most widely used technique for analyzing biopesticides in water is LC. Most biochemical pesticides cannot be analyzed by GC because of high polarity, low volatility, and poor thermal stability [14]. In this technique, the separation of compounds is based on their differential affinity towards the stationary and mobile phases [40]. In the case of biopesticide residue analysis, reversed-phase liquid chromatography (RP-LC) is most commonly used. Thereafter, residue is measured using various types of detectors such as a diode array detector (DAD), a mass spectrometry detector (MSD), a tandem mass spectrometer (MS/MS), or an ultraviolet-visible wavelength detector (UV) [41].

GC is another technique used to detect biopesticide residues in water. In this technique, the sample to be analyzed is first volatilized on the injector, and then the analyte vapor enters the analytical column. Then there will be a gas-liquid interaction between the analyte, which is gaseous, and the liquid stationary phase. Separation is based on the differential partitioning of the gaseous sample with the stationary phase. The stationary phase columns are mostly composed of 100% polydimethylsiloxane, 95% dimethyl—5% diphenylpolysiloxane, 14% cyanopropyl-phenyl—and 86% dimethylpolysiloxane [4]. Detectors to detect the target compounds include an electron capture detector, a flame ionization detector, a nitrogen-phosphorus detector, or a mass spectrometer [42]. Hyphenated techniques such as GC-MS and GC-MS/MS are also used in biopesticides analysis [43]. Despite having high sensitivity and effectiveness, the GC technique has some limitations. Biopesticides commonly have highly complex chemical structures, large molecular weights, and high boiling points, so they are not eligible to be analyzed by GC. Derivatization of the compounds is required in this case. But derivatization takes time, is tedious, and may lead to inaccurate and less reproducible results. That is why LC and immunoassay techniques are used alternatively to analyze biopesticide residues in water [14, 44].

Capillary electrophoresis (CE) is a method of separating biopesticides from aqueous samples [45]. Capillary zone electrophoresis is most commonly used where compounds are separated in a buffer solution. After separation, biopesticidal compounds are detected by detectors such as UV, DAD, or MS [44].

Microbial pesticides and water PIPs are detected using an immunoassay (IA), polymerase chain reaction (PCR), culture, or metabolic technique. Chromatographic techniques are expensive and require tiresome sample preparation. Immunoassay (IA) is a substitute for them and does not require extraction. Moreover, IA is a rapid and sensitive method [46]. Microbial proteins such as Cry1Ab and pyrethrins are detected using enzyme-linked immune-sorbent assay (ELISA) [47, 48]. The genes responsible for executing pesticide activity are isolated, amplified, and quantified in PCR [49]. The measurements of gene amplification are taken during the elongation phase and are called real-time PCR. Quantitative PCR (qPCR) is the most extensively used PCR technique. The qPCR products are mostly fluorescent labeled [50]. This genetic technique can be used to analyze microbes that are not culturable. In the culture method, the water sample is incubated with agar. The microbes present in the water sample are then isolated and identified using the proper technique [49]. In the case of metabolic techniques, a phenotypic identification method, the spent

TABLE 7.4

Techniques Used for Analyzing Biopesticide Residue in Water Samples

Biopesticides	Techniques Used for Extraction and Detection of Biopesticides in Water	Reference
Abamectin	HF-LPME followed by LC-MS/MS	[52]
Azadirachtin	HPLC	[53]
Rotenone	Mini SPE followed by UHPLC-Q-TOF/MS	[54]
DNA of *Choristoneura fumiferana* NPVegt-/lacZ⁺	PCR	[55]
Pyrethroids	UA dispersive LLME	[56]
Cry1Ab gene of *B. thuringiensis kurstaki*	qPCR	[57]
Cry1Ab protein of *B. thuringiensis*	ELISA	[48]

media or the metabolic products of the microorganism are detected using chromatography or electrophoresis [51]. Other techniques involved in microbial pesticides and PIP analysis are whole-cell recognition and surface recognition analysis used to detect cell-surface lipids and proteins [49].

7.7 CONCLUSION AND FUTURE DIRECTIONS

Biopesticides are being used in agriculture and animal farming as an excellent alternative to conventional pesticides, which harm the environment and humans. Biopesticides are relatively safer. However, biopesticides entering the aquatic environment may pose a risk to aquatic life, such as fish, crabs, fauna, flora, etc. Various studies are being conducted on biopesticide residues in water. But more research is needed on the biodegradation, toxicity, and analysis of biopesticides in water. Techniques to reduce the exposure of biopesticides to aquatic systems should be developed. Farmers should also be aware of the proper utilization of biopesticides to avoid any health and environmental hazards. Modern techniques for sample analysis can also be helpful. As biopesticide use is increasing daily to prevent the harmful effects of chemical pesticides, more studies on the treatment of water contaminated with these compounds can be beneficial shortly.

REFERENCES

1. Dhakal R, Singh DN. Biopesticides: A key to sustainable agriculture. Int J Pure App Biosci. 2019;7(3):391–6.
2. Kandpal V. Biopesticides. Int J Environ Res Dev. 2014;4(2):191–6.
3. Nkechi EF, Ejike OG, Ihuoma NJ, Maria-Goretti OC, Francis U, Godwin N, et al. Effects of aqueous and oil leaf extracts of Pterocarpus santalinoides on the maize weevil, Sitophilus zeamais pest of stored maize grains. Afr J Agric Res. 2018;13(13):617–26.

4. Sanganyado E, Munjanja B, Nyakubaya VT. Biopesticide residue in water. In Biopesticides handbook. CRC Press, Boca Raton, 2015:93–117.
5. Rathore HS, Nollet LM. Pesticides: Evaluation of environmental pollution. CRC Press, Boca Raton, FL, 2012.
6. Gupta S, Dikshit A. Biopesticides: An ecofriendly approach for pest control. J Biopesticides. 2010;3:186.
7. Market B. Biopesticides Market by Type (Bioinsecticides, Biofungicides, Bionematicides, and Bioherbicides), Source (Microbials, Biochemicals, and Beneficial Insects), Mode of Application, Formulation, Crop Application, and Region-Global Forecast to 2025.© 2020 [June, 2020]. https://www.marketresearch.com/MarketsandMarkets-v3719/Biopesticides-Type-Bioinsecticides-Biofungicides-Bionematicides-31989580/
8. Hikal WM, Baeshen RS, Said-Al Ahl HA. Botanical insecticide as simple extractives for pest control. Cogent Biol. 2017;3(1):1404274.
9. Sharma S, Malik P. Biopesticides: Types and applications. Int J Adv Pharm, Biol Chem. 2012;1(4):508–15.
10. Seaman D. Trends in the formulation of pesticides—an overview. Pest Sci. 1990;29(4):437–49.
11. Battaglin WA, Sandstrom MW, Kuivila KM, Kolpin DW, Meyer MT. Occurrence of azoxystrobin, propiconazole, and selected other fungicides in US streams, 2005–2006. Water Air Soil Pollut. 2011;218(1):307–22.
12. Chandler D, Bailey AS, Tatchell GM, Davidson G, Greaves J, Grant WP. The development, regulation and use of biopesticides for integrated pest management. Philos Trans R Soc Lond B Biol Sci. 2011;366(1573):1987–98.
13. Jørgensen LF, Kjær J, Olsen P, Rosenbom AE. Leaching of azoxystrobin and its degradation product R234886 from Danish agricultural field sites. Chemosphere. 2012;88(5):554–62.
14. Rodrigues ET, Lopes I, Pardal MÂ. Occurrence, fate and effects of azoxystrobin in aquatic ecosystems: A review. Environ Int. 2013;53:18–28.
15. Konstantinou IK, Hela DG, Albanis TA. The status of pesticide pollution in surface waters (rivers and lakes) of Greece. Part I. Review on occurrence and levels. Environ Pollut. 2006;141(3):555–70.
16. Syafrudin M, Kristanti RA, Yuniarto A, Hadibarata T, Rhee J, Al-Onazi WA, et al. Pesticides in drinking water—a review. Int J Environ Res Pub Heal. 2021;18(2):468.
17. Lewis KA, Tzilivakis J, Warner DJ, Green A. An international database for pesticide risk assessments and management. Hum Ecol Risk Assess. 2016;22(4):1050–64.
18. Gunasekara A. Environmental fate of Pyrethrins. Environmental monitoring branch. Department of Pesticide Regulation, Sacramento, CA, 2014:1–19.
19. Lushchak VI, Matviishyn TM, Husak VV, Storey JM, Storey KB. Pesticide toxicity: A mechanistic approach. Excli J. 2018;17:1101.
20. Cycoń M, Piotrowska-Seget Z. Pyrethroid-degrading microorganisms and their potential for the bioremediation of contaminated soils: A review. Front Microbiol. 2016;7:1463.
21. Mpumi N, Mtei KM, Machunda R, Ndakidemi PA. The toxicity, persistence and mode of actions of selected botanical pesticides in Africa against insect pests in common beans, P. vulgaris: A review. Am J Plant Sci. 2016;7.
22. Hinson D. Rotenone characterization and toxicity in aquatic systems. Principles of Environmental Toxicity. University of Idaho, Moscow, 2000.
23. National Research Council. Soil and water quality: An agenda for agriculture. National Academies Press, Washington, DC, 1993.
24. Ortiz-Hernández ML, Sánchez-Salinas E, Dantán-González E, Castrejón-Godínez ML. Pesticide biodegradation: Mechanisms, genetics and strategies to enhance the process. Biodegradation. 2013:251–87.
25. Wakeil N. Botanical pesticides and their mode of actions. Review article. Gesunde Pflanzen. 2013;65(4):125–49.

26. Yang C, Lim W, Song G. Mediation of oxidative stress toxicity induced by pyrethroid pesticides in fish. Comp Biochem Physiol Part C Toxicol Pharmacol. 2020;234:108758.
27. Pino-Otín MR, Ballestero D, Navarro E, González-Coloma A, Val J, Mainar AM. Ecotoxicity of a novel biopesticide from Artemisia absinthium on non-target aquatic organisms. Chemosphere. 2019;216:131–46.
28. Jiang X, Hansen HCB, Strobel BW, Cedergreen N. What is the aquatic toxicity of saponin-rich plant extracts used as biopesticides? Environ Pollut. 2018;236:416–24.
29. Hasanuzzaman M, Rahman M, Salam M. Identification and quantification of pesticide residues in water samples of Dhamrai Upazila, Bangladesh. Appl Wat Sci. 2017;7(6):2681–8.
30. Bylda C, Thiele R, Kobold U, Volmer DA. Recent advances in sample preparation techniques to overcome difficulties encountered during quantitative analysis of small molecules from biofluids using LC-MS/MS. Analyst. 2014;139(10):2265–76.
31. Wu J, Lu J, Wilson C, Lin Y, Lu H. Effective liquid—liquid extraction method for analysis of pyrethroid and phenylpyrazole pesticides in emulsion-prone surface water samples. J Chromatogr A. 2010;1217(41):6327–33.
32. Clement R, Hao C. Liquid—liquid extraction: Basic principles and automation. 2012.
33. Suffet IH, Faust SD. Liquid-liquid extraction of organic pesticides from water: The p-value approach to quantitative extraction. ACS Publications, Washington, DC, 1972.
34. Ghambarian M, Yamini Y, Esrafili A. Developments in hollow fiber based liquid-phase microextraction: Principles and applications. Microchim Acta. 2012;177(3):271–94.
35. Lavilla I, Bendicho C. Fundamentals of ultrasound-assisted extraction. Water extraction of bioactive compounds. Elsevier, Amsterdam, 2017:291–316.
36. Albero B, Tadeo JL, Pérez RA. Ultrasound-assisted extraction of organic contaminants. TrAC Trends Anal Chem. 2019;118:739–50.
37. Żwir-Ferenc A, Biziuk M. Solid phase extraction technique—trends, opportunities and applications. Polish J Environ Stud. 2006;15(5).
38. Wu W, Wu Y, Zheng M, Yang L, Wu X, Lin X, et al. Pressurized capillary electrochromatography with indirect amperometric detection for analysis of organophosphorus pesticide residues. Analyst. 2010;135(8):2150–6.
39. Björklund E, von Holst C, Anklam E. Fast extraction, clean-up and detection methods for the rapid analysis and screening of seven indicator PCBs in food matrices. TrAC Trends Anal Chem. 2002;21(1):40–53.
40. Bhattu, M, Kathuria, D, Billing, BK, Verma, M. Chromatographic techniques for the analysis of organophosphate pesticides with their extraction approach: A review (2015–2020). Anal Methods. 2022;14(4):322–58.
41. Cequier E, Sakhi AK, Haug LS, Thomsen C. Development of an ion-pair liquid chromatography—high resolution mass spectrometry method for determination of organophosphate pesticide metabolites in large-scale biomonitoring studies. J Chromatogr A. 2016;1454:32–41.
42. Santos F, Galceran M. The application of gas chromatography to environmental analysis. TrAC Trends Anal Chem. 2002;21(9–10):672–85.
43. Reilly TJ, Smalling KL, Orlando JL, Kuivila KM. Occurrence of boscalid and other selected fungicides in surface water and groundwater in three targeted use areas in the United States. Chemosphere. 2012;89(3):228–34.
44. Elbashir AA, Aboul-Enein HY. Separation and analysis of triazine herbcide residues by capillary electrophoresis. Biomed Chromatogr. 2015;29(6):835–42.
45. Chang P-L, Hsieh M-M, Chiu T-C. Recent advances in the determination of pesticides in environmental samples by capillary electrophoresis. Int J Environ Res Pub Heal. 2016;13(4):409.
46. Yao J, Wang Z, Guo L, Xu X, Liu L, Xu L, et al. Advances in immunoassays for organophosphorus and pyrethroid pesticides. TrAC Trends Anal Chem. 2020;131:116022.

47. Watanabe T, Shan G, Stoutamire DW, Gee SJ, Hammock BD. Development of a class-specific immunoassay for the type I pyrethroid insecticides. Anal Chim Acta. 2001;444(1):119–29.
48. Strain KE, Whiting SA, Lydy MJ. Laboratory and field validation of a Cry1Ab protein quantitation method for water. Talanta. 2014;128:109–16.
49. Noble RT, Weisberg SB. A review of technologies for rapid detection of bacteria in recreational waters. J Water Health. 2005;3(4):381–92.
50. Botes M, de Kwaadsteniet M, Cloete TE. Application of quantitative PCR for the detection of microorganisms in water. Anal Bioanal Chem. 2013;405(1):91–108.
51. Plimmer JR. Analysis, monitoring, and some regulatory implications. In Biopesticides: Use and delivery. Springer, Berlin, 1999:529–52.
52. Park JH, Abd El-Aty A, Rahman MM, Choi JH, Shim JH. Application of hollow-fiber-assisted liquid-phase microextraction to identify avermectins in stream water using MS/MS. J Separat Sci. 2013;36(17):2946–51.
53. de Menezes ML, Dalbeto AC, Cruz C. Determination of biopesticide azadirachtin in samples of fish and in samples of water of fish ponds, using cromatography liquid of high performance. CEP. 2004;17033(360):2.
54. Peng L-Q, Ye L-H, Cao J, Chang Y-X, Li Q, An M, et al. Cyclodextrin-based miniaturized solid phase extraction for biopesticides analysis in water and vegetable juices samples analyzed by ultra-high-performance liquid chromatography coupled with quadrupole time-of-flight mass spectrometry. Food Chem. 2017;226:141–8.
55. England L, Trevors J, Holmes S. Extraction and detection of baculoviral DNA from lake water, detritus and forest litter. J Appl Microbiol. 2001;90(4):630–6.
56. Yan H, Liu B, Du J, Yang G, Row KH. Ultrasound-assisted dispersive liquid—liquid microextraction for the determination of six pyrethroids in river water. J Chromatogr A. 2010;1217(32):5152–7.
57. Douville M, Gagné F, Blaise C, André C. Occurrence and persistence of Bacillus thuringiensis (Bt) and transgenic Bt corn cry1Ab gene from an aquatic environment. Ecotoxicol Environ Saf. 2007;66(2):195–203.

8 Biopesticide Residues in Soil

Beatriz Martín-García, Alba Reyes-Ávila, José L. Martínez-Vidal, Antonia Garrido Frenich, and Roberto Romero-González

CONTENTS

8.1 BIOPESTICIDE RESIDUES IN SOILS

8.1.1 INTRODUCTION

More than 99.7% of pesticides are accumulated in the environment, undergoing various physicochemical catabolism and biodegradable processes that are strongly linked to the composition and activity of the soil microbial community.[1] Microorganisms play an important role in different soil processes, such as the decomposition of organic matter and the nutrient cycle, which are greatly involved in the function of soil ecosystems.[2,3] Therefore, the loss of the soil microbial community structure can lead to significant changes in fertility, which is a prerequisite for plant growth.[4] For this reason, in the last few years, biopesticides derived from plants, animals, microbes, and other natural substances have been used for pest control in crops, minimizing the use of chemical pesticides.[5]

Biopesticide registration protocols and dossier requirements differ slightly from country to country, especially in Europe and Organization Economic Cooperation Development (OECD) countries. Different data, such as identification and description of the organism/ingredients, biological qualities, bioefficacies in the laboratory/ screen house and field, safety/ecotoxicity studies, toxicology, and packaging are necessary to register biopesticides globally.[6] In the European Union (EU), biopesticides are not recognized as a regulatory category and are registered as plant protection products (PPPs) under EC 1107/2009. According to this regulation, biopesticides have only been included in PPP when they present a clear benefit for plant production, and they do not pose any negative effects on the health of human or animals or undesirable environment effects.[7] However, there are fewer biopesticide-active substances registered in the EU than in the United States, India, or Brazil because the EU-based biopesticide regulations are more complex.[8]

For agricultural purposes, biopesticides are usually applied in higher concentrations than synthetic ones, which pushes them away from the idea that they are completely harmless. As a result, there is a growing interest in determining their mobility and fate in order to assess their potential to become environmental toxins.[9] Previous studies have evaluated the biocidal effects of certain biopesticides in the soil by using different detection tests, such as polyphasic microbial assays, polymerase chain reaction-denaturing gradient gel electrophoresis (PCR-DGGE), or soil microbial biomass, among others.[4,10,11] These studies have showed that certain biopesticides can exhibit negative effects on the soil bacterial community. Therefore, the detection as well as the evaluation of the toxicity of biopesticides in soil is a precondition to improving the regulation of biopesticides in the near future.[12] This chapter covers the analysis and presence of biopesticides in soils. The first section includes soil pre-treatment to recover these substances by using conventional extraction techniques such as Shoxhlet,[13] solid-liquid extraction (SLE) by shaking,[14–16] headspace-solid phase microextraction (HS-SPME),[17] ultrasound assisted extractions (UAE),[9] and QuEChERS (quick, easy, cheap, effective, rugged, and safe),[18–21] this last being the most employed extraction technique. The second part collects information about the instrumental analysis employed for the detection of these compounds. Most of them are based on gas chromatography (GC) and liquid chromatography (LC), coupled with various detection systems, such as UV, diode array, or mass spectrometry (MS). The last section includes information about the fate of biopesticides in soils by different processes, including adsorption and degradation, as well as their effects on the microbial community.

8.2 SOIL SAMPLE PRE-TREATMENT

Sample pre-treatment is the most important step before the analytical determination because of the low concentration levels of biopesticides in soils, their different chemical properties, and the complexity of the matrix. A high recovery of these analytes is needed in order to improve the sensitivity of the method and the reliability of the results.[22] Extraction of biopesticides from soil is based on differences in their physical and chemical properties, such as solubility, polarity, and volatility. In this sense, selective extraction procedures simplify or completely eliminate the sample purification steps required prior to chromatographic analysis.[23]

Previous studies have reported the use of different techniques for the extraction of biopesticides from soils (Table 8.1). The conventional Soxhlet extraction has been used to extract azadirachtin A from soil. This method required high solvent consumption (200 mL) and a long extraction time (6 hours).[13] Therefore, other, more

TABLE 8.1

Extraction Techniques Used for Biopesticide Determination in Soils [a]

Extraction Technique	Biopesticide	Recovery	Description	Ref
Soxhlet	Azadirachtin A	54–79%	50 g of soil + 200 mL Extraction time: 6 h	13
SLE	Rotenone	95–100%	5 g of soil + 10 mL ethyl acetate Extraction time: 30 min	14,15
	Nicotine, sabadine, veratridine, rotenone, azadirachtin, cevadine, deguelin, spinosad, pyrethrins	1–36%	5 g of soil + 10 mL of ethyl acetate Extraction time: 30 min* Type: rotatory agitator	19
		3–37%	10 g of soil + 30 mL of ethyl acetate/methanol (3:1, v/v) Extraction time: 30 min Type: sonicator	19
SLE (SPE)	Pyrethrins (pyrethrin I and II, cinerin I and II, and jasmolin I and II)	88.1–104%	20 g of soil + 20 mL MeCN + 3 g NaCl Extraction time: 2 min 10 mL supernatant MeCN was dried and redissolved with 1 mL acetone + n-hexane (1:9; v/v) SPE clean-up: Anhydrous Na$_2$SO$_4$ (1 g) was added into Florisil SPE cartridge (preconditioned with 10 mL n-hexane)	16
	Pyrethrin I and II	NP	Air-dried soil samples and biopesticide solutions at 1:5, 1:25, 1:50, and 1:100 g/mL were mixed for 24 h in rotary extractor SPE clean-up: C18 cartridges of 500 mg octadecyl. The cartridges were eluted with 14 mL of methanol	28
PLE	Nicotine, sabadine, veratridine, rotenone, azadirachtin, cevadine, deguelin, spinosad, pyrethrins	3–53%	5 g of soil+ ethyl acetate/methanol (3:1, v/v)	19

(Continued)

TABLE 8.1 (Continued)

Extraction Technique	Biopesticide	Recovery	Description	Ref
HS-SPME	96 xenometabolites	NP	6 g of soil Temperature: 40 °C Extraction time: 30 min Type of fiber coating: 50/30 μm DVB/CAR/PDMS	24
UAE	Rotenone	72–92%	30 g of soil + 50 mL MeCN Extraction time: 30 min Clean-up step: 2 g anhydrous sodium, 5 g Florisil/activated carbon (90/10 w/w), and 2 g anhydrous sodium bottom-up. Elution with 50 mL acetone/petroleum ether (v/v, 50/50)	27
	Cinnamaldehyde and diallyl disulfide	70 and 61%	6 g of soil + 3 mL of ethyl acetate Extraction time: 20 min	9
QuEChERS	Azadirachtin, rotenone, spinosyn A and D	83–104%	5 g of soil + 5 g of water + 50 μL of internal standard solution (isoproturon-D6 at 150 μg mL^{-1}) + 100 μL acetic acid in 10 mL MeCN Extraction time: 5 min Salts and buffers: 0.5 g disodium hydrogen citrate sesquihydrate, 1 g trisodium citrate dihydrate, 4 g anhydrous magnesium sulfate, and 1 g sodium chloride Clean-up step: 150 mg PSA and 150 mg C18	18
	Nicotine, sabadine, veratridine, rotenone, azadirachtin, cevadine, deguelin, spinosad, pyrethrins	30–110%	5 g of soil Hydration step: 2.5 mL of water was added and shaken for 30 min Solvent: 5 mL MeCN with 1% acetic acid Extraction time: 1 min Salts and buffers: 4 g MgSO$_4$, 4 g NaCl, 0.5 g disodium hydrogen citrate sesquihydrate, 1 g trisodium citrate dihydrate No clean-up	19

Extraction Technique	Biopesticide	Recovery	Description	Ref
	Spinosad	98–102%	10 g of soil Hydration: 7 mL of water was added and was vortexed for 25–30 min Solvent: 10 mL MeCN Extraction time: 5–6 min. Clean-up step: 1.5 mL aliquot of supernatant was transferred to a 2-mL C-18 SPE tube	20
	Carvone	95.7–104.2%	5 g + 5 mL of water + 10 mL n-hexane Extraction time: 2 min Salts and buffers: 2 g MgSO$_4$ and 1 g NaCl, Clean-up step: 1 mL of the supernatant was transferred to a 2.5 mL polypropylene tube containing 40 mg C18 and 100 mg Na$_2$SO$_4$	21
	d-limonene oxide isomers, (-)-carveol isomers, and (-)-carvone	71.2 to 114.5%	5 g of soil + 10 mL n-hexane + 5 mL of water Extraction time: 2 min. Salts and buffers: 2 g MgSO$_4$ and 1 g NaCl Extraction time: 1 min Clean-up step: 1 mL of the supernatant with 2.5 mL polypropylene tube containing 40 mg C18 and 5 mg GCB and vortexed for 2 min	26

Notes: [a] Abbreviations: SLE: solid liquid extraction; SPE: solid phase extractions; HS-SPME: headspace-solid phase microextraction; UAE: ultrasound-assisted extractions; QuEChERS: quick, easy, cheap, effective, rugged, and safe; NP: not provided; PSA: primary secondary amine; C18: octadecylsilyl; DVB: divinyl-benzene; CAR: carboxen; PDMS: polydimethylsiloxane; MeCN: acetonitrile efficient techniques have been employed for the recovery of certain biopesticides in soils, such as the SLE technique by shaking. This has been used for the extraction of rotenone, using 10 mL of ethyl acetate as the extraction solvent, and has an extraction time of 30 minutes, obtaining high recoveries (95–100%).[14,15] Feng et al.[16] reported the determination of pyrethrin residues, including pyrethrin I and II, cinerin I and II, and jasmolin I and II in turnips (turnip leaves, turnip tubers, and the whole of the plant) and cultivated soil. The extraction was carried out by vortex agitation of soil with 20 mL of acetonitrile and 3 g of NaCl for 2 minutes. The dry extract was redissolved with 1 mL acetone + n-hexane (1:9; v/v) for solid phase extraction (SPE) clean-up, using 1 g of anhydrous sodium sulfate (Na_2SO_4), which was added into a Florisil cartridge (SPE cartridge, 500 mg, 3 mL), and suitable recoveries were achieved (88–104%).

A recent study used HS-SPME to recover the volatile metabolomics residues (xenometabolites) that were released after the application of *Myrica gale* methanolic extract in soils.[24] This study employed as optimal conditions 5 minutes for incubation time, 30 minutes of extraction time, 40 °C extraction temperature, and 50/30 µm Divinyl-benzene/Carboxen/Polydimethylsiloxane as fiber coating. The extracted compounds were desorbed and introduced in the analytical instrument.

UAE is another fast and efficient technique for extracting biopesticides. The high yield from UAE is due to the cavitation, which allows disruption of the cell wall by ultrasound waves.[25] López-Serna et al.[9] reported the use of UAE during 20 minutes for the extraction of cinnamaldehyde (CAD) and diallyl disulfide (DAD) in soil by using different solvents, achieving the highest recovery rates of 70% and 61% for CAD and DAD, respectively, when ethyl acetate was used.

QuEChERS is an extraction and clean-up technique that has shown higher recoveries for different classes of compounds, including biopesticides, chloroalkanes, phenols, and perfluoroalkyl substances.[22] For instance, Drozdzyński and Kowalska[18] determined azadirachtin, rotenone, spinosyn A, and spinosyn D for the first time by using a citrate-buffered QuEChERS extraction method followed by a clean-up step by dispersive SPE (d-SPE), using primary secondary amine (PSA) and octadecylsilyl (C18). The method employed 5 g of soil, 5 g of water and 0.1% of acetic acid in 10 mL of acetonitrile (MeCN), and the mixture was shaken for 5 minutes. Recoveries higher than 83% were achieved with this extraction method. Another study reported the determination of 15 biopesticides in soil including nicotine, sabadine, veratridine, rotenone, azadirachtin, cevadine, deguelin, spinosad D, pyrethrins, and piperonyl butoxide in agricultural soils.[19] This study evaluated different extraction procedures such as SLE using mechanical shaking, sonication, pressurized liquid extraction (PLE), and modified-citrate QuEChERS. Ethyl acetate was used for SLE sonication and PLE, whereas water and 1% of acetic acid in acetonitrile was used for QuEChERS (Table 8.1). Recovery values were 30–110% when QuEChERS was tested, which were higher than those obtained with SLE (3–37%) and PLE (3–53%) for all compounds as shown in Figure 8.1. Therefore, this approach was chosen as the most suitable extraction technique for the determination of biopesticides.[19] El-Saeid et al.[20] applied a non-buffered citrate QuEChERS method to extract spinosad from soils. For that, 10 g of soil with 7 mL of water were shaken for 25–30 minutes, and

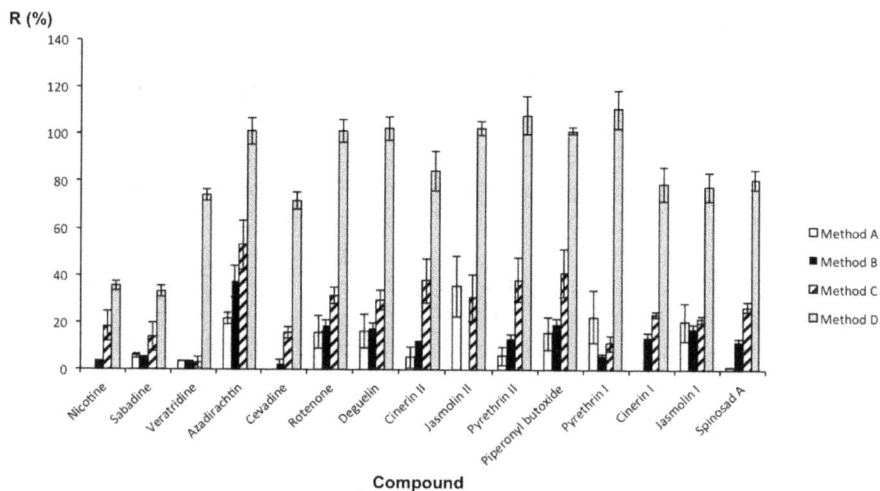

FIGURE 8.1 Comparison of extraction procedures for the determination of biopesticides in soil samples by ultrahigh pressure LC-MS/MS. Method A: Solid-liquid extraction using mechanical shaking; Method B: Solid-liquid extraction using sonication; Method C: Pressurized liquid extraction; Method D: Modified QuEChERS.

Source: Reproduced with permission from John Wiley and Sons (Ref. 19).

then MeCN (10 mL) was added to the samples, and the final mixture was shaken for 5–6 minutes. A clean-up step based on d-SPE, using C18 as sorbent, was utilized, obtaining recoveries from 98 to 102%. Huang et al.[21] extracted and purified carvone from soil using a QuEChERS-based method. Briefly, 5 g of soil was placed into a polypropylene tube with 10 mL of *n*-hexane and 5 mL of ultra-pure water and vortexed for 2 minutes. Then the extract was cleaned up with a polypropylene tube containing 40 mg of C18 and 100 mg of $NaSO_4$ and shaken with vortex for 2 minutes. This method provided recoveries from 95.7 to 104.2%.[21] D-Limonene has also been extracted with the same mixture of solvents as in the previous study, and the extract was purified with C18 and graphitized carbon black (GCB), achieving recoveries from 71.2 to 114.5%.[26]

Comparing the extraction techniques, QuEChERS showed better biopesticide recoveries with less extraction solvent (1/1 and 1/2 (*w/v*)) than other common extraction techniques, such as Soxhlet, SLE, or UAE (1/2–3/5 (*w/v*)).[13–15,18,19,27] Lower values of RSD were also observed for the determination of nicotine, sabadine, veratridine, rotenone, azadirachtin, cevadine, deguelin, spinosad, and pyrethrins by QuEChERS than with SLE and PLE.[19] It is important to note that the extraction of pyrethrins by an assisted agitation SLE with vortex employed acetonitrile in the same solid-to-solvent ratio as QuEChERS (1/1 (*w/v*)), obtaining similar recoveries with a shorter extraction time (1–2 minutes).[16,19] In addition, previous studies reported similar recoveries of 95–100% for rotenone by assisted agitation SLE with ethyl acetate in the same ratio as QuEChERS (1/2 (*w/v*)).[14,15,19] However, QuEChERS needed a rehydration step of the soil, and this can take up to 30 minutes, increasing the overall extraction time.[19] Other studies added water with the extraction solvent without a previous hydration step, reducing the sample preparation.[21,26] In summary, QuEChERS could be the best option for extracting biopesticides from soils. For that purpose, an optimization of extraction parameters, such as the sample-to-solvent ratio, solvent type, and the extraction time is needed to establish the best conditions for the highest recoveries of specific biopesticides from soils. The most employed extraction solvent in QuEChERS was a mixture of 1% of acetic acid in MeCN for the recovery of azadirachtin, rotenone, spinosyn A and D, nicotine, sabadine, veratridine, rotenone, azadirachtin, cevadine, deguelin, spinosad, and pyrethrins,[18–20] whereas *n*-hexane was recently used to extract d-limonene oxide isomers, (-)-carveol isomers, and (-)-carvone.[21,26] The ratio of solid to solvent was from 2/1 to 1/2 (*w/v*), and the time of extraction ranged from 1 to 6 minutes.[18–21,26]

8.3 INSTRUMENTAL ANALYSIS

After suitable sample preparation, instrumental analytical methods are applied to provide key information about the composition of biopesticides in processed soil samples. GC and LC are the most used methods, providing a quantitative analysis with good resolution. Classic detectors provide limited sensitivity and information on the compounds present in the samples. Therefore, MS is commonly coupled with chromatographic techniques, providing more information to identify compounds based on their mass-to-charge (*m/z*) values. Next, the main analytical methods employed for the detection of certain biopesticides in soils are described.

8.3.1 CHROMATOGRAPHIC METHODS

GC has been used for the analysis of biopesticides in soil coupled with MS analysers because most biopesticides have a high volatility.[13] However, certain biopesticides such as rotenone have a low volatility and/or thermolabile characteristics, so they are analysed by LC. Table 8.2 shows an overview of the main characteristics of the GC and LC methods employed for the detection of certain biopesticides in soils. It can be observed that the (5%-phenyl)-methylpolysiloxane phase (HB-5MS or HP-5MS GC columns) was the stationary phase commonly used to separate biopesticides, including cinnamaldehyde and diallyl disulfide, limonene oxide isomers, (-)-carveol isomers, carvone, pyrethrins, and xenometabolites, whereas 5% phenyl polysilphenylsiloxane was the stationary phase used to separate spinosad. Columns with different lengths between 30 and 60 m can be used, with an internal diameter of 0.25 mm and 0.25 μm of film thickness.[9,16,20,21,24,26] These stationary phases provide excellent separation and robustness for analytical applications, delivering excellent inertness with active compounds. In addition, the injection volume ranged from 1 μL to 5 μL, whereas flow rate was usually equal to or lower than 1.5 mL/min when MS was used and 15 mL/min when a flame ionization detector (FID) was applied for the detection of azadirachtin A.[13] The splitless injection mode allowed the detection of several biopesticides that were found in low quantities in soils, improving the sensitivity of the developed method. Nevertheless, split injection with a ratio 5:1 has been shown to be suitable to analyse limonene oxide isomers, carveol isomers, and carvone because their concentration is high enough to be detected (LOD of 2–16 µg/kg and LOQ 6–50 µg/kg).[21,26]

In the case of LC, reversed phase is applied with C18 as stationary phase, utilising columns with lengths ranging from 100 to 250 mm.[14,18,19,27] In general, it should be noted that when reversed phase is used, nonpolar compounds such as pyrethrins are strongly retained, whereas polar compounds such as nicotine are slightly retained. Conventional high-performance liquid chromatography (HPLC) had been used in previous investigations by using stationary phase with particle sizes of 5 µm, providing an analysis time of more than 15 minutes for a single compound (rotenone).[14,15,27] The use of columns filled with sub-2 µm particles (ultra-high performance liquid chromatography [UHPLC]) provided better chromatographic resolution than conventional HPLC, reducing analysis time, obtaining narrower peaks, increasing the signal-to-noise ratio, and improving the sensitivity of the analytical method. Therefore, these properties enabled simultaneous determination of more than one compound in less than 10 minutes.[18,19] The composition of the mobile phase affected the separation of analytes, influencing their retention times, selectivity, and peak efficiency. A mixture of acetonitrile or methanol and water is commonly used in the mobile phase. Some additives can be added to the water, such as ammonium acetate,[18] ammonium formate,[19] or trifluoroacetic acid,[15] to maintain a relatively constant pH upon dilution and improve the ionization, avoiding retention and selectivity changes.[27] The isocratic elution mode has been used for the determination of rotenone (Table 8.2), although a gradient profile has been commonly employed for the simultaneous determination of several biopesticides to reduce analysis time.[18,19,28] Finally, it is important to note that injection volume ranged between 5 µL to 100 µL, whereas the flow rate was 0.3–0.4 mL/min with the MS detection, and it was 1.0–1.5 mL/min with UV or DAD detection (see Table 8.2).

8.3.2 DETECTION

Traditional analytical methods have mainly used UV and DAD detectors, coupled with HPLC, or FID coupled with GC, for the determination of biopesticides in soils. Thus, rotenone was determined by UV detection at 295 or 299 nm[14,27] or using DAD at 295 nm.[14] Other biopesticides, such as pyrethrins, were detected at 230 nm.[28] GC-FID has been used for the determination of azadirachtin A in soils.[13] However, HPLC-UV and GC-FID only provided the retention time at which the compounds eluted from the column; this information was not enough for the reliable identification of the targeted compounds since other substances coextracted from the matrix could be present and eluting at the same retention time. Taking into account this fact, MS must be used for high sensitivity and unambiguous detection, confirmation, and determination of the analytes.

Consequently, conventional detection has been replaced by MS,[18] improving the sensitivity and selectivity of the developed methods. In GC, the most used mass analyser is single quadrupole (Q), equipped with an electron impact (EI) ionization system for the detection of biopesticides in soils (Table 8.2). The main benefits of this analyser are the large dynamic range and high scan frequency and the ability of the EI to obtain reproducible fragmentations for the analysed compounds. A previous

TABLE 8.2

Gas Chromatography Methods Used for the Determination of Biopesticides in Soils[a]

Compounds	Stationary Phase	Carrier Gas/ Mobile Phase	Chromatographic Conditions — Other Conditions	Detection	LOQ	Ref
Azadirachtin A	O-17 column	N_2	Flow rate: 15 mL/min Injection volume: 5 µL	FID	NP	13
Cinnamaldehyde and diallyl disulfide	HP-5MS GC column (30 m, 0.25 mm, 0.25 µm)	He	GC-MS interface, ion source, and quadrupole temperatures: 280, 230, and 150 °C, respectively Flow rate: 1.0 mL/min Injection volume: 2µL Splitless mode Total analysis time: 15.91 min	Q-MS	4.4–16 µg/kg 34–82 µg/kg	9
Pyrethrins	DB-5MS fused-silica (30 m × 0.25-mm, 0.25-µm)	He	Injector temperature: 260 °C Flow rate: 1.2 mL/min Injection volume: 2µL	MSD	50 µg/kg	16
96 xenometabolites	Agilent J&W DB-5MS GC (30 m, 0.25 mm, 0.25 µm)	He	Inlet temperature: 230 °C Flow rate: 1 mL/min Splitless mode Total analysis time: 30 min	EI-Q-MS	NP	24
Spinosad	G.C. Column T.R. ™ 5 MS (30 m × 0.25 mm, 0.25 µm)	He	Transfer line temperature: 280 °C Injector temperature: 230 °C Flow rate: 1.3 mL/min Injection volume: 1µL Splitless mode Total analysis time: 19.2 min	MS/MSTQD	2–13 µg/kg	20

Compounds	Stationary Phase	Chromatographic Conditions		Detection	LOQ	Ref
		Carrier Gas/ Mobile Phase	Other Conditions			
Limonene oxide isomers, (-)-carveol isomers, and (-)-carvone	HP-5MS (60 m x 0.25 mm, 0.25 μm)	He	Inlet, MS interface, Ion source and quadrupole temperatures: 270, 280, 280, and 150 °C, respectively Split ratio: 5:1 Flow rate: 1.5 mL/min Injection volume: 1 μL	Q-MS	6–48 μg/kg	26
Carvone	HP-5MS (60 m x 0.25 mm, 0.25 μm)	He	Ion source, quadrupole, and interface temperatures: 280, 150, and 270 °C, respectively Split ratio: 5:1 Flow rate: 1.5 mL/min Injection volume: 1 μL	Q-MS	10–50 μg/Kg	21

Notes: [a] Abbreviations: LOQ: limit of quantification; Q: single quadrupole; QqQ: triple quadrupole; MS: mass spectrometry; MS/MS: tandem mass spectrometry; NP: not provided; EI: electron impact; MSTQD: spectrometry triple quadrupole; MSD: mass selective detector; FID: flame ionization detector research analysed pyrethrins using this technology by selective ion monitoring (SIM) mode, using Q as the analyser, allowing their identification by monitoring different ions.[16] For instance, cinerin I, jasmolin I, and pyrethrin I have a common fragment at m/z 123 with the highest abundance, but these compounds are distinguished by other fragments ions which are less abundant (m/z 150 and 168 for cinerin I, m/z 69 and 135 for jasmolin I, and m/z 81, 105, and 162 for pyrethrin I). On the other hand, cinerin II and jasmolin II had the same fragment ions (m/z 107, 93, 121, 167) but were distinguished by different retention times because jasmolin had one more methyl group in the molecule and therefore eluted later than cinerin.[16]

Huang et al.[26] reported a rapid and sensitive GC-MS method for the simultaneous determination of d-limonene and its oxidation products (cis-limonene oxide, trans-limonene oxide, cis-(-)-carveol, trans- (-)-carveol, and (-)-carvone in soils. Chromatograms and mass spectra of d-limonene and its oxidation products with their quantifier ions, confirmatory ions, and retention times are showed in Figure 8.2. It can be observed that the isomers cis and trans of limonene oxide have the same fragment ions at m/z 43, which is the most abundant, whereas the less abundant ones are m/z 41 and 67. They can be distinguished by their different retention times, eluting the cis isomer before the trans compound. Nevertheless, cis and trans carveol presented one different fragment with less abundance (m/z 55 and 41, respectively) which allowed the distinction between them.

Recently, GC-Q was also applied for characterising unknown metabolites by fast spectral library search and/or structural elucidation.[24] This research characterized 96 xenometabolites after the application of a *Myrica gale* methanolic extract in soils to study the dissipation of its volatile residues by using a Q mass analyser and spectral library for structural elucidation. Among them, 63 compounds were identified for the

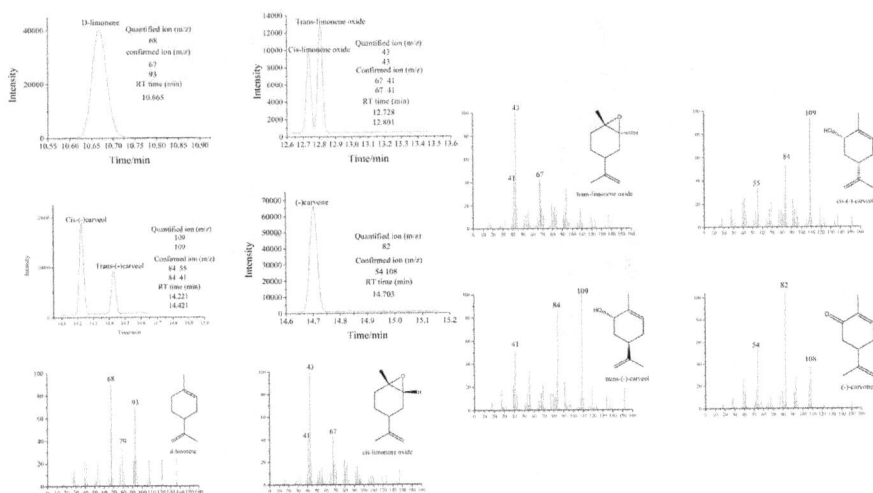

FIGURE 8.2   GC-MS chromatograms and mass spectrums of d-limonene and its oxidation products with their quantifier ions, confirmatory ions, and retention times.

Source: Reproduced with permission from Elsevier (Ref. 27)

first time (47 bioherbicide components and 16 degradation byproducts). Six of the most abundant biopesticides were eucalyptol, L-terpinen-4-ol, α-terpineol, α-terpineol acetate, 3,7(11)-selinadiene, and germacrone. The rest of the xenometabolites were mainly terpenes, aromatic and aliphatic esters, alcohols, and ketones. The fast identification was carried out by using the NIST library for EI-MS fragmentation spectra and the Kovats retention index (RI) calculation. The detection of numerous metabolites was possible by the fragmentation with EI of the parent compounds through this simple spectral library NIST despite the low resolution of the Q analyzer. In addition, Kovats RI calculations ensured higher identification confidence. Therefore, this method has been shown to be a useful tool to study the environmental fate of volatile xenometabolites in emerging complex biopesticides.

Moreover, recent research employed a triple quadrupole analyser (QqQ) for the detection of pesticides, including biopesticides in soils.[20] This method allowed the detection of 14 pesticide residues of organochlorine pesticides (OCPs), organophosphates (OPPs), pyrethroids, carbamates, and biopesticides. Among them, the spinosad biopesticide was detected in these soils. Therefore, the use of GC with QqQ has shown to be a sensitive and effective method for screening and monitoring biopesticide residues. Finally, it should be noted that GC coupled with Q and QqQ analysers allowed the quantification of biopesticides from soils with quantification limits (LOQs) between 6 to 82 μg/kg (Table 8.2).

In the case of LC, the ionisation of biopesticides is carried out by atmospheric pressure ionization (API), applying either electrospray (ESI)[29,30] or atmospheric pressure chemical ionization (APCI).[31] APCI has shown better results for certain compounds such as rotenone and pyrethrins[32] since it was less affected by the matrix components than ESI. However, a lower sensitivity was obtained from APCI interface than from ESI. On the other hand, suitable sensitivity was obtained for the analysis of azadirachtins, salannin, and nimbin from neem extracts using ESI.[33] Schaaf et al.[34] reported the determination of azadirachtin by using several ionisation conditions, observing that APCI provided better results. Nevertheless, recent research used ESI for the ionisation of azadirachtins from seed and leaf extracts of *Azadirachta indica*[30] and pyrethrins in environmental samples, obtaining a high sensitivity.[29] In addition, in the case of analysis of biopesticides in soils, ESI has been the ionization system utilised (Table 8.3), using Q or QqQ as analysers. For instance, Cavoski et al.[14,15] reported the use of LC coupled with both analysers to detect rotenone and its main product of photodegradation (12aβ-hydroxyrotenone) in soils. Rotenone presented m/z 395 [M+H]$^+$ and 436 [M+ H$^+$CH$_3$CN]$^+$ adducts, whereas 12aβ-hydroxyrotenone gave the m/z 393 [M+H-H$_2$O]$^+$ adduct. These were identified and confirmed by LC/MS analysis monitoring in the single-ion mode, the ions 395 and 393 m/z. However, Q was not able to distinguish one analyte from another or interferences with overlapped retention time (t_R) and the same m/z. Consequently, a simultaneous detection was carried out by the QqQ analyser, which allowed the monitoring of two or more different mass transitions (precursor ion > product ion) in many analytes. Currently, "qualifier ions" in addition to the "quantifier ions" are used as an alternative to mass transitions to exclude interference in individual samples based on a typical constant ratio in the number of ions from both transitions.[35] The multiple reaction monitoring (MRM) mode of the QqQ allowed

TABLE 8.3
Liquid Chromatography Methods Used for the Determination of Biopesticides in Soils[a]

Compounds	Stationary Phase	Chromatographic Conditions		Detection	LOQ	Ref
		Carrier Gas/Mobile Phase	Other Conditions			
Pyrethrin I and II	Waters Radial-pak 8MBC1810 (4 μm)	MeCN and water Gradient profile	Column temperature: 25 °C Flow rate: 1.5 mL/min Injection volume: 100 μL	UV (λ = 230 nm)	15–25 ng	28
Rotenone	Acclaim C18 reverse (150 mm x 4.6 mm, 5 μm)	MeCN and water Isocratic (60:40 v/v)	Flow rate: 1.0 mL/min Injection volume: 20 μL	UV (λ = 295 nm)	NP	14
	Waters XTerra C18 (250 mm x 4.6 mm, 5 μm)	MeCN and water Isocratic (60:40 v/v)	Flow rate: 1.0 mL/min Injection volume: 100 μL	DAD (λ = 295 nm)	NP	14
	Waters XTerra MS RP$_{18}$ (250 mm x 2.1 mm, 5 μm)	MeCN and water (0.1 % TFA) Isocratic (60:40 v/v)	Flow rate: 0.4 mL/min Injection volume: 20 μL	Q-MS (ESI)	15 μg/kg	14
	XDB (250 mm x 2.1 mm, 5 μm)	MeCN and water Isocratic (75:25 v/v)	Flow rate: 0.4 mL/min Injection volume: 20 μL	QqQ-MS/MS (ESI)	NP	15
	Zorbax TC-C18 (250 mm x 4.6 mm, 5 μm)	MeCN and water Isocratic (70:30 v/v)	Flow rate: 1.0 mL/min Injection volume: 10 μL	UV (λ = 299 nm)	20.3 μg/kg	27
Azadirachtin, rotenone, spinosyn A and D	BEH C18 (100 mm x 2.1 mm, 1.7 μm)	Water and MeOH with 0.1% ammonium acetate Gradient profile	Column temperature: 30 °C Flow rate: 0.3 mL/min Injection volume: 5 μL	QqQ-MS/MS (ESI)	6–9 μg/kg	18
Nicotine, sabadine, veratridine, rotenone, azadirachtin, cevadine, deguelin, spinosad, pyrethrins	BEH C18 (100 mm x 2.1 mm, 1.7 μm)	MeOH and aqueous solution of ammonium formate (5 mM). Gradient profile	Column temperature: 30 °C Flow rate: 0.3 mL/min Injection volume: 5 μL	QqQ-MS/MS (ESI)	1–10 μg/kg	19

Notes: [a] Abbreviations: MeCN: acetonitrile; DAD: diode array detector; ESI: electrospray ionization; LOQ: limit of quantification; MeOH: methanol; Q: single quadrupole; QqQ: triple quadrupole; MS: mass spectrometry; MS/MS: tandem mass spectrometry; NP: not provided; UV: ultraviolet; MSTQD: spectrometry triple quadrupole; TFA: trifluoroacetic acid the selection and detection of target pesticides with an improvement in the selectivity.[36] Drożdżyński et al.[18] reported a method for the simultaneous analysis of three organic farming bioinsecticides (azadirachtin, rotenone, spinosyn A and D) in soil samples by UHPLC-MS/MS. This research performed the selection of specific MRM transitions for each analyte. Therefore, this proposed methodology allowed the selective determination of selected biopesticide residues at trace levels with a great analytical performance. In addition, Prestes et al.[19] employed UHPLC-MS/MS for the simultaneous analysis of more than ten biopesticide residues in soil samples. It is important to note that the precursor ion of azadirachtin corresponded to the sodium adduct ion (m/z 743) $[M+Na]^+$.[37] In the case of spinosyn A and B, these had a precursor adduct $[M+H]^+$ (m/z 732 and 746), and they had the same product ions at m/z 142.[18] It is also worth emphasising that, in the case of pyrethrins, class II had the $[M+H]^+$ and $[M+H+CH_3CN]^+$ adducts, whereas those in class I also provided the adduct $[M+H_2O]^+$.[38] Furthermore, each pair of pyrethrins had the same two main product ions (m/z 161 and 133 for pyrethrin I and II, m/z 149 and 107 for cinerin I and II, and m/z 163 and 107 for jasmolin I and II) when $[M+H]^+$ was selected as precursor ion.[19] However, certain compounds such as rotenone and deguelin had the same precursor ion (m/z 395) and product ion spectra; thus, their determination was carried out considering their different retention times when simultaneous determination was performed.[19] Finally, it should be mentioned than when LC was combined with MS or MS/MS, low LOQs ranging from 1 to 15 µg/kg were obtained (Table 8.3).

8.4 FATE AND MOBILITY OF BIOPESTICIDES IN SOILS

The fate and mobility of biopesticides in soils imply complex mechanisms that are influenced by a variety of processes, including volatilisation, leaching, adsorption/desorption, and degradation by physical, chemical, and biological processes, which are provided when biopesticides are released to the environment.[39,40] Among them, adsorption and degradation are the key processes to predict the fate of biopesticides in soils. The soil adsorption coefficient (K_d) is a measurement of the quantity of chemical substance adsorbed onto soil per amount of water. Because adsorption is mostly accomplished by partition into soil organic matter, K_d is commonly normalized to the soil's organic carbon content, and the distribution coefficient is expressed as K_{oc}. This is known as the organic carbon-water partition co-efficient (Equation 8.1).

$$K_{oc} = (Kd \times 100)/\% \text{ Organic carbon} \qquad \text{(Equation 8.1)}$$

K_{oc} is commonly estimated considering the octanol-water partition coefficient (K_{ow}) and water solubility. Generally, it has been found that adsorption of biopesticides is positively correlated with octanol-water partition coefficient and negatively correlated with their water solubility.[20,39] Therefore, K_{oc} will be a measure of the soil adsorption, and it is useful to predict the mobility of organic soil compounds. For that reason, larger K_{oc} values indicate that biopesticides are strongly bound to the soil. Table 8.4 presents a summary of K_{oc} as well as the time for 50% disappearance (DT_{50}) of the most important biopesticides.

TABLE 8.4

Physical and Chemical Properties of Biopesticides

Compound	K_{oc} (L/kg)	DT_{50} (Days)	Ref
Rotenone	10000	5 hours–2.76	14,27
Azadirachtin A	20.6–875.1	19.8–43.9	42,43
Azadirachtin B	Not available	20.8–59.2	42,43
Spinosyn A	35024	6.5–46.3	48,49
Spinosyn D	Not available	11.3–62.6	48
Sabadine	1.8×10^5	Not available	64
Veratridine	2.3×10^6	Not available	64
Cevadine	9.7×10^4	Not available	64
Sabadinine	6.1×10^4	Not available	64
Nicotine	100	< 0.5	53
Pyrethrin I	26915	1.8	56
Pyrethrin II	2042	73.2	56
Cinerin I	9332	2.7	56
Cinerin II	700	97.2	56
Jasmolin I	21380	1.9	56
Jasmolin II	1622	36.8	56
Carvone	111	0.2–5	59

Biopesticides are non-persistent under the field conditions, and most of them are mainly degraded by light and temperature.[41] In this sense, transformation products are generated by degrading biopesticides through a series of complex chemical reactions. In this sense, rotenone was widely studied, and its degradation depends on photolysis, soil properties including organic matter and clay concentration, and temperature.[14,15,27] A study reported a fast initial degradation of rotenone with DT_{50} varying from 5 to 7 hours by photolysis reaction.[14] Another study reported the half-lives of rotenone, which ranged from 1.98 to 2.76 days in soil.[27] Different reactions, such as O-demethylation, epimerization, epoxidation, hydroxylation, and dehydration, are involved as a result of the photolysis. In addition, a higher degradation of rotenone is provided when the soils contain a higher organic matter.[14] Moreover, Cavoski et al.[15] revealed that an increase of 10 °C in temperature provided a decreasing in the DT_{50} value by a factor of 4.2 for rotenone, its main degradation product being 12 aβ-hydroxyrotenone.[14]

In the case of azadirachtins, azadirachtin A has a low to very high mobility (K_{oc} = 20.6–875.1 L/kg), but there is no information on K_{oc} for azadirachtin B, although it is believed that azadirachtin B has similar adsorption/desorption endpoints to azadirachtin A.[42] Their degradation followed first-order kinetics with different half-lives, depending on several factors, such as temperature. The persistence of azadirachtin A and B was determined at two different temperatures (15 and 25

°C) after application of the commercial neem insecticide to soil, observing that temperature affects the degradation rates. The DT_{50} for azadirachtin A was 19.8–43.9 and 20.8–59.2 days for azadirachtin B at 15 °C and 25 °C, respectively.[43] Moreover, two unknown degradation products were commonly observed in soil although they were not identified by HPLC-UV at $\lambda = 215$ nm. Other studies evaluated the effect of azadirachtin in soil, observing that the population of bacteria, actinomycetes, and diazotrophs was not affected because of the addition of azadirachtin to culturable soils.[44] Other studies reported that azadirachtin had no significant negative effects on arbuscular mycorrhiza populations.[45] Additionally, Suciu et al.[46] also evidenced that trifloxystrobin and azadirachtin did not have adverse effects on soil microbial functions, even at high dose rates. Recent studies have revealed that although azadirachtin is considered environmentally safe due to its biological origin, it has adverse effects on rhizospheric bacterial and fungal communities at different plant growth stages, similar to synthetic pesticides.[11,47] Furthermore, as demonstrated by its genes and transcripts, azadirachtin has detrimental impacts on plant growth promotion, the nitrogen-fixing bacterial community, and nitrification. Therefore, the content of azadirachtin in soils should be regulated due to its adverse effects.[11]

Spinosyn A and D are transformed to spinosyn B (a metabolite of spinosyn A) and N-demethylated spinosyn D (a metabolite of spinosyn D) in soil. DT_{50} values ranged from 6.5–46.3 days for spinosyn A and 11.3–62.6 days for spinosyn D, depending on the soil conditions.[48] Furthermore, half or more of the spinosad was adsorbed to the interior of soil particles and was not available for photodegradation.[48] Moreover, spinosyn A and D and spinosyn B presented low to no mobility in soils, whereas the N-demethyl spinosyn D metabolite exhibited medium to no mobility. It was concluded that the adsorption of spinosyn A and D and its metabolites was not pH dependent.[48] Moreover, spinosyn A had a value of $K_{oc} = 35024$ L/kg, whereas a K_{oc} value was not available for spinosyn D, although it is assumed that this had the same sorption characteristics as spinosyn A.[49] The dissipation of spinosad in soil was evaluated after the application of a spinosad formulation (Tracer 45.5 % SC) sprayed in the field at two doses of 51.0 and 102.0 g active ingredient (a.i.)/ha in 500 L water at the 50% fruiting stage. The DT_{50} in soil was 6.36 and 6.91 days for the recommended dose and the double one, respectively. In addition, a second spray of similar treatment was done 15 days after first spray, obtaining a DT_{50} between 5.49 and 6.76 days.[50] The dissipation of spinosad was 98.1 and 76.9% by 15 days for the recommended and double dose, respectively. A recent study reported the effects of sub-lethal doses of two insecticides, one biologically derived (spinosad) and one synthetic organophosphate (chlorpyrifos), on earthworm *Eisenia foetida* and microorganisms in organic soil.[51] Early DNA damage was estimated in earthworms exposed to chlorpyrifos while the impact of spinosad was only significant at the end of the toxicity test.

In relation to sabadilla alkaloids (cevadine and veratridine), their degradation was faster under photolysis (hours) than under hydrolysis, since hydrolysis is expected to occur at much slower rates (days to years).[52] K_{oc} values ranged from $6.1 \cdot 10^4$ (sabadinine) to $2.3 \cdot 10^6$ L/kg (veratridine). In addition, nicotine usually degraded to cotinine and presented DT_{50} and K_{oc} values in soils of 0.5 days and 100 L/kg,[53] respectively. Bulenga Lisuma et al.[54] indicated that nicotine sorption isotherms fit a Freundlich model, revealing their adsorption was based on the soil depths (0–50 cm), ranging from 2.81 to 4.61 mg/kg in sandy loam and sandy soils, respectively. The nicotine desorption ranged from 0.89 to 1.12 mg/kg in loamy sand and sandy loam soils, respectively.

Pyrethrins were strongly adsorbed into soil surfaces and were commonly considered not mobile, with a DT_{50} between 1.9 to 97.2 days.[55,56] It was observed that pyrethrins degraded very quickly upon exposure to sunlight, and they did not persist in the environment beyond a few weeks.[16] The half-life of pyrethrin I and II was less than two hours in field conditions, whereas under dark conditions, there was little degradation over time.[57] Moreover, the sorption of pyrethrins into soil increased with raising soil organic matter content. Thus, it was found that soils containing twice the organic matter content as native soils absorbed more pyrethrins, and their mobility was reduced by humic acids.[28] Feng et al.[16] revealed that the pyrethrins' dissipation was 1.0–1.3 days. This study showed a faster degradation of pyrethrins in the greenhouse than in the open field.

Regarding the monoterpenoid biopesticide compounds, such as α-terpineol, limonene, thymol, menthol, carvone, eucalyptol, and perillaldehyde, they are aromatic and volatile at near room temperature.[21] Monoterpenoid biopesticides were susceptible to oxidation, cyclization, isomerization, dehydrogenation, and other

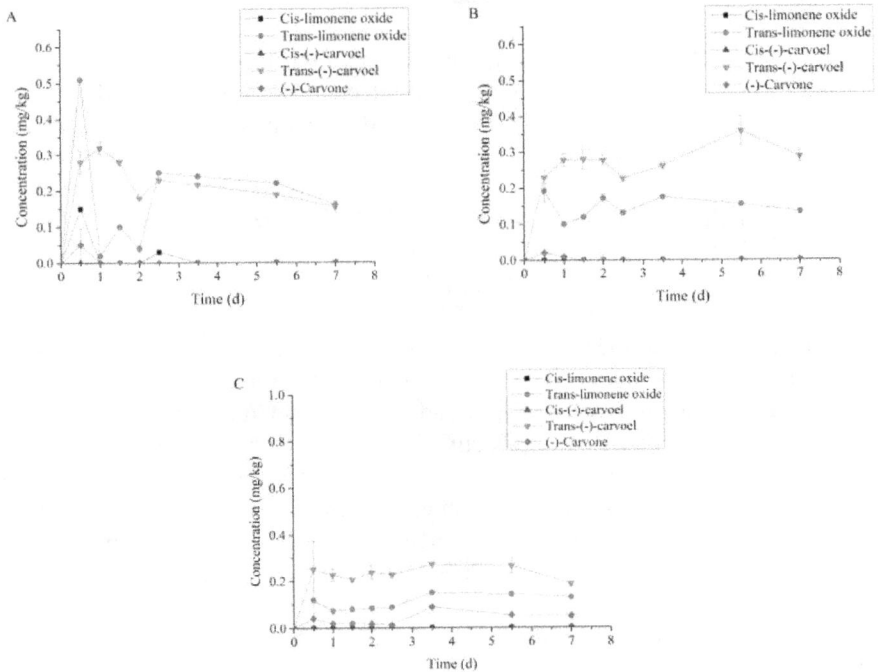

FIGURE 8.3 Dynamic of d-limonene oxidation productions in non-autoclaved S1(A), S2(B), S3(C) at the 10 mg/kg spiked levels.

Source: Reproduced with permission from Elsevier (Ref. 27)

breakdown reactions, and as a result of these features, they were not environmentally stable.[21,58] For instance, carvone, limonene oxide, carveol, and limonene hydroperoxide were all formed when the limonene monoterpenoid was oxidized in the environment.[21] Carvone had a high mobility in soil of 111 L/kg, and it was not persistent, with a DT_{50} ranging from 0.2–5 days.[59] Huang et al.[26] reported the production of d-limonene oxidation products under indoor simulated conditions (Figure 8.3.), which was quickly dissipated within seven days after its application in soils, and some oxidation products were generated. The production of *trans*-compound isomers was higher than their *cis* isomers in three soils, Jilin (S1), Jiangxi (S2), and Sichuan (S3). Monoterpenoids were degraded mostly by microbes in soil and water, with a small percentage lost by leaching and evaporation.[60] Environmental factors, such as soil type, pH, temperature, humidity, and precipitation, will influence this degradation process.[61] Previous research had discovered that the decomposition of monoterpenoids produced by *Myrtis communis* in the soil was accelerated during periods of high microbial activity.[21,62] Furthermore, some of these monoterpenoids can help pesticide degradation by biochemical reactions in contaminated soils.[63]

8.5 CONCLUSIONS

The use of biopesticides is still limited due to expensive manufacturing processes, inadequate storage stability, vulnerability to environmental conditions, effectiveness issues, etc., compared to synthetic pesticides. About 75% of biopesticide products consist of microbial biopesticides. Moreover, plant extracts and essential oil–based pesticide products could be an excellent alternative to synthetic chemicals. These natural components should be controlled in soil to determine their possible toxic effects, as well as their mobility and persistence. Previous studies have determined biopesticides in soils, including cinnamaldehyde, diallyl disulfide, azadirachtin, rotenone, spinosyn A and D, nicotine, sabadine, veratridine, cevadine, deguelin, pyrethrins, and limonene and its oxidation products. It must be considered that an extraction step prior to their analyses is needed to obtain a good recovery. QuEChERS was the most used extraction technique for most of these biopesticides, obtaining higher recoveries than SLE, SPE, and PLE. Subsequently, LC and GC coupled with MS have been the most employed analytical techniques for their determination, obtaining low LOD and LOQ values.

On the other hand, adsorption and degradation are the most significant processes that can predict the fate of biopesticides in soils. Nevertheless, there is a little information about their degradation products from parent compounds, and these can be more active than the original biopesticides, or their toxicity could be higher. Therefore, other analytical techniques based on high-resolution mass spectrometry (HRMS) analysers, such as TOF or Orbitrap, could be suitable tools for simultaneous detection of target, non-target, and unknown biopesticides in one single run. They must be used to achieve precise mass measurements of known biopesticide metabolites or transformation products through various degradation processes as well as for the identification of unknown biopesticides. Thus, the combination of efficient extraction procedures and analysis by GC and LC, coupled with HRMS analysers, could be a good strategy for the determination of several families of biopesticides in a short time. Finally, this type of instrumentation will be needed to get comprehensive results when degradation studies are developed.

8.5.1 ACKNOWLEDGMENTS

The authors gratefully acknowledge the University of Almeria (UAL-FEDER 2014–2020); the Regional Ministry of Economy, Knowledge, Business; and the University of the Junta de Andalucía for financial support (Project Ref. UAL2020-FQM-B1943). BMG is also grateful for personal funding through the Juan de la Cierva Program (funded by MCIN/AEI/10.13039/501100011033 and European Union NextGenerationEU/PRTR).

REFERENCES

1. Kumar N, Mukherjee I, Sarkar B, Paul RK. Degradation of tricyclazole: Effect of moisture, soil type, elevated carbon dioxide and Blue Green Algae (BGA). *J Hazard Mater.* 2017;321:517–527. doi:10.1016/j.jhazmat.2016.08.073
2. Bowles TM, Acosta-Martínez V, Calderón F, Jackson LE. Soil enzyme activities, microbial communities, and carbon and nitrogen availability in organic agroecosystems across an intensively-managed agricultural landscape. *Soil Biol Biochem.* 2014;68:252–262. doi:10.1016/j.soilbio.2013.10.004
3. Campos EVR, Proença PLF, Oliveira JL, Bakshi M, Abhilash PC, Fracetoa LF. Use of botanical insecticides for sustainable agriculture: Future perspectives. *Ecol Indic.* 2019;105:483–495.
4. Shao H, Zhang Y. Non-target effects on soil microbial parameters of the synthetic pesticide carbendazim with the biopesticides cantharidin and norcantharidin. *Sci Rep.* 2017;7(1). doi:10.1038/s41598-017-05923-8
5. Kumar J, Ramlal A, Mallick D, Mishra V. An overview of some biopesticides and their importance in plant protection for commercial acceptance. *Plants.* 2021;10(6):1–15. doi:10.3390/plants10061185
6. Smith CJ, Perfetti TA. A comparison of the persistence, toxicity, and exposure to high-volume natural plant-derived and synthetic pesticides. *Toxicol Res Appl.* 2020;4. doi:10.1177/2397847320940561
7. Frederiks C, Wesseler JHH. A comparison of the EU and US regulatory frameworks for the active substance registration of microbial biological control agents. *Pest Manag Sci.* 2019;75(1):87–103. doi:10.1002/ps.5133
8. Damalas CA, Koutroubas SD. Current status and recent developments in biopesticide use. *Agriculture.* 2018;8(1). doi:10.3390/agriculture8010013
9. López-Serna R, Ernst F, Wu L. Analysis of cinnamaldehyde and diallyl disulfide as eco-pesticides in soils of different textures—a laboratory-scale mobility study. *J Soils Sediments.* 2016;16(2):566–580. doi:10.1007/s11368-015-1249-5
10. Selim S, Martin-Laurent F, Rouard N, Gianinazzi S, van Tuinen D. Impact of a new biopesticide produced by Paenibacillus sp. strain B2 on the genetic structure and density of soil bacterial communities. *Pest Manag Sci.* 2007;63:269–275. doi:10.1002/ps
11. Singh S, Gupta R, Sharma S. Effects of chemical and biological pesticides on plant growth parameters and rhizospheric bacterial community structure in Vigna radiata. *J Hazard Mater.* 2015;291:102–110. doi:10.1016/j.jhazmat.2015.02.053
12. Romero-González R, Frenich AG, Martínez-Vidal JL. Biopesticides residues in soil. In: *Biopesticides handbook,* 2015:123–135, Boca Raton, FL: CRC Press.

13. Agyarko K, Kwakye PK, Bonsu M, Osei BA, Asare Donkor N, Amanor E. Breakdown of azadirachtin A in a tropical soil amended with neem leaves and animal manures. *Pedosphere*. 2006;16(2):230–236. doi:10.1016/S1002-0160(06)60048-9

14. Cavoski I, Caboni P, Sarais G, Cabras P, Miano T. Photodegradation of rotenone in soils under environmental conditions. *J Agric Food Chem*. 2007;55(17):7069–7074. doi:10.1021/jf0708239

15. Cavoski I, Caboni P, Sarais G, Miano T. Degradation and persistence of rotenone in soils and influence of temperature variations. *J Agric Food Chem*. 2008;56(17):8066–8073. doi:10.1021/jf801461h

16. Feng X, Pan L, Wang C, Zhang H. Residue analysis and risk assessment of pyrethrins in open field and greenhouse turnips. *Environ Sci Pollut Res*. 2018;25(1):877–886. doi:10.1007/s11356-017-0015-1

17. Vilela ADO, Faroni LRDA, Rodrigues AAZ, et al. Headspace solid-phase microextraction: Validation of the method and determination of allyl isothiocyanate persistence in cowpea beans. *ACS Omega*. 2020;5(34):21364–21373. doi:10.1021/acsomega.0c01385

18. Drożdżyński D, Kowalska J. Rapid analysis of organic farming insecticides in soil and produce using ultra-performance liquid chromatography/tandem mass spectrometry. *Anal Bioanal Chem*. 2009;394(8):2241–2247. doi:10.1007/s00216-009-2931-5

19. Prestes OD, Padilla-Sánchez JA, Romero-González R, Grio SL, Frenich AG, Martínez-Vidal JL. Comparison of several extraction procedures for the determination of biopesticides in soil samples by ultrahigh pressure LC-MS/MS. *J Sep Sci*. 2012;35(7):861–868. doi:10.1002/jssc.201101057

20. EL-Saeid MH, Alghamdi AG. Identification of pesticide residues and prediction of their fate in agricultural soil. *Water Air Soil Pollut*. 2020;231(6). doi:10.1007/s11270-020-04619-6

21. Huang C, Zhou W, Bian C, Wang L, Li Y, Li B. Degradation and pathways of carvone in soil and water. *Molecules*. 2022;27:1–14.

22. Pszczolinska K, Michel M. The QuEChERS approach for the determination of pesticide residues in soil samples: An overview. *J AOAC Int*. 2016;99(6):1403–1414. doi:10.7540/jaoacint.16-0274

23. Hassaan MA, El Nemr A. Pesticides pollution: Classifications, human health impact, extraction and treatment techniques. *Egypt J Aquat Res*. 2020;46(3):207–220. doi:10.1016/j.ejar.2020.08.007

24. Ghosson H, Raviglione D, Salvia M-V, Bertrand C. Online headspace-solid phase microextraction-gas chromatography-mass spectrometry-based untargeted volatile metabolomics for studying emerging complex biopesticides: A proof of concept. *Anal Chim Acta J*. 2020;1134:58–74.

25. Pratiwi FA, Utami TS, Arbianti R. Using ultrasonic assisted extraction to produce a bioinsecticide from cigarette butt waste and green solvent to control armyworm infestation. *Int J Technol*. 2020;11(7):1329–1336. doi:10.14716/ijtech.v11i7.4474

26. Huang C, Bian C, Wang L, Zhou W, Li Y, Li B. Development and validation of a method for determining d-limonene and its oxidation products in vegetables and soil using GC-MS. *Microchem J*. 2022;179:107470. doi:10.1016/j.microc.2022.107470

27. Zhou Y, Zhang N, Wang K, Li W, Li H, Zhang Z. Dissipation and residue of rotenone in cabbage and soil under field conditions. *Bull Environ Contam Toxicol*. 2013;91(2):251–255. doi:10.1007/s00128-013-1040-5

28. Antonious GF, Patel GA, Snyder JC, Coyne MS. Pyrethrins and piperonyl butoxide adsorption to soil organic matter. *J Environ Sci Heal—Part B Pestic Food Contam Agric Wastes*. 2004;39(1):19–32. doi:10.1081/PFC-120027436

29. Song L, Wang J, Gao Q, et al. Simultaneous determination of five azadirachtins in the seed and leaf extracts of Azadirachta indica by automated online solid–phase extraction coupled with LC-Q-TOF-MS. *Chem Cent J*. 2018;12(85):1–9. doi:10.1186/s13065-018-0453-y

30. Ccanccapa-Cartagena A, Masiá A, Picó Y. Simultaneous determination of pyrethroids and pyrethrins by dispersive liquid-liquid microextraction and liquid chromatography triple quadrupole mass spectrometry in environmental samples. *Anal Bioanal Chem.* 2017. doi:10.1007/s00216-017-0422-7

31. Peruga A, Hidalgo C, Sancho J V, Hernández F. Development of a fast analytical method for the individual determination of pyrethrins residues in fruits and vegetables by liquid chromatography—tandem mass spectrometry. *J Chromatogr A.* 2013;1307:126–134. doi:10.1016/j.chroma.2013.07.090

32. Caboni P, Sarais G, Angioni A, Garau VL, Cabras P. Fast and versatile multiresidue method for the analysis of botanical insecticides on fruits and vegetables by HPLC/DAD/ MS. *J Agric Food Chem.* 2005;53:8644–8649.

33. Barrek S, Paisse O. Analysis of neem oils by LC-MS and degradation kinetics of azadirachtin-A in a controlled environment characterization of degradation products by HPLC-MS-MS. *Anal Bioanal Chem.* 2004;378:753–763. doi:10.1007/s00216-003-2377-0

34. Schaaf O, Jarvis AP, Esch SA Van Der, Giagnacovo G, Oldham NJ. Rapid and sensitive analysis of azadirachtin and related triterpenoids from Neem (Azadirachta indica) by high-performance liquid chromatography—atmospheric pressure chemical ionization mass spectrometry. *J Chromatogr A.* 2000;886:89–97.

35. Schuster C, Habler K, Vogeser M. Averaging of results derived from different, simultaneously acquired mass transitions in ID-LC-MS/MS—potential impact on measurement imprecision. *Clin Mass Spectrom.* 2020;17:1–3. doi:10.1016/j.clinms.2020.06.001

36. Shin Y, Lee J, Park E, Lee J, Lee HS, Kim JH. A quantitative tandem mass spectrometry and scaled-down quechers approach for simultaneous analysis of pesticide multiresidues in human urine. *Molecules.* 2019;24(7). doi:10.3390/molecules24071330

37. Turnipseed SB, Storey JM, Wu IL, Andersen WC, Madson MR. Extended liquid chromatography high resolution mass spectrometry screening method for veterinary drug, pesticide and human pharmaceutical residues in aquaculture fish. *Food Addit Contam Part A Chem Anal Control Expo Risk Assess.* 2019;36(10):1501–1514. doi:10.1080/19440049. 2019.1637945

38. Wan J, He P, Chen Y, Zhu Q. Comprehensive target analysis for 19 pyrethroids in tea and orange samples based on LC-ESI-QqQ-MS/MS and LC-ESI-Q-ToF/MS. *LWT.* 2021;151:112072. doi:10.1016/j.lwt.2021.112072

39. Copaja SV., Gatica-Jeria P. Effects of clay content in soil on pesticides sorption process. *J Chil Chem Soc.* 2021;65(1):5086–5092. doi:10.4067/S0717-97072021000105086

40. Su W, Hao H, Wu R, Xu H, Xue F, Lu C. Degradation of mesotrione affected by environmental conditions. *Bull Environ Contam Toxicol.* 2017;98(2):212–217. doi:10.1007/ s00128-016-1970-9

41. Ujváry I. *Pest control agents from natural products.* Volume 1, 3rd ed. Elsevier Inc., 2010. doi:10.1016/B978-0-12-374367-1.00003-3

42. Arena M, Auteri D, Barmaz S, et al. Peer review of the pesticide risk assessment of the active substance azadirachtin (Margosa extract). *EFSA J.* 2018;16(9). doi:10.2903/j. efsa.2018.5234

43. Stark JD, Walter JF, Stark JD, Walter JF. Persistence of azadirachtin a and b in soil: Effects of temperature and microbial activity. *J Environ Sci Heal Part B.* 1995;30(5):685–698. doi:10.1080/03601239509372960

44. Gopal M, Gupta A, Arunachalam V, Magu SP. Impact of azadirachtin, an insecticidal allelochemical from neem on soil microflora, enzyme and respiratory activities. *Bioresour Technol.* 2007;98(16):3154–3158. doi:10.1016/j.biortech.2006.10.010

45. Wang MT, Rahe JE. Impact of azadirachtin on Glomus intraradices and vesicular-arbuscular mycorrhiza in root inducing transferred DNA tranformed roots of Daucus Carota. *Environ Toxicol Chem.* 1998;17:2040–2050.

46. Suciu N, Vasileiadis S, Puglisi E, et al. Azadirachtin and trifloxystrobin had no inhibitory effects on key soil microbial functions even at high dose rates. *Appl Soil Ecol.* 2019;137:29–38. doi:10.1016/j.apsoil.2019.01.016

47. Walvekar VA, Bajaj S, Singh DK, Sharma S. Ecotoxicological assessment of pesticides and their combination on rhizospheric microbial community structure and function of Vigna radiata. *Environ Sci Pollut Res.* 2017;24(20):17175–17186. doi:10.1007/s11356-017-9284-y

48. Arena M, Auteri D, Barmaz S, et al. Peer review of the pesticide risk assessment of the active substance spinosad. *EFSA J.* 2018;16(5). doi:10.2903/j.efsa.2018.5252

49. UE. Directive 98/8/EC concerning the placing biocidal products on the market Spinosad (PT 18) Assessment report Finalised in the Standing Committee on Biocidal Products at its meeting on May 27, 2010 in view of its inclusion in Annex I to Directive 98/8/EC CO. 2010;18. https://circabc.europa.eu/sd/a/6f5e6afd-a508-45cf-9074-0d8620c21813/Spinosad Assessment Report post SCclean0211.pdf.

50. Adak T, Mukherjee I. Dissipation kinetics of spinosad from tomato under sub-tropical agroclimatic conditions. *Environ Monit Assess.* 2016;188(5). doi:10.1007/s10661-016-5291-6

51. De Bernardi A, Marini E, Casucci C, et al. Ecotoxicological effects of a synthetic and a natural insecticide on earthworms and soil bacterial community. *Environ Adv.* 2022;8:100225. doi:10.1016/j.envadv.2022.100225

52. Reregistration eligibility decision exposure and risk assessment on lower risk pesticide chemicals sabadilla alkaloids. *Environ Prot Agency.* 2004.

53. Seckar JA, Stavanja MS, Harp PR, Yi Y, Garner CD, Doi J. Environmental fate and effects of nicotine released during cigarette production. *Environ Toxicol Chem.* 2008;27(7):1505–1514. doi:10.1897/07-284.1

54. Lisuma JB, Mbega ER, Ndakidemi PA. Nicotine release at the tobacco rhizosphere and their adsorption capacities in different soil textures. *Rhizosphere.* 2020;15:100210. doi:10.1016/j.rhisph.2020.100210

55. *Reregistration Eligibility Decision for Pyrethrins,* 2007.

56. Crosby DG. Environmental fate of pyrethrins. In *Pyrethrum flowers; production, chemistry, toxicology, and uses* (Casida JE and Quistad GB, eds.). Oxford University Press, 1995.

57. Grdiša M, Gršić K. Botanical insecticides in plant protection. *Agric Conspec Sci.* 2013;78(2):85–93.

58. De Carvalho CCCR, Da Fonseca MMR. Carvone: Why and how should one bother to produce this terpene. *Food Chem.* 2006;95(3):413–422. doi:10.1016/j.foodchem.2005.01.003

59. Arena M, Auteri D, Barmaz S, et al. Peer review of the pesticide risk assessment of the active substance carvone (substance evaluated d-carvone). *EFSA J.* 2018;16(8). doi:10.2903/j.efsa.2018.5390

60. Fenner K, Canonica S, Wackett LP, Elsner M. Evaluating pesticide degradation in the environment: Blind spots and emerging opportunities. *Science (80-).* 2013;341(6147):752–758. doi:10.1126/science.1236281

61. Gámiz B, Facenda G, Celis R. Nanoengineered sorbents to increase the persistence of the allelochemical carvone in the rhizosphere. *J Agric Food Chem.* 2019;67(2):589–596. doi:10.1021/acs.jafc.8b05692

62. Gámiz B, Hermosín MC, Celis R. Appraising factors governing sorption and dissipation of the monoterpene carvone in agricultural soils. *Geoderma.* 2018;321:61–68. doi:10.1016/j.geoderma.2018.02.005

63. Han D, Yan D, Wang Q, et al. Effects of soil type, temperature, moisture, application dose, fertilizer, and organic amendments on chemical properties and biodegradation of dimethyl disulfide in soil. *L Degrad Dev.* 2018;29(12):4282–4290. doi:10.1002/ldr.3177

64. Reregistration eligibility decision exposure and risk assessment on lower risk pesticide chemicals sabadilla alkaloids. *Environ Prot Agency.* 2004.

Section 3

Classes of Biopesticides

9 Biochemical Pesticides

*Zareeka Haroon, Hyda Haroon, Irzam Haroon,
Showkat Rasool Mir, and Yawar Sadiq*

CONTENTS

9.1 INTRODUCTION

Unconventional pesticides like biochemical pesticides and biological pesticides are made from things like bacteria, plants, animals, and certain minerals. These include plant-incorporated protectants, or pesticidal substances, which plants produce from additional genetic material (such as corn genetically modified to produce *Bacillus thuringiensis toxins*); microbial pesticides made of bacteria, fungi, viruses, etc., as the active ingredient; and biochemical pesticides made of naturally occurring substances that control pests by non-toxic mechanisms (such as pheromones or some insect growth regulators). These are naturally occurring chemicals that use non-toxic ways to manage pests, such as plant extracts, fatty acids, or pheromones. Contrarily, conventional insecticides are synthetic substances that usually render pests inactive or dead. Plant growth regulators, pheromones, and other chemicals that repel or attract pests are examples of biochemical pesticides. Biochemical pesticides also include chemicals that interfere with growth or mating. If a pesticide fits the requirements for a biochemical pesticide, it can often be challenging to identify whether a natural pesticide controls the pest by a non-toxic mode of action. Because of the sensitivity of their olfactory systems, insects can only be controlled with extremely small doses of semiochemicals. The codling moth (*Cydia pomonella*), which is currently being controlled by the use of pheromones in apple orchards (using a method of mating disruption through confusion), serves as an example of this. Only around 1 g/ha of pheromone is required to control insects using the confusion technique, while an insect dispenser releases roughly 1 g/ha of pheromone per hectare. Contrarily, in this instance, 1 kg/ha of pesticide would be required for a typical treatment. Furthermore, on ecological and environmental grounds, the use of nonselective insecticides

DOI: 10.1201/9781003265139-12

is being questioned. A pheromone product called Oriental Beetle MD is used to manage the oriental beetle, a turf- and container-grown plant pest in some areas. It is one of the few substances accessible to ornamental plant growers for controlling insects through mating disruption (P. Witzgall et al.,1999) (T.G.T. Jaenson et al., 2006).

- **Sex pheromones**

Biochemical insecticides known as insect sex pheromones have long been used in IPM systems as mass-trapping and monitoring instruments. Commercially accessible sex pheromone lures include insects like cucumber moths (*Diaphania indica*), common army worms, beet army worms, and tomato fruit worms.

An integrated pest management (IPM) technique based on sex pheromones has been created and promoted by AVRDC for controlling the eggplant fruit and shoot borer in South Asia (Alam et al., 2003; Alam et al., 2006). According to Alam et al. (2006), Bangladesh reduced its use of pesticides by 70% after using an IPM method to combat eggplant fruit and shoot borer. With the help of natural enemies, this IPM technique decreased the overuse of pesticides in eggplant production systems. According to Naresh et al. (1986) and Bangladeshi research, *Trathala flavo orbitalis* is an efficient parasitoid of the eggplant fruit and shoot borer (Alam and Sana, 1964). However, its role in pest management has rarely been studied and does not seem to be substantial. The original findings from Bangladesh indicated that just 10% of people were parasitized. However, following one year of eggplant production without pesticide spraying, the average amount of parasitism increased roughly thrice. The parasitism rate was significantly higher (39.3–48.9%) during the period of heavy production. It would be possible to attack the insect population on a sustainable basis and eliminate the need for pesticides if this level of parasitism could be maintained over greater areas all year long (Alam et al., 2003).

A synthetic sex pheromone was created for the legume pod borer that attracted male moths in Benin and Ghana while (E,E)-10,12-hexadecadienal alone was more successful in Burkina Faso (Downham et al., 2003, 2004). (Downham, 2006). According to unpublished data from the AVRDC, neither pheromone worked in Taiwan or Southeast Asia, but a different combination did in India (Hassan, 2007). The application of trap-based pest monitoring is hampered in several significant subsistence legume crop regions of the world by the geographical diversity in the mix of the legume pod borer. A network has been established to improve the sex pheromones of the legume pod borer and create an IPM strategy based on them (Srinivasan et al., 2007).

- **Aggregation pheromones**

For the purpose of controlling the striped flea beetle (*Phyllotreta striolata*) on brassica plants, efforts are being made to develop an IPM technique based on aggregation pheromones. Male striped flea beetles that are actively eating release an aggregation pheromone. Previously, a sesquiterpene from the congeneric species *P. cruciferae* was discovered to be the male aggregation pheromone (Soroka et al., 2005). The aggregation pheromone of the striped flea beetle has seven male-specific sesquiterpenes that have been discovered. However, (+)-(6R,7S)-himachala- 9,11-diene was shown to be the active substance in male-specific tissues. When combined with the volatile from the host plant allyl isothiocyanate, this synthetic pheromone's activity attracted a lot of *P. striolata* in a controlled environment (Beran et al., 2011).

9.2 AGENTS FOR BIOCHEMICAL PEST CONTROL

Reduced-risk biochemical pesticides with various broad biologically functional groups are approved for use on both crops and ornamentals in addition to vegetables. The following are succinct descriptions of significant biochemical pest control agents:

- **Hormones**

These are biochemical substances that are created in one area of an organism and then transported to another area where they have an impact on behaviour, control, or regulation. The description of natural and synthetic substances capable of interfering with the processes of development and reproduction of the target insects has provided new methods for creating insect control agents. Ecdysteroids and juvenile hormones are two novel types of insect growth and developmental hormones that are mimicked by pesticides. There is potential to use novel control in the biosynthesis, function, and metabolism of neuropeptide structures (Hoffmann and Lorenz, 1998).

- **Enzymes**

Protein molecules that act as enzymes are the means by which genes are expressed and biochemical processes are catalysed. Enzymes that impede digestive functions in the insect gut are one way that plants defend themselves from insect herbivores. The evolutionary history of these enzymes, their location in the plant kingdom, and the methods by which they function in the protease-rich environment of the animal digestive tract are all poorly understood (J. Chen et al., 2007). Additionally they are considered a potential defence mechanism against insects in the transgenic production of insecticidal proteins, including α-amylase and protease inhibitors (T.H. Schuler et al., 1998).

- **Feeding deterrents**

A substance known as a feeding deterrent prevents an insect pest from feeding, which leads to its eventual starvation and death. Inhibition of crop damage results in the insect starving to death over time. The root bark of the Chinese medical plant *Dictamnus dasycarpus* is a strong feeding deterrent against two stored-product insects, according to a screening for insecticidal principles from many Chinese medicinal herbs (*Tribolium castaneum* and *Sitophilus zeamais*). Two feeding inhibitors have been identified from methanol extract using bioassay-guided fractionation. Based on their spectroscopic data, the chemicals have been identified as fraxinellone and dictamnine. It has been shown that the drugs fraxinellone and dictamnine have the ability to prevent *S. zeamais* and *T. castaneum* adults and larvae from feeding on them (S.A.A. Amer et al., 2001; Z.L. Liu et al., 2002).

- **Semiochemicals**

These are substances that are released by both plants and animals that alter the behaviour of receptor organisms of the same or different types. The terms *pheromones* (acting between members of the same species) and *allelochemicals* (acting between members of different species) are frequently used to refer to various semiochemicals (chemical signals). The four types of allelochemicals are allomones (beneficial to the transmitter), kairomones (beneficial to the recipient), synomones (beneficial to both), and apneumones (from non-living sources). Pheromones are compounds that a member of one species releases into the environment which alter the behaviour of other members of the same species. Chemicals called allomones are released by one species and change the behaviour of another species in the emitting species's favour. Kairomones are substances released by one species that influence another species's behaviour in favour of the receptor species. Insect life conditions, including feeding, mating, and egg laying, are determined by semiochemicals (ovipositing). Semiochemicals are thus prospective tools for the targeted elimination of pest insect populations. Pheromones or kairomones can be employed for biological control to detect and monitor insect populations. For the effective application of conventional or non-conventional pesticides, monitoring is crucial. Pheromone-based mating disruption is a promising and frequently effective pest management method (confusion technique). Another tactic is to use semiochemicals as feeding inhibitors. Attracting, trapping, and killing the pest insects is the most popular semiochemical control method (T. Norin, 2007).

- **Plant growth regulators**

Simply put, substances used to modify a plant's or plant part's growth are referred to as plant growth regulators, growth regulators, or plant hormones. Plant growth regulators are categorised as pesticides from the standpoint of regulatory control. Plants naturally create compounds that are poisonous, inhibitive, or stimulating or have other modifying effects on the same type of plants or other plants. Plant hormones or phytohormones are names of several of them. Through a number of means, including rupturing plant tissue's cell membranes, preventing the synthesis of amino acids, and preventing the development of photosynthesis-related enzymes, some plant oils can function as efficient contact herbicides. Products with cotton seed, clove, garlic, cedar, and rosemary oils as active components are examples of low-risk insecticides (K.I.M. Jeong-Kyu et al., 2005).

9.3 METHYL SALICYLATE (MESA) AS BIOCHEMICAL PESTICIDE

Volatile semiochemicals are increasingly seen as potential additions to integrated pest management plans for insects (Smart et al., 2014). Semiochemicals might contain substances that directly deter pests, draw in predatory animals, or act as "elicitors" to activate defence mechanisms that give the host plant resilience (Maffei et al., 2012). Some semiochemicals may be capable of acting in more than one of these ways. Although it is usually believed that plants' salicylic acid biochemical pathway serves as a defence against biotrophic diseases, it also appears to work against some herbivorous arthropods, especially those with a piercing/sucking feeding style (Aerts et al., 2021). MeSA may have different modes of action that can be exploited for pest management (James, 2003; Ninkovic et al., 2003; Byers et al., 2021). It is attractive to a range of natural enemies of arthropod pests (Mallinger et al., 2011; Orre Gordon et al., 2013) and possibly birds (Rubene et al., 2019) and has shown repellency against aphids (Glinwood and Pettersson, 2000; Prinsloo et al., 2007; Digilio et al., 2012). MeSA is often reported as a plant volatile that is induced by insect feeding and may play a role in defence signaling within (Heil and Ton, 2008) or between plants (Shulaev et al., 1997). Thus, the compound can act as a defence elicitor (Heil and Ton, 2008), providing an additional mode of action against pests. MeSA could be especially effective against piercing/sucking pests like aphids. In fact, several researchers (Pettersson et al., 1994; Ninkovic et al., 2003; Prinsloo et al., 2007; Xu et al., 2018) have shown that MeSA can reduce aphid populations in crops. Outside of luring in natural enemies, the compound's method of action isn't always obvious, especially when it comes to the relative weight of direct and plant-mediated effects. This is especially true for the *Rhopalosiphum padi* L. (cherry-oat aphid), which has been effectively controlled by the release of MeSA in cereal fields (Pettersson et al., 1994; Ninkovic et al., 2003). *Prunus padus* L. serves as *R. padi's* winter host, and grasses and cereals serve as its summer hosts. Furthermore, previous research (Ninkovic et al., 2003; Glinwood et al., 2007) demonstrated that cereal plants exposed to MeSA can decrease the host acceptance of *R. padi*, indicating that MeSA may have an impact on this aphid both directly as an ecological cue in the lifecycle and indirectly by inducing unfavourable changes in the host plant. Aphids find, pick, and colonise host plants in a methodical manner (Pettersson et al., 2007). Attraction is first mediated by volatile chemical and/or visual cues. The choice to start feeding may be influenced by cues from the plant surface after landing. The stylet probes the plant tissues and moves mostly between cells before piercing the phloem sieve components to start feeding. In this highly developed feeding process, MeSA has been demonstrated to function as a mobile signal for systemic acquired resistance (SAR) by the plant's conversion of MeSA to salicylic acid (SA) (Park et al., 2007). SA activates defences and is essential for a plant's resistance to phloem-feeding insects, such as aphids (Kaloshian and Walling, 2005; Goggin, 2007; Smith and Boyko, 2007; Spoel and Dong, 2012; van Dam et al., 2018). Aphid colonisation and feeding behaviour in wheat have been found to be disrupted by exogenous application of the SA analogue benzothiadiazole (BTH), and population growth rates on susceptible and resistant tomato cultivars have also been shown to be reduced (Li et al., 2002; Cooper et al., 2004).

9.4 BENEFITS OF BIOCHEMICAL PESTICIDES

Certain biochemical insecticides cause crops to manufacture and accumulate greater amounts of specific proteins and other substances that prevent the growth of fungi and other pests. Plant extracts have a variety of special advantages for producers when used as biopesticides. In general, plant-based substances break down quickly, lowering the possibility of residues on meals. The pre-harvest period for many of these products is extremely brief. Due to their diverse mode of action, the majority of the products have large windows of crop safety, and resistance to them does not form as quickly as it does to synthetic pesticides. Many plant extracts that are employed as pesticides have an immediate effect, stopping insect feeding and further crop damage. Additionally, many plant extract pesticides are more selective to insect target and safer to beneficial insects because they work on the gut of the insect and disintegrate quickly in the environment (Taverne, 2001; Gilrein, 2014).

Further benefits include:

- Safe for humans, the environment, and natural enemies.
- Resistance to certain insecticides is less likely to develop.
- Cost effective, renewable, biodegradable, and user friendly
- Active components quickly deteriorate, making them more tolerable.
- They are stable and can be kept in storage for a long time.
- Particularly in the case of semiochemicals, they are insect specific.

9.5 DISADVANTAGES OF BIOCHEMICAL PESTICIDES

- The basic formulations of the insecticide necessitate a larger quantity.
- Slow motion is used.
- Higher specificity and demand that the infection or pest be precisely identified.
- Inconsistent efficacy.

9.6 CONCLUSION

This chapter provides information about biochemical pesticides, which are an environmentally beneficial kind of insect control. Chemical regulation is the quick way to reduce the number of insect pests. However, the wide spread of indiscriminate use of highly harmful synthetic chemical pesticides had led to ecological imbalance in addition to their damaging effects on living things, including people. Consequently, it is necessary to create materials and producers in an eco-friendly environment. There are a variety of interesting pest control options that can be predicted, including the use of biochemical pesticides, all of which need to be carefully investigated and tapped for use in integrated pesticide management programmes. Products used to protect public health and agricultural safety must have proof that they are successful in controlling their intended pests. Additional criteria for products used on food plants include minimum hazards and the use of instructions along with the percentage of each ingredient's concentrations. In conclusion, a class of substances known as biochemical pesticides manage and improve natural plant protection mechanics in order to more efficiently meet the needs of the global food supply. In general, all chemical insecticides must be registered with the appropriate authorities before being used in agriculture. During the registration process, their safety and effectiveness will be carefully evaluated, and with correct application, these registered pesticides will leave little residue and pose a negligible risk to food safety. Recent studies on biochemical star chemicals and their uses for pest control have demonstrated that even relatively common and structurally simple compounds can act as significant chemical signals and exhibit biological activity. In the wide variety of species with one compound having multiple functions depending on the species, the chemical methyl salicylate is a good example of this class. Additionally, it is obvious that multi-discipline research in the field of logical chemistry is effective and will offer instruments for long-term control of several pest insects and other creatures.

REFERENCES

Aerts, N., Pereira Mendes, M., and Van Wees, S. C. M. 2021. Multiple levels of crosstalk in hormone networks regulating plant defense. Plant J. 105, 489–504. doi: 10.1111/tpj.15124.

Alam, A. Z., and Sana, D. L. 1964. Biology of Leucinodes Orbonalis Guenee in East Pakistan. In: Review of Research, Division of Entomology, 1947–64. Dhaka: Agriculture Information Service, Department of Agriculture, 192–200.

Alam, S. N., Hossain, M. I., Rouf, F. M. A., Jhala, R. C., Patel, M. G., Rath, L. K., Sengupta, A., Baral, K., Shylesha, A. N., Satpathy, S., Shivalingaswamy, T. M., Cork, A., and Talekar, N. S. 2006. Implementation and promotion of an IPM Strategy for control of eggplant fruit and shoot borer in South Asia. AVRDC publication number 06–672. AVRDC—The World Vegetable Center, Shanhua, Taiwan. Tech Bull. 36, 74.

Alam, S. N., Rashid, M. A., Rouf, F. M. A., Jhala, R. C., Patel, J. R., Satpathy, S., Shivalingaswamy, T. M., Rai, S., Wahundeniya, I., Cork, A., Ammaranan, C., and Talekar, N. S. 2003. Development of an integrated pest management Strategy for eggplant fruit and shoot borer in South Asia, AVRDC—The World Vegetable Center, Shanhua, Taiwan. Techn Bull. 28, 66.

Amer, S. A. A., Refaat, A. M., and Momen, F. M. 2001. Repellent and oviposition-deterring activity of rosemary and sweet marjoram on the spider mites tetranychus urticae and eutetranychus orientalis acari: tetranychidae. Acta Phytopathol Entomol Hungarica. 36(1–2), 155–164.

Beran, F., Mewis, I., Srinivasan, R., Svoboda, J., Vial, C., Mosimann, H., Boland, W., Büttner, C., Ulrichs, C., Hansson, B. S., and Reinecke, A. 2011. Male Phyllotreta Striolata (F.) produce an aggregation pheromone: identification of male-specific compounds and interaction With host plant volatiles. J Chem Ecol. 37, 85–97.

Byers, J. A., Maoz, Y., Cohen, B., Golani, M., Fefer, D., and Levi-Zada, A. 2021. Protecting avocado trees from ambrosia beetles by repellents and mass trapping (push—pull): experiments and simulations. J Pest Sci. 94, 991–1002. doi: 10.1007/s10340-020-01310-x.

Chen, J., Hua, G., Jurat-Fuentes, J. L., Abdullah, M. A., and Adang, M. 2007. Synergism of Bacillus thuringiensis toxins by a fragment of a Toxin-binding cadherin. Proc Nat Acad Sci USA. 104, 13901–13906.

Cooper, W. C., Jia, L., and Goggin, F. L. 2004. Acquired and r-gene-mediated resistance against the potato aphid in tomato. J Chem Ecol. 30, 2527–2542. doi: 10.1007/s10886-004-7948-9.

Digilio, M. C., Cascone, P., Iodice, L., and Guerrieri, E. 2012. Interactions between tomato volatile organic compounds and aphid behaviour. J Plant Interact. 7, 322–325. doi: 10.1080/17429145.2012.727104.

Downham, M. C. A. 2006. Maruca vitrata pheromone trapping in West Africa. www.nri.org/maruca/.

Downham, M. C. A., Hall, D. R., Chamberlain, D. J., Cork, A., Farman, D. I., Tamo, M., Dahounto, D., Datinon, B., and Adetonah, S. 2003. Minor components in the sex Pheromone of legume Podborer: Maruca vitrata Development of an attractive blend. J Chem Ecol. 29, 989–1012.

Downham, M. C. A., Tamo, M., Hall, D. R., Datinon, B., Adetonah, S., and Farman, D. I. 2004. Developing pheromone traps and lures for Maruca vitrata in Benin, West Africa. Entomol Exp Appl. 110, 151–158.

Gilrein, D. 2014. Pesticide Advancements Rise While Risks Drop: Features-Pest Control. Riverhead, NY: Cornell University Cooperative Extension of Suffolk County.

Glinwood, G. T., Karpinska, B., Ahmed, E., Jonsson, L. M. V., and Ninkovic, V. 2007. Aphid acceptance of barley exposed to volatile phytochemicals differs between plants exposed in daylight and darkness. Plant Signal Behav. 2, 321–326. doi: 10.4161/psb.2.5.4494.

Glinwood, R. T., and Pettersson, J. 2000. Change in response of Rhopalosiphum padi spring migrants to the repellent winter host component methyl salicylate. Entomol Exp Appl. 94, 325–330. doi: 10.1046/j.1570-7458.2000.00634.x.

Goggin, F. L. 2007. Plant—aphid interactions: molecular and ecological perspectives. Curr Opin. Plant Biol. 10, 399–408. doi: 10.1016/j.pbi.2007.06.004.

Hassan, M. N. 2007. Re-investigation of the female sex Pheromone of the legume pod-borer, Maruca vitrata (Lepidoptera: Crambidae). PhD thesis, University of Greenwich, 244p.

Heil, M., and Ton, J. 2008. Long-distance signalling in plant defence. Trends Plant Sci. 13, 264–272. doi: 10.1016/j.tplants.2008.03.005.

Hoffmann, K. H., and Lorenz, M. W. 1998. Recent advances in hormones in insect pest control. Phytoparasitica. 26(4), 323–330.

Jaenson, T. G. T., Garboui, S., and Palsson, K. 2006. Repellency of oils of lemon eucalyptus, geranium, and lavender and the mosquito repellent MyggA® natural to ixodes ricinus (Acari: Ixodidae) in the laboratory and field. J Med Entomol. 43(4), 731–736.

James, D. G. 2003. Synthetic herbivore-induced plant volatiles as field attractants for beneficial insects. Environ Entomol. 32, 977–982. doi: 10.1603/0046-225X-32.5.977.

Kaloshian, I., and Walling, L. L. 2005. Hemipterans as plant pathogens. Annu Rev Phytopathol. 43, 491–521. doi: 10.1146/annurev.phyto.43.040204.135944.

Kim, J.-K., Kang, C.-S., Lee, J.-K., Kim, Y.-R., Han, H.-Y., and Yun, H.-K. 2005. Evaluation of repellency effect of two natural aroma mosquito repellent compounds, citronella and citronellal. Entomol Res. 35(2): 117–120.

Li, X., Schuler, M. A., and Berenbaum, M. R. 2002. Jasmonate and salicylate induce expression of herbivore cytochrome P450 genes. Nature. 419, 712–715. doi: 10.1038/nature01003.

Liu, Z. L., Xu, Y. J., Wu, J., Goh, S. H., and Ho, S. H. 2002. Feeding deterrents From Dictamnus dasycarpus Turcz against two stored-product insects. J Agric Food Chem. 50(6), 1447–1450.

Maffei, M. E., Arimura, G. I., and Mithöfer, A. 2012. Natural elicitors, effectors and modulators of plant responses. Nat Prod Rep. 29, 1288–1303. doi: 10.1039/c2np20053h.

Mallinger, R. E., Hogg, D. B., and Gratton, C. 2011. Methyl salicylate attracts natural enemies and reduces populations of soybean aphids (Hemiptera: Aphididae) in soybean agroecosystems. J Econ Entomol. 104, 115–124. doi: 10.1603/EC10253.

Naresh, J. S., Malik, V. S., and Balan, J. S. 1986. Estimation of Fruit damage and larval population of brinjal fruit borer, Leucinodes orbonalis Guen. And its parasitization by Trathala sp. On brinjal. Bull Entomol (India), 27: 44–47.

Ninkovic, V., Ahmed, E., Glinwood, R., and Pettersson, J. 2003. Effects of two types of semiochemical on population development of the bird cherry oat aphid Rhopalosiphum padi in a barley crop. Agric For Entomol. 5, 27–34. doi: 10.1046/j.1461-9563.2003.00159.x.

Norin, T. 2007. Semiochemicals for insect pest management. Pure Appl Chem. 79(12), 2129–2136.

Orre Gordon, G. U. S., Wratten, S. D., Jonsson, M., Simpson, M., and Hale, R. 2013. "Attract and reward": combining a herbivore-induced plant volatile with floral resource supplementation—multi-trophic level effects. Biol Control. 64, 106–115. doi: 10.1016/j.biocontrol.2012.10.003.

Park, S. W., Kaimoyo, E., Kumar, D., Mosher, S., and Klessig, D. F. 2007. Methyl salicylate is a critical mobile signal for plant systemic acquired resistance. Science. 318, 113–116. doi: 10.1126/science.1147113.

Pettersson, J., Pickett, J. A., Pye, B. J., Quiroz, A., Smart, L. E., Wadhams, L. J., et al. 1994. Winter host component reduces colonization by bird-cherry-oat aphid, Rhopalosiphum padi (L.) (homoptera, aphididae), and other aphids in cereal fields. J Chem Ecol. 20, 2565–2574. doi: 10.1007/BF02036192.

Pettersson, J., Tjallingii, W. F., and Hardie, J. 2007. Host-plant selection and feeding. In: Aphids as Crop Pests, eds E. H. F. Van and R. Harrington. Wallingford, OX: CABI, 173–195. doi: 10.1079/9780851998190.0087.

Prinsloo, G., Ninkovic, V., van der Linde, T. C., van der Westhuizen, J., Pettersson, J., and Glinwood, R. 2007. Test of semiochemicals and a resistant wheat variety for Russian wheat aphid management in South Africa. J Appl Entomol. 131, 637–644. doi: 10.1111/j.1439-0418.2007.01213.x.

Rubene, D., Leidefors, M., Ninkovic, V., Eggers, S., and Low, M. 2019. Disentangling olfactory and visual information used by field foraging birds. Ecol Evol. 9, 545–552. doi: 10.1002/ece3.4773.

Schuler, T. H., Poppy, G. M., Kerry, B. R., and Denholm, I. 1998. Environmental risk assessment of transgene products using honey Bee (Apis mellifera) larvae. Trends Biotechnol. 16, 168–175.

Shulaev, V., Silverman, P., and Raskin, I. 1997. Airborne signalling by methyl salicylate in plant pathogen resistance. Nature. 385, 718–721. doi: 10.1038/385718a0.

Smart, L. E., Aradottir, G. I., and Bruce, T. J. A. 2014. Role of semiochemicals in integrated pest management. In: Integrated Pest Management: Current Concepts and Ecological Perspective. Amsterdam: Elsevier, 93–109. doi: 10.1016/B978-0-12-398529-3.00007-5.

Smith, C. M., and Boyko, E. V. 2007. The molecular bases of plant resistance and defense responses to aphid feeding: Current status. Entomol Exp Appl. 122, 1–16. doi: 10.1111/j.1570-7458.2006.00503.x.

Soroka, J. J., Bartelt, R. J., Zilkowski, B. W., and Cosse, A. A. 2005. Responses of flea beetle Phyllotreta cruciferae to Synthetic aggregation pheromone components and host Plant volatiles in field trials. J Chem Ecol. 31, 1829–1843.

Spoel, S. H., and Dong, X. 2012. How do plants achieve immunity? Defence without specialized immune cells. Nat Rev Immunol. 12, 89–100. doi: 10.1038/nri3141.

Srinivasan, R., Tamo, M., Ooi, P. A. C., and Easdown, W. 2007. IPM for Maruca vitrata on food legumes in Asia and Africa. Biocontrol News Inf. 28, 34–37.

Taverne, J. 2001. Malaria on the Web and the mosquito-repellent properties of basil. Trends Parasitol. 17(6), 299–300.

van Dam, N. M., Wondafrash, M., Mathur, V., and Tytgat, T. O. G. 2018. Differences in hormonal signaling triggered by two root-feeding nematode species result in contrasting effects on aphid population growth. Front Ecol Evol. 6, 88. doi: 10.3389/fevo.2018.00088.

Witzgall, P., Backman, A. C., Svensson, M., Koch, U. T., Rama, F., El-Sayed, A., Brauchli, J., Arn, H., Bengtsson, M., and Lofqvist, J. 1999. Behavioral observations of codling moth, Cydia pomonella, in Orchards permeated with synthetic pheromone. BioControl. 44, 211–327.

Xu, Q., Hatt, S., Lopes, T., Zhang, Y., Bodson, B., Chen, J., et al. 2018. A push—pull strategy to control aphids combines intercropping with semiochemical releases. J Pest Sci. 91, 93–103. doi: 10.1007/s10340-017-0888-2.

10 Microbial Biopesticides

Zareeka Haroon, Hyda Haroon, Asfia Shabbir,
Irzam Haroon, and Yawar Sadiq

CONTENTS

10.1 INTRODUCTION

Soil sustains almost all forms of life and hence can be regarded as a life-ensuring or life-sustaining component. Soil microbes, co-evolution with plants, and biomineralisation are some of the features that clearly shows the dynamic nature of soil. The extensive use of fertilisers and pesticides has degraded the quality of soil, altering the physical (structure, porosity, aeration, water infiltration, etc.) as well as the chemical (concentrations of specific chemicals, pH, cation exchange capacity, anion exchange capacity, base saturation, salinity, sodium adsorption ratio, etc.) nature of soil. Although fertilisers and pesticides are considered a necessary evil for ensuring global food security, these have a long persistence in soil and affect a wide range of soil functions (nutrient cycling, soil fertility, improving plant productivity through enhanced availability of limited nutrients, and decomposition of organic as well as inorganic matter) and properties. This persistence of fertilizers and pesticides also has an impact on the soil microbes that results in the degradation of soil health and fertility. As far as soil microbes are concerned, they have a strong influence on the

DOI: 10.1201/9781003265139-13

physical nature of soil and help it pursue eco-friendly practices indicating soil health and activity and, hence, is regarded as a bio indicator. Ever since the practices of farming and cultivation began, man has been continuously depleting and degrading the soil health in numerous ways. Hence, a need is felt for sustainable agriculture to ensure safer and more productive agriculture practices.

Microbial pesticides in general enlist various microorganisms like bacteria, protozoa, microsporidia, fungus, virus, and their bioactive-compounds as the active ingredients in the pest management agents. They tend to provide a better alternative to chemical pesticides. There is a deep connection between soil and microorganisms. Various bacteria have been strained for their metabolites that can be used against various pests for application on useful plants like fruits, vegetables, and some ornamentals. It has been seen that a handful of industries have produced a limited number of marketing products to control pests using microbes (Lacey et al., 2015). However, there is a challenge in transferring these microbial inoculants from laboratories to the fields successfully.

Currently, various studies aimed at investigating the role of plant growth promoters have found that various microorganisms colonise plant tissues and vessels from the root to the shoot system of the plants. Most of them are regarded as growth enhancers which interfere with plant life by producing metabolites. Various microbes have higher efficacy in crop productivity by enhancing nutrient uptake, regulating phytohormones, and increasing the tolerance of plants to various biotic and abiotic stresses. There is a symbiotic relationship between various microbes in general and bacteria in specific, and different plant parts like roots, fruits, stems, leaves, etc. that are often associated with plant health and increased crop productivity. Thus, these microbes can be regarded as biocontrol agents or growth promoters (Shimizu et al., 2009).

10.2 WHAT ARE MICROBIAL PESTICIDES?

Infections caused by agriculture pests can be controlled by various natural and biologically occurring compounds known as biopesticides. Different types of biopesticides have been developed from different sources. Based on living microorganism or natural products, microbial pesticides are known to provide the best possible pest management. For a long time, crops were protected from disease by chemical pesticides and fungicides, and fertilisers are still used to enhance the fertility of the soil, which, in turn, pollutes our groundwater and environment and ultimately results in biomagnification. An urgent need to think about alternate means is felt.

Biocontrol agents seem to show promising results in this regard without damaging the environment further. Moreover, they tend to be an important component in sustainable agriculture. The microbial pesticides not only protect the agricultural crops but also provide nutrients to the plants by enhancing various metabolites and metabolic activities. *Microbial pesticides* is a common term used for any biocontrol agent, such as microorganisms including viruses, protozoa, fungi, bacteria and their bioactive compounds, microbial herbicides, bioinsecticides, etc. Microbial pesticides are used nowadays on larger scale as they are cost efficient and less costly than the other means. Microbial pesticides or biopesticides get easily multiplied in soil as the rate of reproduction of these microbes is tremendous, and they leave no residual problem. Handling these microbial pesticides is very easy and requires no special training. This practice is safer for the person who applies them. Application of microbial pesticides not only controls disease, but, in some cases, it also enhances various growth-related phenomena that result in increased crop yields. Microbial pesticides can be regarded as a safer way sustainable agriculture can be attained with social acceptability, economic productivity, and environmental stewardship. The main characterisation of microbial pesticides is suppression of the growth and proliferation of the pest population. Microbial pesticides achieve this by diverse mechanistic actions.

Different chemical pesticides add heavy metal to soil, which has adverse effects on it. Cadmium (Cd) has extremely high mobility in soil, and it affects the essential microorganisms and tends to absorb soil's organic matter, which is important for plant growth and development. Likewise, soil's pH and sorption can be affected by the accumulation of lead (Pb), which can even damage the DNA and adversely affect essential physiological processes like reduction in photosynthetic rate, leading to the death of the plant. Copper (Cu), zinc (Zn), and many other elements are also responsible for the toxicity of the soil. These pose a serious problem to both plant and animal health. This problem can be overcome by moving towards a sustainable practice of agriculture, which can be attained by the greater use of microbial pesticides, which tend to enhance the developmental aspects of plants without adding these toxic heavy metals to the soil.

These microbial and biochemical pesticides comprise 5% of the global pesticide market, with microbial pesticides owning the top position (Pathma et al., 2021). It was seen in a study that a complete preference for these microbial pesticides is somewhere hindered by the shortage of the supply against the high demand by farmers and, to a greater extent, the slow action of most of these microbial pesticides (Verma et al., 2021). Despite these shortcomings, microbial pesticides are preferred because their toxicity, if any, is tolerable. Microbial pesticides are usually action specific and biodegradable, and importantly, the pest resistance issues caused by chemical pesticides can be countered by different microbial pesticides (Mishra et al., 2020). Cost-effective technology, judicious use of resources, productivity, and environmental safety can be regarded as the key concerns of microbial pesticides.

10.3 PROPERTIES OF MICROBIAL PESTICIDES

Due to their superiority to chemical insecticides and reduced impact on non-targeted organisms, microbial pesticides are a better solution. Since they are not closely related to the target pest, the organisms utilised in microbial pesticides are essentially harmless and nonpathogenic to humans, animals, and other organisms. The major advantage of microbial insecticides is the safety they provide. Most microbial insecticides do not directly impact beneficial insects (including predators or parasites of pests) in treated areas because the toxic effect of microbial insecticides is frequently specific to a single group or species of insects. The majority of microbial insecticides can be used in conjunction with synthetic chemical insecticides as appropriate because, for the most part, they do not cause harm the microbial product or render it inactive. They also promote the growth of beneficial soil microorganisms, which helps the roots of plants. They contribute to the increase in crop yield in this way (Usta, 2013). To effectively reduce the pathogen in the host, it is essential to understand the biocontrol mechanisms in microbial pesticides. A single mechanism is covered by several strains, and others combine them. Examples of the usage of cyclo-lipopeptides, such as fengycins generated by *Bacillus subtilis*, in the prevention of disease include the defence of injured apple fruits against the grey mould disease brought on by *Botrytis cinerea* (Ongena et al., 2005). Numerous hyperparasites, particularly yeast and fungi like *Pichia* and *Trichoderma*, interact directly and degrade the fungal cell, or they fight back using antimicrobial substances, become hyper-parasitic, directly attach to the pathogen cells, disrupt the pathogen's signals, or make the plant host resistant (Harman, 2006). Finally, some biocontrol agent (BCAs) can prevent pathogens from infecting plants by degrading the chemical signal messengers required for quorum sensing, such as acyl homoserine lactones, which the pathogen uses to initiate infection (Molina et al., 2003). Gram-positive bacteria like Bacillaceae and gram-negative bacteria like Pseudomonadaceae are where the majority of the strains of microorganisms used to treat pome fruit tree illnesses are found, as well as numerous yeasts and fungi, particularly those belonging to the Basidyomicota order (Montesinos and Bonaterra, 2009).

10.4 INGESTION AND SORPTION OF MICROBIAL PESTICIDES

Because pesticide molecule uptake by microbes is a prerequisite for aerobic transformation, molecular diffusion from a boundary layer surrounding the microbe to its interior via the microbial surface becomes a critical process (Aksu, 2005). When pesticides are taken up, either energy-mediated active transport or passive sorption to microbial surfaces can be proposed. It is convenient to distinguish these processes not only by comparing uptake between living and dead cells but also by reversibility in sorption. The plot of the uptake rate by living microbes versus substrate concentration reveals saturation kinetics in the biodegradation of four simple organic compounds, including m-cresol (Pfaender and Bartholomew, 1982). Phenol was incorporated into the cellular macromolecules of living microbes using similar kinetics (Chesney et al., 1985). Lal and Saxena (1982) reported rapid species-dependent uptake of organochlorine pesticides by microbes, most likely as a result of the sorption-metabolism balance. The pesticides were mostly absorbed by microbes regardless of cell viability, indicating a passive process. The pesticide sorption coefficient to microbes is commonly estimated using linear and Freundlich isotherms, and it is known to correlate well with the n-octanol/water partition coefficient (Kow), as reported for algae. Pesticide sorption to fungi has been studied in relation to biosorbents' removal of contaminants from water. Several pesticide sorption isotherms could not be explained by assuming a monolayer surface coverage, according to *Rhizopus anhizus* studies (Bell and Tsezos, 1988). Furthermore, pesticide sorption was not fully explained by pesticide partitioning to fungal cell walls, and the involvement of cytoplasmic components leaking from the cells was highly suspect (Lièvremont et al., 1998; Tsezos and Bel, 1989).

10.5 ROLE OF BACTERIA AS MICROBIAL BIOPESTICIDE

Due to the prominence of *Bacillus thuringiensis* (*Bt*) products for controlling caterpillars, bacteria were among the first microbiological control agents (MCAs) to be produced and continue to dominate the microbial pesticide business. Bacteria tend to be produced in huge biofermenters and include a combination of spores, crystal proteins, and inert carriers (Sanahuja et al., 2011).

10.5.1 *PAENIBACILLUS POPILLIAE*

Currently, a product with *Paenibacillus* (Bacillus) *popilliae* spores that cause milky disease in *Popillia japonica*, a Japanese beetle species, and closely-related scarab larvae is approved for use on turf. The first bacteria to be registered as an MCA in the United States was *Paenibacillus popilliae*, which has been extensively employed to control *P. japonica* (Klein and Kaya, 1995).

10.5.2 *BT GALLERIAE* AND *BT TENEBRIONIS*

For the control of beetles, *Bt galleriae* and *Bt tenebrionis* strains are registered. On vegetables, legumes, fruits, and grasses, the former is indicated for the management of specific beetles in the Buprestidae, Scarabaeidae, and Curculionidae families. Both the Colorado potato beetle (*Leptinotarsa decemlineata*), which attacks potatoes and tomatoes, and the elm leaf beetle (*Pyrrhalta luteola*), which attacks elm trees, are controlled by the latter, according to its labelling.

10.5.3 *PASTEURIA* SPP. AND *BACILLUS FIRMUS*

The bacteria *Pasteuria* spp. form endospores and mycelium, and they are obligate parasites of some phytoparasitic nematodes. Another bacterium, *B. firmus*, colonises the rhizosphere and may produce metabolites toxic to some nematodes, preventing nematode infection of roots. In recent years, a number of bionematicides aimed at a variety of phytoparasitic nematodes have been developed as a result of the antagonistic action of both of these bacteria (Wilson and Jackson, 2013). Nortica 5WG, marketed for turf grass (Crow, 2014) and Clariva, used as a seed treatment for soybeans and other field crops, are examples of products sponsored by major international corporations (Mourtzinis et al., 2017).

10.5.4 *BT ISRAELENSIS* (*BTI*) AND *LYSINIBACILLUS SPHAERICUS*

For a long time, especially in environmentally sensitive locations, *Bti* and *Lysinibacillus* (Bacillus) *sphaericus* have been utilised for vector and nuisance fly management in public health (Lacey, 2007). Four *Bti* strains are currently recognised for the control of fungus gnats (Diptera: Sciaridae) in greenhouses and interiorscapes, as well as mosquito and blackfly larvae in aquatic habitats. A different *Bti* strain (SUM-6218) is being examined by the EPA right now. For residual control of mosquito larvae, one product, VectoLex FG, a granular formulation of *L. sphaericus* 2362, is currently available.

10.6 ROLE OF FUNGI AS MICROBIAL BIOPESTICIDE

Eleven entomopathogenic and nematophagous fungal strains are actively registered as microbial pesticides. Thrips, whiteflies, aphids, and other field, greenhouse, and nursery pests, as well as numerous plant-parasitic nematodes, are the main targets of these products. Fungi are the only microbial agents capable of directly penetrating the cuticle and targeting the majority of sap-feeding pests due to their contact action.

10.6.1 *Purpureocillium lilacinum* and *Myrothecium verrucaria*

Currently, bionematicides made from the soil fungus *Purpureocillium lilacinum* and *Myrothecium verrucaria* are sold commercially (MeloCon and DiTera). These products, which contain spores or fermentation solids, respectively, naturally parasitize or, in the case of the latter, are toxic to a variety of plant-parasitic nematodes, such as false root knot nematodes, cyst nematodes, root lesion nematodes, root knot nematodes, and burrowing nematodes. These products can be used as partial substitutes for chemical soil fumigants and are registered for a number of crops (Dong et al., 2015; Baidoo et al., 2017).

10.6.2 *Isaria fumosorosea*

Whiteflies, aphids, thrips, leafminers, plant bugs, and a number of soil pests are controlled by the Apopka strain 97 of *Isaria fumosorosea* (also known as *Paecilomyces fumosoroseus*, PFR-97, and Ancora). Spider mites and mealybugs are the principal pests targeted by PFR-97 treatments on crops cultivated outdoors, whereas whiteflies, thrips, and aphids are the main pests targeted in greenhouses. This product is provided as a dried blastospore formulation, allowing for quick germination under ideal circumstances (Avery et al., 2013). For the control of whiteflies in ornamental and vegetable crops, the strain FE 9901 is marketed as NoFly in Europe and was once offered under this name in the US (Arthurs and Dara, 2019).

10.6.3 *Metarhizium brunneum*

The F52 strain of *M. brunneum* is indicated for the control of thrips, whiteflies, mites, and weevils in a variety of fruits, vegetables, and ornamental crops, as well as ticks in turf. A granular formulation of Met52 EC is also available for use against soil pests such the black vine weevil and thrips pupae, and it can also be used to treat pests that are foliar (Arthurs and Dara, 2019). In spite of this, due to industry mergers, this strain is not now available on the market. In the past, BioBlast, a termite control product, was sold alongside the ESF1 strain (Wright et al., 2005).

10.7 ROLE OF BACULOVIRUSES AS MICROBIAL BIOPESTICIDES

Eight baculoviruses are currently approved for use to manage *Lepidoptera* larvae in field- and greenhouse-grown vegetables, ornamentals, tree fruit, forestry, and storage facilities. Although baculoviruses generated in vivo from viral occlusion bodies have been utilised for a long time, fresh interest may have been sparked by markets for integrated and organic insecticides (Szewczyk et al., 2006). Baculoviruses like *Bt* require the host to consume them in order to be lethal. They seem to be more host specific than the *Bt*s that are active against Lepidoptera, but they are most effective against younger instars and have minimal residual activity (Grzywacz, 2016).

10.7.1 GRANULOVIRUSES (GV)

One of the most commonly utilised baculoviruses in the United States, the codling moth granulovirus (CpGV-M), has a number of commercially successful products for controlling the codling moth, *Cydia pomonella*, on apples and other pome fruits. The oriental fruit moth, *Grapholita molesta*, in stone fruits is another pest identified by the V22 isolate of CpGV (Madex HP). In the early 2000s, many CpGV products were introduced in the United States and quickly became well liked by organic farmers (Arthurs and Lacey, 2004; Arthurs and Dara, 2019). When compared to the 100,000 ha in Europe where the Carpovirusine formulation is utilised, CpGV was employed on more than 4,000 ha annually in both conventional and organic orchards within five years (Lacey et al., 2008). To combat the Indian meal moth, *Plodia interpunctella*, in stored goods, another PiGV-based product has been employed (Lord et al., 2007).

REFERENCES

Aksu, Z. 2005. Application of biosorption for the removal of organic pollutants: A review. Process Biochem., 40(3–4), 997–1026.

Arthurs, S., Dara, S.K. 2019. Microbial biopesticides for invertebrate pests and their markets in the United States. J Invertebr Pathol., 165, 13–21.

Arthurs, S.P., Lacey, L.A. 2004. Field evaluation of commercial formulations of the Codling moth granulovirus (CpGV): Persistence of activity and success of seasonal Applications against natural infestations in the Pacific Northwest. Biol Control., 31, 388–397.

Avery, P.B., Pick, D.A., Aristizábal, L.F., Kerrigan, J., Powell, C.A., Rogers, M.E., Arthurs, S.P. 2013. Compatibility of *Isaria fumosorosea* (Hypocreales: Cordycipitaceae) blastospores with agricultural chemicals used for management of the Asian citrus psyllid, *Diaphorina citri* (Hemiptera: Liviidae). Insects., 4(4), 694–711.

Baidoo, R., Mengistu, T., McSorley, R., Stamps, R.H., Brito, J., Crow, W.T. 2017. Management of root-knot nematode (*Meloidogyne incognita*) on *Pittosporum tobira* Under greenhouse, field, and on-farm conditions in Florida. J Nematol., 49(2), 1330–1339.

Bell, J.P., Tsezos, M. 1988. The selectivity of biosorption of hazardous organics by microbial biomass. Water Res, 22(10), 1245–1251.

Chesney, R.H., Sollitti, P., Rubin, H.E. 1985. Incorporation of phenol carbon at trace concentrations by phenol-mineralizing microorganisms in fresh water. Appl Environ Microbiol., 49(1), 15–18.

Crow, W.T. 2014. Effects of a commercial formulation of Bacillus firmus I-1582 on golf Course bermudagrass infested with *Belonolaimus longicaudatus*. J Nematol., 46(4), 331–335.

Dong, H., Zhou, X.G., Wang, J., Xu, Y., Lu, P. 2015. Myrothecium verrucaria strain X-16, a Novel parasitic fungus to *Meloidogyne hapla*. Biol Control, 83, 7–12.

Grzywacz, D. 2016. Basic and applied research: Baculovirus. In: Lacey, L.A. (Ed.), Microbial Control of Insect and Mite Pests: From Theory to Practice. Academic, Amsterdam, pp. 27–46.

Harman, G.E. 2006. Overview of mechanisms and uses of trichoderma spp. Phytopathology, 96(2), 190–194.

Klein, M.G., Kaya, H. 1995. *Bacillus* and *Serratia* species for scarab control. Mem I OsCr., 90(1), 87–95.

Lacey, L.A. 2007. *Bacillus thuringiensis serovariety israelensis* and *Bacillus sphaericus* for Mosquito control. In: Floore, T.G. (Ed.), Biorational control of mosquitoes. Am Mosq Control Assoc Bull., 7, 133–163.

Lacey, L.A., Grzywacz, D., Shapiro-Ilan, D.I., Frutos, R., Brownbridge, M., Goettel, M.S. 2015. Insect pathogens as biological control agents: Back to the future. J Invertebr Pathol., 132, 1–41.

Lacey, L.A., Thomson, D., Vincent, C., Arthurs, S.P. 2008. Codling moth granulovirus: A Comprehensive review. Biocontrol Sci Tech., 18, 639–663.

Lal, R., Saxena, D.M. 1982. Accumulation, metabolism, and effects of organochlorine insecticides on microorganisms. Microbiol Rev., 46(1), 95–127.

Lièvremont, D., Seigle-Murandi, F., Benoit-Guyod, J.L. 1998. Removal of PCNB from aqueous solution by a fungal adsorption process. Wat Res., 32(12), 3601–3606.

Lord, J.C., Campbell, J.F., Sedlacek, J.D., Vail, P.V. 2007. Application and evaluation of Entomopathogens for managing insects in stored products. In: Lacey, L.A., Kaya, H.K. (Eds.), Field Manual of Techniques in Invertebrate Pathology. Springer, Dordrecht, pp. 677–693.

Mishra, J., Dutta, V., Arora, N.K. 2020. Biopesticides in India: Technology and sustainability linkages. 3 Biotech., 10, 1–12.

Molina, L., Constantinescu, F., Michel, L., Reimmann, C., Duffy, B., Défago, G. 2003. Degradation of pathogen quorum-sensing molecules by soil bacteria: A preventive and curative biological control mechanism. FEMS Microbiol Ecol., 45(1), 71–81.

Montesinos, E., Bonaterra, A. 2009. Microbial pesticides. In: Schaechter M (Ed.), Encyclopedia of Microbiology. Elsevier, New York, pp. 110–120.

Mourtzinis, S., Marburger, D., Gaska, J., Diallo, T., Lauer, J., Conley, S. 2017. Corn and Soybean yield response to tillage, rotation, and nematicide seed treatment. Crop Sci., 57, 1704–1712.

Ongena, M., Jacques, P., Touré, Y., Destain, J., Jabrane, A. and Thonart, P. 2005. Involvement of fengycin-type lipopeptides in the multifaceted biocontrol potential of Bacillus subtilis. Appl Microbiol Biotechnol., 69(1), 29–38.

Pathma, J., Kennedy, R.K., Bhushan, L.S., Shankar, B.K., Thakur, K. 2021. Microbial biofertilizers and biopesticides: Nature's assets fostering sustainable agriculture. In: Recent Developments in Microbial Technologies. Springer, Singapore, pp. 39–69.

Pfaender, F.K., Bartholomew, G.W. 1982. Measurement of aquatic biodegradation rates by determining heterotrophic uptake of radiolabeled pollutants. Appl Environ Microbiol, 44(1), 159–164.

Sanahuja, G., Banakar, R., Twyman, R.M., Capell, T., Christou, P. 2011. *Bacillus thuringiensis*: A century of research, development and commercial applications. Plant Biotechnol J., 9(3), 283–300.

Shimizu, M., Yazawa, S., Ushijima, Y.A. 2009. Promising strain of endophytic Streptomyces sp. For biological control of cucumber anthracnose. J Gen Plant Pathol., 75, 27–36.

Szewczyk, B., Hoyos-Carvajal, L., Paluszek, M., Skrzecz, I., De Souza, M.L. 2006. Baculoviruses—re-emerging biopesticides. Biotechnol Adv., 24(2), 143–160.

Tsezos, M., Bell, J.P. 1989. Comparison of the biosorption and desorption of hazardous organic pollutants by live and dead biomass. Water Res., 23(5), 561–568.

Usta, C. 2013. Microorganisms in biological pest control—a review (bacterial toxin application and effect of environmental factors). Curr Prog Biol Res., 13, 287–317.

Verma, D.K., Guzmán, K.N.R., Mohapatra, B., Talukdar, D., Chávez-González, M.L., Kumar, V., et al. 2021. Recent trends in plant-and microbe-based biopesticide for sustainable crop production and environmental security. In: Prasad, R., Kumar, V., Singh, J., Upadhyaya, C.P. (Eds.), Recent Developments in Microbial Technologies. Springer, Singapore, pp. 1–37.

Wilson, M.J., Jackson, T.A. 2013. Progress in the commercialisation of bionematicides. Biocontrol., 58(6), 715–722.

Wright, M.S., Raina, A.K., Lax, A.R. 2005. A strain of the fungus *Metarhizium anisopliae* For controlling subterranean termites. J Econ Entomol., 98(5), 1451–1458.

11 Plant-Incorporated Protectants

*Faraat Ali, Kumari Neha, Hasan Ali,
and Arvind Kumar Sharma*

CONTENTS

11.1 INTRODUCTION

Biopesticides are live organisms (natural enemies) or their products (phytochemicals, microbial products) or byproducts (semiochemicals) that can be employed to control plant-harming pests (**Dayan & Duke, 2014**). Microbial pesticides, biochemicals generated from microorganisms, and other natural sources are all examples of biopesticides (**Sarwar, 2015**). They play a significant role in crop protection, but they are usually used in conjunction with other substances, such as chemical pesticides, as part of bio-intensive integrated pest management. Typically, these are made by growing and concentrating naturally existing organisms and their metabolites, such as bacteria and other microorganisms, fungi, nematodes, proteins, and so on. These are frequently regarded as critical components of integrated pest management (IPM) systems and have garnered considerable practical attention as alternatives to synthetic chemical plant protection solutions (**Sarwar, 2015**). Biopesticides are pesticides relying on microorganisms or natural products, and they include naturally produced pest-controlling compounds (biochemical pesticides), pest-controlling microorganisms (microbial pesticides), and pesticidal compounds manufactured by plants with extra genetic material (plant-incorporated protectants) (**Sarwar, 2015**). Pesticides derived from natural substances such as animals, plants, microorganisms, and minerals are referred to as biopesticides.

DOI: 10.1201/9781003265139-14

11.2 PLANT-INCORPORATED PROTECTANTS

Plant-incorporated protectants (PIPS) are pesticidal compounds produced by agricultural plants from genetic material that has been added to the plant genome by biotechnology, providing protection against specific pests or other chemical stresses (**Meshram et al., 2022**; **Sudakin, 2003**; **Razaq & Shah, 2022**) (Figure 11.1). Because their nuclear material is stably combined with genetic material from a naturally occurring bacterium, these plants are genetically modified and are often known as transgenic plants. As a result, transgenic plants exhibit typical traits of the additional genetic material, which they express to kill pests. So far, many PIPs have been produced (**Razaq & Shah, 2022**). For example, scientists can transfer the gene for a specific *Bt* pesticidal protein into a plant's genetic code. The plant then produces a pesticidal protein, which suppresses the pest when it feeds on the plant. Herbicide-tolerant crops and plant-incorporated protectants are two types of genetically altered crops. Crops are also genetically modified or "stacked" to express numerous features, such as resistance to multiple herbicides or resistance to herbicides and insecticides. In all these examples, the features are accomplished by modifying the organism's genetic composition by inserting certain genes. The first *B. thuringiensis* (*Bt*) plant-incorporated protectant has registered 11 *Bt* plants, 5 of which are no longer active. Corn and cotton *Bt* cultivars are released, and a *Bt* soy variety is registered. Despite industry assurances that transgenic plants will reduce pesticide use, insects have developed resistance to the genetically modified crops. This calls into question the usefulness of natural *Bt* utilized in organic food production, as well as the loss of a crucial tool (**Sarwar, 2015**). PIPs that have been registered to date include plants that have been genetically altered to create a *B. thuringiensis* toxin (*Bt*-PIPs) and plants that have been designed to produce plant virus coat proteins (PVCP-PIPs). *Bt*-PIPs and PVCP-PIPs, respectively, protect plants from insect pests and harmful viruses. The *Bt* pesticidal protein gene is inserted into a plant's genetic code. The plant then produces the chemical that kills the bug instead of the *Bt* bacteria. All prototype biopesticides arose from the bacteria *B. thuringiensis* (*Bt*), which produces a toxin (*Bt* toxin) that binds to insecticides and attaches to insect gut receptor protein, disrupting the gut when consumed. They are already in the environment and do not endanger human health (**Meshram et al., 2022**). Canada, Japan, Australia, New Zealand, and Argentina have implemented science-based regulatory review procedures for the application of biotechnology products such as plant-incorporated protectants, which rely on a balancing of benefits and risks in regulatory decision-making. European Union countries have taken a more careful approach to biotechnology legislation, particularly plant-incorporated protectants, and there is presently a moratorium on the approval of new genetically modified organisms (**Sudakin, 2003**).

Transgenic plants that express insecticidal endotoxins derived from soil bacteria (*Bt* plants) and the expression of these toxins provide protection against insect crop loss. *Bt* endotoxins' lethality is heavily dependent on the alkaline environment of the insect gut, which ensures that these toxins are not active in vertebrates, including humans. In general, *Bt* toxins kill insect larvae by acting on the stomach cells of the larvae eating the toxin. Several members of the class Insecta, notably those belonging to the insect orders Lepidoptera and Coleoptera, are highly vulnerable to *Bt* toxin.

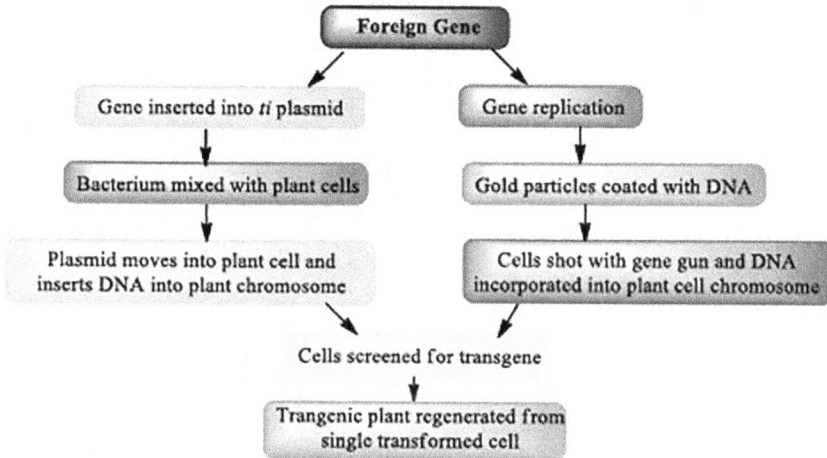

FIGURE 11.1   Gene insertion strategies into plant cell.

When feeding, insects absorb *Bt* toxin from the transgenic plant host, which enters the insect stomach; the alkaline nature of the gut denatures insoluble *Bt* crystals into soluble ones. In the presence of gut protease enzymes, the soluble crystals release active poisons. These active poisons enter the cellular membranes, adhere, and begin generating spores. This activity causes gut paralysis, which causes insects to stop eating and starve until they die (Razaq & Shah, 2022). These proteins have been produced commercially to target the principal pests of cotton, tobacco, tomato, potato, corn, maize, and rice, providing for higher coverage by reaching places on plants that foliar sprays cannot reach. There are multiple *Bt* strains, each with a unique Cry protein, and over 60 Cry proteins have been found. The majority of *Bt* maize hybrids express the Cry1Ab protein, while a handful also express the Cry1Ac or Cry9C proteins, all of which are directed against the corn borer, *Ostrinia nubilalis* (Hubner) (Lepidoptera) (Sarwar, 2015). Transgenic plants with the Cry gene (which produces toxins) from the *Bt* bacteria, *B. thuringiensis*, are the most common examples. Several agricultural crops have been genetically modified to include *B. thuringiensis* traits. This transformation, in which a simple plant becomes a transgenic plant, is carried out via manipulation of biological or physical means. For example, for transgenic *Bt*, crop plants that are to be modified are injected with the corresponding gene from the *Bt* bacterium by the use of recombinant DNA technology with the aid of *Agrobacterium tumefaciens*. This technology was first used in genetic modification of tomato plants in 1987, and after its success, its application was extended to benefit many other economically important agricultural crops, including cotton, tobacco, etc., (Razaq & Shah, 2022).

PIPs interest and implementation have skyrocketed in recent decades, with more acceptability in poorer nations where chemical pesticide hazards are high, and consumer health and safety are important priorities. Apart from the United States, where PIPs are classified as biopesticides, many other nations' regulatory agencies do not classify PIPs as biopesticides due to consumer concerns about GM crops. Frequent consumption of GM crops may result in health issues (hormonal and neurological disturbance) in their consumers (**Razaq & Shah, 2022**).

When plants are genetically engineered to create pesticides by this method, the Environmental Protection Agency (EPA) regulates them as pesticides (NPIC, 2017). As a result, the pesticide generated by such a plant, as well as the genetic material transferred, are collectively referred to as PIPs. Cry proteins are the first-generation insecticidal PIPs that GM crops carrying transgenes from the soil bacterium *B. thuringiensis* were permitted; next-generation double-stranded ribonucleic acid (dsRNA) PIPs have just been authorized. The use of either Cry protein or dsRNA PIPs, like traditional synthetic pesticides, results in their release into recipient habitats. This section emphasizes the information gaps and obstacles discovered while researching the environmental fate of Cry protein. PIPs enhance the state of the research required for the continuing assessment of the environmental destiny of dsRNA PIPs. In June 2017, the EPA approved the first PIP, SmartStax Pro, to assist US farmers in controlling maize rootworm, a destructive corn pest that has gained resistance to multiple conventional pesticides. It was created using RNAi technology and was designed to target the western corn rootworm (*Diabrotica virgifera virgifera*) (Niu et al., 2017). The RNAi mechanism operates at the messenger RNA level, employing a sequence-dependent method of action that distinguishes it from traditional agrochemicals in terms of potency and selectivity. Moreover, the use of RNAi in agricultural protection can be accomplished not only through the application of PIPs via plant modification but also through nontransformative approaches such as the utilization of sprayable RNA formulations as direct control agents, resistance factor repressors, or developmental disruptors (**Parween & Jan, 2019**).

11.3 PLANT-INCORPORATED PROTECTANTS IN VARIOUS CROPS

1. *Bacillus thuringiensis*

   Bacillus bacteria produce parasporal, proteinaceous, and crystal inclusion bodies (Cry toxins), which are pesticides against over 150 kinds of arthropod insects. Several toxins, including a-exotoxin, b-exotoxin, c-exotoxin, d-endotoxin, louse factor exotoxin, mouse factor exotoxin, water-soluble toxin, Vip3A, and enterotoxin, have been extracted and reported from *Bt* strains. Each toxin has unique pesticidal properties based on its strength and receptor specificity (**Razaq & Shah, 2022**). *B. thuringiensis* is a worldwide organism that can be found in soils and on foliar surfaces. *B. thuringiensis* is usually encountered in nature as a spore, and during the sporulation process, the organism releases endotoxin, an insecticidal crystal protein expressed on bacterial plasmids. *B. thuringiensis* can also produce a non-protein exotoxin with differential toxicodynamic effects mediated by RNA polymerase suppression. *B. thuringiensis*–protected crops, such as maize, potato, and cotton, are the most extensively researched application of plant-incorporated protectants. These crops include genetic material encoding several insecticidal crystal proteins, which give protection against agriculturally important insect pests when expressed at low (ppm) quantities in target plant tissue. In several studies, the development of *B. thuringiensis*–protected crop types has been linked to effective pest management, higher agricultural yields, and lower use of traditional chemical pesticides. Recent research suggests that *B. thuringiensis*–protected crops may be less susceptible to mycotoxin contamination than conventionally produced crops due to reduced insect damage (which serves as a source of entry for toxigenic fungi) (**Sudakin, 2003**).

2. DvSnf7 RNA

   The first commercial RNAi insecticidal PIP was MON 87411. MON 87411 maize, which expresses DvSnf7 RNA, was created to provide an alternative mechanism of action for corn rootworm protection (*Diabrotica* spp.). The Monsanto Company has produced MON 87411, a GE maize that protects against corn rootworm (CRW) (*Diabrotica* spp.) using RNAi as the insecticidal method. The DvSnf7 RNA produced in MON 87411 is made up of a 968-nucleotide sequence with 240 base pair dsRNA components and a poly A tail designed to target the Snf7 gene of the western corn rootworm (*Diabrotica virgifera virgifera*) (WCR) (DvSnf7) (**Bachman et al., 2016**).

3. 4-hydroxyphenylpyruvate dioxygenase enzyme of *Pseudomonas fluorescens*

   4-Hydroxyphenylpyruvate dioxygenase (HPPD) is an Fe(II)-dependent, non-heme oxygenase that catalyzes the conversion of 4-hydroxyphenylpyruvate to homogentisate (Moran, 2005, Dufourmantel et al., 2007). This reaction involves decarboxylation, substituent migration, and aromatic oxygenation in a single catalytic cycle. HPPD is a member of the α-keto acid–dependent oxygenases that typically require an α-keto acid (almost exclusively α-ketoglutarate) and molecular oxygen to either oxygenate or oxidize a third molecule.

   The transformation catalyzed by HPPD has both agricultural and therapeutic significance. HPPD catalyzes the second step in the pathway for the catabolism of tyrosine, which is common to essentially all aerobic forms of life. In plants, this pathway has an anabolic branch from homogentisate that forms essential isoprenoid redox cofactors such as plastoquinone and tocopherol. Naturally occurring multi-ketone molecules act as allelopathic agents by inhibiting HPPD and preventing the production of homogentisate and hence required redox cofactors. This has been the basis for the development of a range of very effective herbicides that are currently used commercially.

11.4 ENVIRONMENTAL PROTECTION AGENCY

All plants have innate pest defense mechanisms, which frequently contain defensive chemicals. When these protective compounds are claimed to provide plant defense or plant growth regulation in the United States, they are classified as pesticides under the Federal Insecticide, Fungicide, and Rodenticide Act (FIFRA) and may be regulated as such by the United States Environmental Protection Agency (EPA).

PIPs are controlled as pesticides under FIFRA and require a tolerance exemption or exemption from the requirement of a tolerance under FFDCA when expressed in a food or feed crop. Under FIFRA Section 3, the EPA registers PIPs for sale and distribution, resulting in regulations under 40 CFR. The EPA examines each PIP application to see if the proposed use would have unreasonably negative effects on people and the environment. To avoid potentially unjustified adverse effects, the agency may impose (and has imposed) limitations on PIP registration (e.g., conditions to slow or eradicate insect resistance) (https://www.federalregister.gov/documents/2001/12/14/01-30820/pesticide-labeling-and-other-regulatory-revisions). FFDCA Section 408 also applies to PIP crops and human food and animal feed items generated from them when the PIP-expressing plant enters the food or feed stream. The EPA's strategy for delaying insect resistance to *Bt* PIPs has been to demand a public-interest insect resistance management (IRM) program in order to enhance the environmental advantages of these products. Such a proactive approach to resistance management is unprecedented in pesticide regulation (https://www.federalregister. gov/documents/2001/12/14/01-30820/pesticide-labeling-and-other-regulatory-revisions). The EPA's risk assessment of the evolution of insect resistance to *Bt* PIPs is based on genetic, biological, and operational considerations, as well as simulation modeling. All scientific problems surrounding IRM for *Bt* PIPs have been peer reviewed scientifically by EPA's SAP in an open and transparent procedure. Simulation models have played an important role in the development and evaluation of IRM strategies and will continue to do so in the future. Each program includes the following basic components: a field operational refuge strategy, grower agreements that contractually bind growers to follow IRM requirements, a resistance monitoring plan, a remedial measure, a grower education program, a compliance assurance program, and annual reporting. The refuge strategy, resistance monitoring program, and corrective action programs are all based on the unique *Bt* PIP-pest-crop combination. In response to significant breakthroughs in understanding the evolution of resistance and the registration of innovative *Bt* PIP products with varying levels of resistance risk, the EPA has made adjustments to IRM standards for registered *Bt* PIPs (Wozniak & Hughen, 2012).

11.5 CONCLUSION

We believe that future research on PIPs will be most successful if it includes researchers from a variety of scientific disciplines, such as environmental chemistry (to assess environmental processes influencing PIP fate and develop analytical approaches), agricultural sciences (to understand farming practices that influence PIP release), agricultural biotechnology (to elucidate PIP production in and release from GM crops), biochemistry, and molecular biology. Scientists who want to further this transdisciplinary effort will need adequate funding to perform independent research that is valuable to academic, governmental, and industrial groups alike. We anticipate that additional PIPs (e.g., non-Cry protein-based PIPs) will be created in addition to the currently developing dsRNA PIPs. Understanding the environmental fate of existing Cry protein and dsRNA PIPs will establish the groundwork for investigating the fate of new PIPs when they develop. Environmental biomacromolecular chemistry advances will add to the body of evidence currently supporting PIP ERAs and contribute to a fairer and educated assessment of the advantages and potential dangers connected with certain PIPs.

REFERENCES

Bachman, P. M., Huizinga, K. M., Jensen, P. D., Mueller, G., Tan, J., Uffman, J. P., & Levine, S. L. (2016). Ecological risk assessment for DvSnf7 RNA: A plant-incorporated protectant with targeted activity against western corn rootworm. *Regulatory Toxicology and Pharmacology, 81*, 77–88.

Dayan, F. E., & Duke, S. O. (2014). Natural compounds as next-generation herbicides. *Plant Physiology, 166*(3), 1090–1105.

Dufourmantel, N., Dubald, M., Matringe, M., Canard, H., Garcon, F., Job, C., . . . Tissot, G. (2007). Generation and characterization of soybean and marker-free tobacco plastid transformants over-expressing a bacterial 4-hydroxyphenylpyruvate dioxygenase which provides strong herbicide tolerance. *Plant Biotechnology Journal, 5*(1), 118–133.

Meshram, S., Bisht, S., & Gogoi, R. (2022). Current development, application and constraints of biopesticides in plant disease management. In *Biopesticides* (pp. 207–224). Woodhead Publishing.

Moran, G. R. (2005). 4-Hydroxyphenylpyruvate dioxygenase. *Archives of Biochemistry and Biophysics, 433*(1), 117–128.

Niu, X., Kassa, A., Hu, X. et al. (2017). Control of Western corn rootworm (*Diabrotica virgifera virgifera*) reproduction through plant-mediated RNA interference. *Scientific Reports, 7*, 12591. https://doi.org/10.1038/s41598-017-12638-3

Parween, T., & Jan, S. (2019). Chapter 1—Pesticides and environmental ecology. *Ecophysiology of Pesticides, 2019*, 1–38. https://doi.org/10.1016/B978-0-12-817614-6.00001-9

Razaq, M., & Shah, F. M. (2022). Biopesticides for management of arthropod pests and weeds. In *Biopesticides* (pp. 7–18). Woodhead Publishing.

Sarwar, M. (2015). Biopesticides: An effective and environmental friendly insect-pests inhibitor line of action. *International Journal of Engineering and Advanced Research Technology, 1*(2), 10–15.

Sudakin, D. L. (2003). Biopesticides. *Toxicological Reviews, 22*(2), 83–90.

Wozniak, C. A., & McHughen, A. (Eds.). (2012). *Regulation of agricultural biotechnology: The United States and Canada.* Springer Science & Business Media. https://archive.epa.gov/pesticides/news/web/html/npic.html

12 Applications of Pheromones in Crop Protection

Faraat Ali and Leo M.L. Nollet

CONTENTS

DOI: 10.1201/9781003265139-15

12.1 BIOCHEMICAL PESTICIDES

Biochemical pesticides are naturally occurring compounds that control pests by non-poisoning mechanisms while chemical crop protection is profit-induced poisoning of the environment. Biochemical pesticides can be classified in the following biologically distinct functional classes: semiochemicals, plant extracts and oils, plant growth regulators, and insect growth regulators.

12.1.1 SEMIOCHEMICALS

When you type "biopesticides pheromones" in Google Scholar you get 192.000 hits. Only the essence of pheromones is discussed here.

Semiochemicals (Gk. *semeon*, a signal) mediate interaction between organisms. Thus, these chemicals are controlling insect behavior and are also known as behavioral control pesticides. Semiochemicals belong to one of of four categories: pheromones, allomones, kairomones, and allomones-kairomones. The following paragraphs describe these substances in relation to insect pest control.

12.1.1.1 Pheromones

A pheromone is a chemical or a mixture of chemicals released by an organism (insect) to the outside (the environment) that causes a specific reaction in a receiving organism of the same species. Therefore, they are also reported in the literature as social hormones. It is known that most insects communicate in some manner with chemicals. They release traces of highly specific compounds that are volatile in nature, so they permeate the environment quickly and are detected by insects of the same species. Thus, it is an effective chemical means of communication. Pheromones are strong insect sex attractants. Being exocrine in origin (secreted outside the body), these chemicals were earlier called ectohormones. In 1959, German chemists Karlson and Butenandt named them pheromones (Gk., *pherein*, to carry, and *hormone*, to excite or stimulate). Probably the most physiologically potent active molecules known today are insect pheromones. For some species, laboratory-synthesized pheromones have been developed for trapping purposes (e.g., Johnson et al., 2020) [1].

Table 12.1 [2, 3] provides some female moth mating pheromones that have been approved as pesticides.

TABLE 12.1
Some Female Moth Mating Pheromones Approved as Pesticides

Chemical Name of Pheromone (OPP Chemical Code)	Pests Controlled	Use Sites
(1) (Z)-11-Hexadecenal (120001)	Artichoke plume moth	Artichokes
(2) (Z,E)-9–12-Tetradecadienyl acetate (117203) and (Z)-9-tetradecen-1-ol (119409)	Beet armyworm	Alfalfa, cotton, strawberries, vegetables, and tobacco
(3) (Z)-11-Tetradecenyl acetate (128980)	Blackheaded fireworm	Cranberries and fruits
(4) (Z,Z,E)-7,11,13-Hexadecatriennial (029000)	Citrus leafminer moth (CLM) and *Phyllocnistis citrella*	Ornamental and agricultural crops
(5) Myristyl alcohol (001509) (001510) (E,E)-8,10-Dodecadien-1-ol (129028) and (Z)-11-tetradecenyl acetate (128980)	Codling moth (*Cydia pomonella*)	Fruits and nuts
(6) (E,E)-8,10-Dodecadien-1-ol (129028)	Codling moth (*Cydia pomonella*)	Fruits, nuts, ornamental tree/shrubs, and uncultivated agricultural areas
(7) n-Tetradecyl acetate (-)	Codling moth (*Cydia pomonella*)	Pome fruits, stone fruits, and tree nuts
(8) (Z)-11-Hexadecenyl acetate (129071)	Diamondback moth	Manufacturing use
(9) (E,Z)-2,13-Octadecadien-1-yl acetate (117242)	Dogwood borer (DWB) (*Synanthedon scitula*)	Pome fruits, stone fruits, tree nuts, and ornamental nursery crops
(10) (Z)-6-Heneicosen-11-one (129060)	Douglas fir tussock moth	Douglas fir trees
(11) (E)-9-Dodecen-1-ol acetate (119004)	Eastern pine shoot borer	Forest trees and woodland trees
(12) (Z)-11-Tetradecenyl acetate (128980) (Z)-9-dodecenyl acetate (129004)	Grape berry moth	Grapes and vine fruits
(13) cis-7,8-Epoxy-2-methyloctadecane (114301)	Gypsy moth	Forest trees, ornamental evergreen trees, and shrubs

(Continued)

TABLE 12.1 (Continued)

Chemical Name of Pheromone (OPP Chemical Code)	Pests Controlled	Use Sites
(14) (E,E)-8,10-dodecadien-1-ol (129028), lauryl alcohol (001509), myristyl alcohol (001510), (E,E)-8,10- dodecadien-1-ol (129028), (Z)-11-tetradeny acetate (128980)	Hickory shuckworm	Fruits, nuts, and uncultivated agriculture areas
(15) (Z)-8-Dodecen-1-yl acetate (128906), (E)-8- dodecen-1-yl acetate (128907), and (Z)-S-dodecen- 1-ol (128908)	Koa seed worm	Fruits and nuts
(16) (Z)-11-Tetradecenyl acetate (128980)	Leaf rollers	Cranberries and various fruits
(17) 9,11-Tetradecadien-1-ol,1- acetate (12800)	Light brown apple moth (LBAM) and Epiphyas postvittana	Orchards, ornamental nurseries, and vineyards
(18) (Z)-8-Dodecen-1-yl acetate (128906), (E)-8-dodecen-1-yl acetate (128907), and (Z)-8-dodecen-1-ol (128908)	Macadamia nut borer	Fruits and nuts
(19) (Z,Z)-11,13-Hexadecadienal (000711)	Navel orange worm	Oranges
(20) Lauryl alcohol (001509), myristyl alcohol (001510), (E,E)-8,10-dodecadien-1-ol (129028), and (Z)-11- tetradecenyl acetate (128980)	Oblique banded leafroller (Choristoneura rosaceana)	Fruits
(21) (E)-11-Tetradecenyl acetate (129019), (Z)-11-tetradecenyl acetate (128980)	Omnivorous leafroller	Fruits (deciduous), grapes, kiwi, and nuts
(22) (E)-11-Tetradecenyl acetate (129019), (Z)-11-tetradecenyl acetate (128980)	Oriental fruit moth	Fruits and nuts
(23) (E)-5-Decenyl acetate (1117703), (E)-5-decen-1-ol (078038), (Z)-8-Dodece-1-yl acetate (128906), (E)-8-dodecen-1-yl acetate (128907), and (Z)-8- dodecen-1-ol (128908)	Oriental fruit moth	Fruits and nuts.
(24) Lauryl alcohol (001509), myristyl alcohol (001510), (E,E)-8,10-dodecadien-1-ol (129028), and (Z)-11- Tetradecenyl acetate (128980)	Pandemis leafroller (Pandemis pyrusana)	Fruits

Chemical Name of Pheromone (OPP Chemical Code)	Pests Controlled	Use Sites
(25) (E)-5-Decenyl acetate (117703), (E)-5-decen-1-ol (078038), (Z)-8-dodecen-1-yl acetate (128906), (E)-8-dodecen-1-yl acetate (128907), and (Z)-8- dodocen-1-ol (128908)	Peach twig borer	Fruits and nuts
(26) (E)-5-Decen-1-ol acetate (117703) and (E)-decen-1-ol (078038)	Peach twig borer	Fruits, nuts, and agricultural crops
(27) 7,11-Hexadecadien-1-yl acetate (114103)	Pink bollworm	Cotton
(28) (Z,E)-7,11-Hexadecadien-1-yl acetate (114101) and (Z,Z)-7,11-hexadecadien-1-yl acetate (114102)	Pink bollworm	Cotton
(29) (E)-11-Tetradecen-1-ol acetate (129019)	Sparganothis fruitworm	Cranberries
(30) (E,Z,Z)-3,8,11-Tetradecatrienyl acetate (011472)	Tomato leaf miner and Tuta absoluta	Ornamental and agricultural crops, specifically tomato
(31) (Z)-4-Tridecen-1-yl acetate (121901) and (E)-4-tridecen-1-yl acetate (121902)	Tomato pinworm	Eggplant, tomato, and vegetables
(32) (E,Z)-3,13-Octadecadien-1-ol (-) and (Z,Z)-3,13- octadecadien-1-ol (-)	Western poplar clearwing moth	Poplars, white birch willows, and locust
(33) (Z)-9-tetradecen-1-yl acetate (129109)	Coding moth (Cydia pomonella), oblique banded leafroller (Choristoneura rosaceana), pandemis leafroller (Pandemis pyrusana), fruit tree leafroller (Pandemis limitata) and European leafroller (Archips rosanus)	Apples, pears, quince, and other pomes fruits; peaches, prunes, plums, nectarines, cherries, and other stone fruits; walnuts, pecans, and other nut crops

Some well-known commercially available synthetic insect sex pheromones include (name (trade name) species attracted):

1. Disparlure (Disparmone) gypsy moth
2. Grandlure (Grandamone) boll weevil
3. Gossyplure (Pherocon PBW) pink bollworm
4. Hexalure (Hexamone) pink bollworm (obsolete)
5. Looplure (Cabblemone-Pherocon CL) cabbage looper
6. Muscalure (Muscamone) house fly
7. Codlelure (Codlemone) codling moth
8. Virelure (None) codling moth
9. None (Z-11) red-banded leafroller, European corn borer, oblique banded leafroller, and smartweed borer

See also https://www.fmc.com/en/innovation/biologicals/pheromones.

To find information on biopesticides and more specific information on pheromones explore the database of the University of Hertfordshire (http://sitem.herts.ac.uk/aeru/bpdb/).

12.1.1.2 Classification of Pheromones

Pheromones, on the basis of the effects they produce, are divided into two groups: primer effect pheromones and releaser effect pheromones.

12.1.1.2.1 Primer Effect Pheromones

Primer effect pheromones operate through the gustatory (taste) sensilla and trigger a chain of physiological changes in the body. In insects—mainly social insects like ants, bees, wasps, and termites—they regulate caste determination and reproduction. So they are of no practical value in insect pest control.

12.1.1.2.2 Releaser Effect Pheromones

Releaser effect pheromones operate through the olfactory (smell) sensilla and regulate the behavior of insects. The pheromones in this category are of the following types:

- Sex pheromones
- Aggregation pheromones
- Alarm pheromones
- Trail pheromones,
- Dispersion pheromones
- Ovipositor pheromones
- Specialized colonial behavior pheromones

The first three types are used in insect pest control.

Sex pheromones: They offer great potential for insect control. A great number of female and male insects are known to produce sex hormones. These sex hormones represent a varied assemblage of compounds. The sex hormones produced by males and females possess different properties as well as modes of action. For example, female sex hormones are species specific and excite males to copulate at longer range than male sex hormones, which only lower the mating resistance of females at a short range. Since they have a short range, other communication stimuli (visual and auditory) may come into play.

The male sex pheromone is of lesser value than the female sex pheromone in insect pest control. The pheromonal communication system consists of three steps: release of the pheromone (exocrine gland), the traveling medium (air or water) to reach the receptor, and the receptor. The mechanism of pheromones' action (reception and transduction) is not yet fully understood. However, it is well known that insect olfaction is more sensitive than that of a man-made electronic detection system (e.g., 30 grams of disparlure, a synthetic sex pheromone attractant, is enough to bait some 60,000 traps per year for the coming 50,000 years) [4].

Aggregation pheromones: These induce aggregation or congregation of insects for protection, reproduction, feeding, or all the three. These pheromones are found mostly in coleopterans and also in pentatomid bugs (*Eurydema rugosa*) and dictyoptera (*Periplaneta americana*). The pertinent examples are the female bark beetle, *Dendroctonus frontalis* and the male phloem beetle, *Ips confusus*. The *Ips* males burrow into the phloem tissue of a suitable host tree by making galleries. Then they secrete the pheromones into the lumen of the hindgut, where they get incorporated into the fecal pellets voided in the galleries. The pheromone volatilizes to permeate the air outside, stimulating the flying population of both males and females. The females are stimulated to enter the galleries, and the males are stimulated to construct new galleries. After mating or continued feeding for some time, pheromone production stops. This may be to avoid overcrowding due to the continuous release of pheromone, or it may be the result of the production of an anti-aggregation pheromone. Well-known examples of aggregation pheromones are frontalin (*Dendroctonus*), ipsenol (*Ips*), periplanone (*Periplaneta*), and dimethyldecanol (*Tribolium*).

Alarm pheromones: This pheromone is produced by several organs: namely, a pair of cornicles or siphunculi near the tip of the abdomen in aphids; abdominal terga in *Dysdercus*; cephalic glands in termites; sting apparatus and mandibular glands in worker bees; and anal, mandibular, Dufour's, and poison glands in ants. They have been reported in Homoptera, Isoptera, and Hymenoptera. In fact, alarm hormones are an anti-predator device, a warning to cospecifics (members of the same species) about the presence or attack of enemy (mostly a predator). This warning deduces different signals in different insects (e.g., dispersion or escape in aphids and bugs, dispersion when foraging or feeding away from the nest, and aggression when inside the nest in ants); attraction towards the source of emission in wasps and worker bees; and aggression in termite soldiers. Worker bees generally attack in sting autotomy that guarantees marking of the enemy by the presence of the odoriferous sting apparatus fortified by the alarm pheromone. This pulls other bees out of the colony to identify and attack the enemy by stinging or biting, in either case releasing the alarm pheromones. Chemically, alarm pheromones are terpenes (aphids), aldehydes (hemipterans), alkyl and alkenyl acetates (honeybee sting glands), 2-heptanone (honeybee mandibular glands), ethyl and methyl ketones and formic acid (ants), and monoterpene hydrocarbons (termite soldiers).

Trail pheromones: The functioning of the pheromones described previously is through air while the terrestrial trail pheromones are laid in the form of intermittent or continuous lines on solid substrate. These lines are sensed by the antennae of trail followers as leading to a destination (a mate or a source of food discovered by the trail maker). These pheromones also maintain the cohesion and social integration of the colony. The organs producing trail pheromones are sterna glands in termites; Dufour's glands, Pavan's glands, poison glands, rectal glands, and tibial glands in ants (depending on the species); and labial glands in caterpillars. In ants, the glands open near the base of the sting, and the pheromone comes down to be drawn on the ground like the nib of a pen drawing a line. Trail pheromones are multicomponent. The chemicals present in some of these pheromones are caproic acid (*Zootermopsis* spp.); 3, 6, 8-dodecartrienal (*Reticulitermes* spp.); and hexanoic, heptanoic, decanoic, and nonanic acids.

Pheromones are chemicals that are produced by the insect and are used for intra-specific communication within the same species. Other classes of chemicals are produced by one species of insects and utilized by another for inter-specific communication. These are called allomones, kairomones, and allomones-kairomones.

12.1.2 ALLOMONES AND KAIROMONES

12.1.2.1 Allomones

An allomone is a chemical or a mixture of chemicals produced by one species of organism and applied to another species in order to induce advantageous response: e.g., defensive secretions of insects and plants that are poisonous or repugnant to attacking predators. One of the terpenes that is known as citral is a mandibular secretion of many ants and bees that acts as a defensive allomone for ants and an attractant for bees.

12.1.2.2 Kairomones

These are chemicals or mixtures of chemicals released by one species of insect and received by a different species to induce a response that is advantageous to the recipient. In this case, the releaser has been found to be the loser, and the recipient (parasite or predator) is a gainer. Thus, kairomones are regarded as allomones or pheromones that evolutionarily backfire (boomerang) on the releaser: i.e., originally a kairomone was meant to be advantageous to the releaser, but in the course of evolution, the recipient, which was to be deterred by the chemical, developed tolerance to it and started using it as a signal to locate the releaser for predation or parasitization. For example, kairomone is the male sex pheromone of the bug *Nezara viridula*, which acts as an attractant to its parasite, the tachinid fly, *Trichopoda pennipes*.

12.1.2.3 Allomones-Kairomones

Allomones-kairomones, or synomones, are chemicals or mixtures of chemicals (endocrine secretions of hymenopterans) frequently functioning as both allomones and kairomones at the same time. For example, the parasitic wasp *Aphytis* is attracted by the sex pheromone of its host, *Aonidiella* spp. In this case, the sex pheromone acts as an allomone and a kairomone simultaneously: i.e., an allomone for host (emitter), which utilizes it for mating, and a kairomone for the parasite (recipient), which takes its help in locating the host. In the same fashion, the ant *Camponotus*, when attacked, ejects formic acid for its own defense (as an allomone), but the acid is utilized by another ant, *Myrmecia gulosa*, to locate the workers of *Camponotus* on which it feeds.

12.2 MECHANISM OF PHEROMONE ACTION

Regnier and Law [5] have reviewed the evidence of intraspecies chemical communication in insects, with emphasis on those studies in which known organic compounds have been implicated. They have concluded that there are two distinct types of pheromones: releaser and primer. Pheromones affect the central nervous system in two different ways. One class of compounds, the releaser pheromones, causes an immediate behavioral response upon reception, as in the case of sex substances. The second group of compounds, the primer pheromones, has a delayed effect on behavior. A typical example is the blockage of ovum implantation in female mice that have been exposed to the odor of an alien male immediately after mating. These delayed responses are the result of physiological changes induced by the pheromones.

12.3 CHEMICAL NATURE OF PHEROMONES

The chemistry of several sex pheromones, particularly female sex pheromones, is known. Some pheromones produced by insects consist of

1. Hydrocarbons, such as linolene, terpinolene, undecane, and tridecane
2. Alcohols, such as octanol and 2,6-dimethyl-5-hepten-1-ol
3. Aldehydes, such as hexanal, 2-hexanal, 4-oxo-2-octenal, 2,6-dimethyl-5-heptenal, citral, and citronellal
4. Ketones such as 2-heptanone, 4-methyl-2-hexanone, 2-methyl-4-heptanone, 4-methyl-3-heptanone, 6-methyl-5-heptene-2-one, 3-octanone, 2-nonanone,3-nonanone, 6-methy-3-octanone, and 2-tridecanone
5. Miscellaneous, such as isoamyl acetate, formic acid, dimethyl disulfide, and dimethyl trisulfide

The following groups of sex pheromones are widely studied: Lepidoptera, Coleoptera, Hymenoptera, Diptera, Hemiptera, and Orthoptera.

The first female sex pheromones (produced by the silk moth *Bombyx mori*) were chemically characterized by Karlson et al. in 1959 [6]. It was obtained in traces (12 mg) by extracting half a million virgin females. It was named as 10, 12-hexadecadienol, a primary alcohol $(CH_3 (CH_2)_2(CH=CH)_2(CH_2)_8CH_2OH$. The pheromone gyplure is produced by the female gypsy moth, *Porthetria dispar*. It is 10-acetoxy-cis-7-hexa-decanol, $CH_3(CH_2)_5CH(OC(O)CH_3)CH_2(CH=CH)(CH_2)_5CH_2OH$. The pheromone Gossyplure (gyptol) is produced by the female pink bollworm, *Pectinophora gossypiella*. It is chemically known as 10-propyl-trans-5, 9-tridecadienal acetate, $[CH_3(CH_2)_2]_2O=CH(CH_2)_2CH=CH(CH_2)_3CH_2OC(O)CH_3$. The pheromone looplure is produced by the female cabbage looper, *Trichoplusia ni*, which is chemically cis-7-dodecenyl acetate $CH_3(CH_2)3CH=CH(CH_2)6OC(O)CH_3$. The pheromone queen substance is produced by the mandibular glands of the honeybee queen. For the drones, it acts as a sex attractant, but for the worker bees in the colony, it is an ovarian inhibitor that does not allow the ovaries to mature. It is 9-oxo-2-decenoic acid, $CH_3C (O) (CH_2)_5CH=CHCOOH$.

A few male pheromones have been chemically characterized. The pheromonal component of the male queen butterfly, *Danaus gilippus berenice*, is an alkaloidal ketone, 2, 3-dihydro-7-methyl-1H-pyrrolizin-1-one, and the male cabbage looper, *Trichoplusia ni*, is 2-phenylethanol. The sex pheromone of the male boll weevil, *Anthonomus grandis* (Coleopetra), contains two terpene alcohols and two terpene aldehydes, and that of the male *Ceratitis capitata* (Diptera) contains methyl-6-nonenoate and methyl-6-nonenol.

The pheromones are specific because of the distinctive shape of the longer molecule (higher molecular weight, 180–3000 Daltons): i.e., high specificity and potency both depend on a reasonably long chain of carbon atoms (10–20). The other two characteristics of pheromones, volatility and diffusibility, decrease with increasing molecular weight or carbon chain length. Thus, there is a need for a reasonable-length carbon chain. The biological activity of pheromones for male attraction (sexual excitement or short-range orientation behavior) is determined by field tests; on the other hand, electroantennogram (EAG) responses can be determined in the laboratory. A few pheromones such as *Bombyx mori*, *Porthetria*, and *Trichoplusia ni* are monocomponent. Their species specificity depends on minor structural variations in the functional moiety of the pheromone, such as the number, position, and configuration of double bonds and carbon atoms. But the majority of pheromones, such as *Adoxophyes* and *A. orana* (two sibling tortricid moths), are muticomponent systems. The species specificity of multicomponent pheromones is due to the variable ratio of a number of different components present. For example, *Adoxophyes orana* and *Clepsis spectrana* both possess (Z)-9-and (Z)-11-tetradecenyl acetates. However, the former has these compounds in the ratio of 90:10 and the latter 25:75.

See also Abd El-Ghany [7].

12.4 SYNTHESIS OF PHEROMONES

Pheromones are species specific and safe to use. Unfortunately, pheromones to control all types of pest are not known so far. It is also obvious that synthetic pheromones have gained prominence in recent years. Therefore, their synthesis has attracted considerable attention and played a vital role in their overall development in integrated pest management. Recently, a considerable amount of work has been published. A few papers are briefly summarized here.

Yadav and Ready [8] have worked out a number of new methodologies in order to use various approaches successfully for the preparation of pheromones.

A critical review was published on isolation, synthesis and identification of pheromones and techniques for their use in plant protection in India [9]. Studies were carried out largely on *Spodoptera litura*, *Pectinophora gossypiella*, the codling moth, the gypsy moth, *Heliothis armigera*, *Earias insulana*, and *Chilo auricilius*.

Millar et al. [10] describe a stereo-specific synthesis of the pheromone, (9Z)-9, 13-tetradecadien-11-ynal of the avocado seed moth, *Stenoma catenifer*. They report that the highly unsaturated aldehyde (9Z)-9, 13-tetradecadien-11-ynal and the corresponding alcohol were identified as possible sex pheromone components. The aldehyde as a single component attracted more male moths than caged virgin female moths, and the addition of the analogous alcohol and/or acetate decreased attraction.

See further references [11–16].

12.5 PLANT PROTECTION BY PHEROMONES

Among different types of pheromones, sex pheromones offer the greatest potential for insect control. The following example is cited to show their potential use. In the Texas high plains, pheromone traps using live male boll weevils captured sufficient overwintered weevils to suppress the population until the late-summer migration of a large number of weevils overpowered the action.

The sex pheromones of pest moths have been studied in detail. After 30 years of study, the gypsy moth sex pheromone was isolated, identified, and synthesized in the laboratory in 1960. Since then, huge quantity of the synthetic female gypsy moth sex pheromone, disparlure, has used in male-trapping programs for this forest pest.

12.6 ANALYSIS OF PHEROMONES

From the 1960s to date, several reports and research articles and a few books and monographs have been published on the detection, extraction, separation, estimation, and characterization of pheromones and their complex mixtures (Beyond Discovery, 2003). Several technical improvements in analytical procedures and techniques have given new dimensions to the productivity of pheromone research. Among the improvements is the use of electroantennography (EAG), gas chromatography (GC), mass spectrometry (MS), simple micro- analytical methods, and the transcriptomic approach. These analytical tools have also been used in combination: i.e., hyphenated techniques such as GC-EAG, GC-MS, and GC-EAG-MS. By coupling GC with EAG, researchers could detect which components in their insect preparations prompted an electrical response.

For more information, see references [18–19].

12.7 TECHNOLOGIES OF PHEROMONE APPLICATION

Considerable research has been published in this area. As pheromones are volatile compounds, efforts have been made to produce a regulated release of these substances in order to allow them to remain longer and at concentrations sufficient to affect the insects. The initially developed methods of pheromone release from dishes or wicks have been replaced by microencapsulation, hallow fiber, pheromone baited traps, and pheromone dispenser methods.

1. Microencapsulation method: In this method, the pheromone is enclosed in small plastic capsules (ca. 50 µm), which are dispensed with conventional spray techniques in order to provide uniform spreading.
2. Hollow fiber method: This technique keeps the pheromone in hollow plastic fibers (capillaries), which are cut into small pieces and scattered by an aircraft. The fibers may also be shaped into hoops (coils) and tied to the upper portion of the stems.
3. Pheromone-baited trap method: The traps are specially devised structures of various shapes and sizes that can be suspended on trees or elevated objects. Several types of traps are now available commercially. Widely known traps are Pherocon R, Sectar XC-26, and Sectar 1.
4. Pheromone dispensers: This recent technology can release the pheromone at a precisely calculated rate. The Hercon controlled-release dispenser is commonly used in India. It is made of plastic laminates; its middle portion, embedded with pheromone (or anylure), acts as a reservoir, and the outer layer acts as a barrier in order to regulate the rate of release of the pheromone. The dispensers are available in different shapes: squares (0.5–2.5 cm side), ribbons, and flakes (confetti). The flakes are scattered by aircraft over dense vegetation (forests, orchards, etc.) while squares and ribbons are placed manually. The flake is equipped with a sticker to stick the particles to the crop foliage. About 10,000 to 20,000 sex pheromone flakes can mask the odor of a "calling" female in a one-acre area, thereby causing mating disruption. A few hundred squares can cover the same in crop fields of the same size.
5. New technologies of pheromone application: Cornell entomologists [20] have unlocked an evolutionary secret as to how insects evolve into new species. This discovery has major implications for the control of insect populations through the disruption of mating. The methods based on this theory are effective as, over time, current eradication methods could become ineffective, similar to the way insects develop chemical pesticide resistance. They made the discovery while examining ways to keep European corn bores from mating, multiplying, and then chewing up farmers' fields. They discovered the existence of a previously undetected gene, beta-14, that can regulate the attractant chemicals produced in the sex pheromone glands of female borers. The gene can be suddenly switched on, changing the pheromone components that females use to attract males for mating.

Cork et al. [21] developed several resin-controlled release formulations for pheromones for use in mating disruption in plant protection. This technique involves male attraction, which invariably requires less pheromone than mating disruption and so is potentially more cost effective. The PVC-resin formulation of sex pheromones also shows a significant reduction in damage relative to that in plots treated with insecticides. The formulation was tested for mating disruption of yellow rice stem borer, *Scirpophaga incertulas*, in replicated 1 ha plots in India over two seasons. Even at an application rate of 10 g active ingredient (a.i.)/ha/season, there was a significant reduction in damage relative to that in plots treated with insecticides. However, application rates of 30–40 g a.i./ha/season are recommended for general use by farmers.

Advantages and drawbacks of the newly devised methods, the polymer nano-encapsulated devices, are discussed for smart release of plant protection agents (CPAs) [22]. CPA release methods are designed for the use of biopesticides at nano levels in plant protection, and the application of the required micronutrients is synchronized by the nature or intensity of plant root exudates. Advances in nano-scale release of costly biopesticides for crop protection herald a "small revolution" for the benefit of sustainable agriculture.

Krupke et al. [23] conducted a case study of a pheromone-based attract-and kill management strategy for the codling moth, *Cydia pomonella*, to examine key behavioral factors mitigating the possible effectiveness of this strategy. Last Call CM is a newly registered attracticide product that combines the primary component of codling moth sex pheromones with the insecticide permethrin. Studies of competition between pheromone point sources in caged trees showed individual attracticide droplets were significantly more attractive to male moths then calling females. They also examined the hypothesis that traditional broadcast insecticides kill a fixed proportion of the population regardless of density while attractant-and-kill methods may be highly inversely density dependent due to competition with natural point sources. In addition, the possible mechanisms, strengths, and weaknesses of this new technology were also studied.

For more information on technologies of pheromones applications, see references [24–29].

12.8 PEST MANAGEMENT WITH PHEROMONES

Insect pest control with sex pheromones may be done in two ways.

12.8.1 POPULATION SURVEY

Synthetic analogues of the sex pheromones of many pests are known, and the number is increasing day by day. They are being used to determine the population density of pests in a given area. The population density helps determine level of pest; either it is under the limit of tolerance, or it has assumed an economically dangerous level that requires control measures. This knowledge saves the unnecessary use of hazardous and costly pesticides. It also helps determine the time of pest buildup and thus the correct time for the application of control measures and the most suitable technology for control.

12.8.2 Behavioral Manipulation

This helps in determining the degree of harm caused by the pest to crops. Two strategies are known to manipulate insect behavior with sex pheromones: stimulation of the normal approach response (i.e., attraction) and disruption of chemical communication between the sexes.

12.8.2.1 Stimulation of Normal Approach Response

Normally, an insect responds to sex pheromone by attraction and approaches the source of the pheromone (i.e., the releaser). This response may be exploited to lead to death in the following three ways.

12.8.2.1.1 Use of Pheromone-Baited Traps

Pheromone-baited traps can be suspended from the branch of a tree or placed in a field. When the pests have gathered in large numbers, they may be killed by chemical pesticides. Several publications have investigated the most suitable characteristics of the trap, such as design, color, shape, entrance, height, etc., in order to make the technique more efficient. When traps are used, these technologies should be kept in mind, assuming that the insects are very sensitive to them. Other important factors are the concentration of the pheromone applied to the trap and the vegetation in which it is to be deployed. This technique is also known as "the male annihilation method" because the female sex pheromone is employed to attract the male. As the wild female releases its pheromone at specific times of the day and night, this technique has the advantage of attracting and snaring the males continuously until the wild females start releasing the pheromone. Then the insects captured by the traps may also be infected with entomogenous pathogens (bacteria, viruses, fungi, etc.), and, if released, they will mate with females to spread the disease in the community.

12.8.2.1.2 Orientation to a Source of Chemosterilant or Insecticide

The insects may be guided/misguided by the use of pheromones to a source of chemosterilant or insecticide, only to be sterilized or get killed. Sterilization has its own advantages. First, the insects sterilized are as good as killed, and secondly, the sterilized insects may sterilize other insects by mating. It is also known that a chemically sterilized male becomes a little insensitive to the concentration of the pheromone produced by the wild female and therefore is unable to compete with the wild male for mates. This difficulty can be overcome by a topical application of the chemosterilant since it has been found that sterilization by ingestion of chemosterilants raises the threshold of the male responsiveness while topical application does not alter it.

12.8.2.1.3 Orientation to a Combination of Light and Pheromones

Nocturnal male Lepidoptera are more strongly attracted to light than to pheromones. This is commonly true for most nocturnal insects. Therefore, if the light and pheromones are used together, a large number of males will be attracted and destroyed by insecticides.

12.8.3 Disruption of Chemical Communication

Natural pheromones can be masked by introducing synthetic pheromones into the environment. Thus, the normal phenomenal communication can be disrupted between the pest species. This disruption will cause a failure among insects to locate their mates, thereby preventing mating. The disruption involves three factors: sensory adaptation, central nervous system (CNS) habituation, and confusion. In sensory adaptation, the olfactory receptors stop functioning (i.e., stop reporting to the CNS) if an animal is exposed to a certain level of an odorant. This occurs only for a brief period (a few seconds to minutes), during which the animal cannot perceive that odorant (pheromone). When the same condition occurs in the CNS itself, it is referred to as CNS habituation or adaptation. In both kinds of adaptations, the male fails to detect the natural pheromone produced by the wild female and thus fails to locate her for mating. Confusion is the result of competition between the synthetic and natural pheromones. If the source of the synthetic pheromone releases more concentration than the wild insect, the insect is more likely to approach the synthetic pheromone than the natural one. Thus both technologies prevent mating and control the population of insects.

12.8.4 Disruption by Parapheromones and Antipheromones

Parapheromones are non-pheromonal chemicals that are structurally and functionally similar to natural pheromones: e.g., caproic acid, which can produce the same response in honeybees as the queen bee substance. This type of pheromonal substances can be employed to mask the effect of natural substances, so they are also known as biological maskers. Some impurities present in synthetic pheromones or in extracted pheromones may act as biological maskers. Pertinent examples are the impurities present in crude extract of gyplure; in technical grade gyplure, for example, methylene chloride alcohol has been found to mask the effects of the gypsy moth's natural pheromone.

Antipheromones are substances produced by certain insects that act as inhibitors of the males of closely related species to prevent wrong mating: i.e. antipheromones help in maintaining reproductive isolation between closely related species. Thus parapheromones and antipheromones are both employed to control pest populations by disruption of pheromonal communication or preventing mating.

12.8.5 Control with Aggregation Pheromones

Aggregation pheromones are used in insect pest control by inducing insects to aggregate and then attack the wrong host plants. Logs infested with bark beetle, *Dendroctonus* spp., are tied to unsuitable host trees. The pheromone produced by the infested insects concentrates the flying population of these beetles on such trees to attack them. Ultimately, the insects die of starvation because these trees are not their real hosts. Similarly, the phloem beetle, *Ips*, can be induced to attack and bore into a gum tree, an inappropriate host, if an air stream is directed towards its trunk from a box containing beetle-infested logs. In this fashion, properly designed pheromone-baited traps incorporating the necessary visual cues may be used as efficiently as resistant strains of trees to destroy these tree-burrowing beetles.

12.8.6 Control with Alarm Pheromones

Many species of aphids release alarm pheromones under the influence of predators. These alarm pheromones induce nearby aphids to stop feeding and move away from the site. The aphids, in their frenzy to escape, may even drop from the plants. The action of alarm pheromones is very effective. Therefore, alarm pheromones may be sprayed in the fields in order to keep aphids off the crops.

12.8.7 Control with Oviposition-Marking Pheromones

The females of the economically important *Rhagoletis* flies, such as apple fly and maggot fly (*R. pomonella*), walnut husk fly (*R. complete*), and cherry fruit fly (*R. cingulata*) drag their ovipositor on the fruit surface immediately after laying eggs to deposit a marking pheromone that deters subsequent attempts to lay eggs in the parasitized host. This marking behavior disperses eggs among available fruits, preventing crowding and thus ensuring adequate food supply to the developing larvae. The marking pheromone is generally water soluble and can be washed off the fruits. The solution obtained from the washing has been found to be effective in deterring egg laying when applied to the fruits. Investigations are in progress to characterize and synthesize this pheromone so it can be used in fruit fly control programs. The flies themselves can be captured by leaving some untreated trees in the orchard with sticky spheres suspended to act as traps.

12.8.8 Control with Trail Pheromones

Trail pheromones have been found to be of particular interest in controlling termites and ants. In fact, trail pheromones can be used to attract pest insects to poison bait, which could have a ravaging effect if brought back to the nest by the foraging workers. Some preliminary experimental work on the leaf cutting ants, *Atta texana*, and pharaoh ants, *Monomorium pharaonis*, which are a great nuisance in buildings, bakeries, and hospitals, has shown that the addition of trail pheromones increases the chances of the poisoned bait being found during the foraging activity of the workers, which, in turn, increases the amount of bait transported to the nest.

12.9 MONITORING OF PEST POPULATION

The normal role of pheromones is monitoring and regulating the pest population. Generally, attention is drawn when the pest population has already built up to an economically harmful level. At this dangerous stage, pest control is uneconomical as larger amounts of costly insecticide are needed, but it is also unsuccessful because the pests will have caused too much damage by the time they can be controlled. Farm practices therefore include several preventive (prophylactic) sprays at regular intervals before the actual arrival of the pest to make sure that the pest is killed early.

These difficulties can be disposed of by a system that can survey and monitor the developing pest population. The desired system has been developed utilizing pheromones. This system can indicate the initiation step of the pest flight (first arrival), coordinate it with the appearance of larvae, and predict egg hatches and population density. This newly developed device has been found to be helpful in the detection of insect populations and new areas of infestation at an early stage, reducing the quantity of costly and hazardous pesticides.

12.10 APPLICATIONS AND LIMITATIONS

12.10.1 APPLICATIONS

(1) A large number of insects can be attracted and killed using a small quantity of pheromones, so the methods are affordable; (2) these are nonhazardous and ecofriendly devices; (3) as these are highly species specific, non-targets are spared; (4) pheromonal devices can collect a large number of insects from long distances for their abatement and destruction, so they are labor saving; (5) it is an easy technique for monitoring the buildup of the pest population; (6) the pheromonal technique is versatile and can keep surveillance on pest entry from long distances through the ports and airports; and (7) they provide knowledge of foreign or new pests to aid in developing new and fast methods of prevention and control.

12.10.2 LIMITATIONS

Pheromone-based technology has the following weaknesses and drawbacks: (1) pheromones are not known for all the pests; (2) pheromones are species specific: i.e. one sex pheromone can attract one sex, and the other sex could still be there to do harm; (3) the pheromone technique requires skilled operators with knowledge and expertise, which has not been developed by most farmers so far; only government agencies can provide them; (4) instant results like those of chemical pesticides cannot be obtained by pheromones, so they cannot be used for short-term control measures; they are best suited for integrated pest management (IPM); (5) pheromone technology is sophisticated and sensitive (to temperature, pressure, wind speed, rain, etc.); thus, great care is required in fabrication, storage, transportation, and usage of these techniques.

12.11 NEW DEVELOPMENTS

Pheromone companies need to compete with the price of insecticides, so the amount of active ingredient or the number of compounds formulated in mating disruption dispensers is critical. (Z)-11-tetradecenyl acetate is a pheromone component of many leaf rollers, which are important, especially in North America [30, 31] and this compound is easier to synthesize than (Z)-9-tetradecenyl acetate. It is obviously more economical to use one formulation all over and to include single compounds rather than complex blends into dispenser formulations.

Among developing Asian countries, a thorough survey has been made regarding the scope of pheromone crop protection in Bangladesh [32]. The empirical data for the study were collected from farmers as the object was to explore the major factors influencing the extent of their practice of pheromone technologies. This author found that the education level of users has significant association with knowledge of pheromone science and technology in crop protection. At this stage, full achievement cannot be made due to inadequate availability of pheromones in the market and insufficient help from the agricultural sector to create motivation and understanding in financially weak farmers. Therefore, government agencies should take necessary steps to promote pheromone crop protection in Bangladesh.

The importance of pheromone-based methods is accentuated in view of increasing problems associated with the use of conventional insecticides. However, for a more widespread use of pheromones, application techniques must become more reliable and more economical. The key to further development is closer communication and collaboration between academic research institutions, the plant protection industry, and extension services. There is data in support of both pessimistic and optimistic views of the future. The use of pheromones will continue to depend on their cost and efficacy compared to other methods. Failure of mating disruption at high population densities, elevated cost, and the occurrence of other insects that can only be controlled with insecticides are serious obstacles. Only continued goal-oriented research will overcome these obstacles and lead to more reliable and widespread application.

For information on the latest developments, see references [33] and [34].

12.12 CONCLUSIONS

1. Pheromone crop protection is a newly developed method. This new weapon uses chemical substances generated by the insects themselves.
2. Unlike conventional pesticides, pheromones do not damage other animals, nor do they pose health risks to people. Pheromones specifically disrupt the reproductive cycles of harmful insects.
3. They also can be used to lure pests into traps to help farmers track insect population growth and stages of development. In this way, farmers can reduce the amount of insecticide they need to spray, and only do so when the insects are at a vulnerable stage or when their numbers exceed certain levels.

REFERENCES

1. Johnson, N. S., Lewandoski, S. A., Alger, B. J., O'Connor, L., Bravener, G., Hrodey, P., . . . Siefkes, M. J. (2020). Behavioral responses of sea lamprey to varying application rates of a synthesized pheromone in diverse trapping scenarios. *Journal of Chemical Ecology*, 46(3), 233–249.
2. Technical Evaluation Report: Pheromones Crops. (2012, March 27). Identification of Petitioned Substances. OMIR 2012 and NPIRS 2012. Electronic Distribution.
3. Ware, G. W. (1978). *The pesticide book*. W. H.Freeman and Company.
4. Fernandes, R. A., Chandra, N., & Gangani, A. J. (2020). Three decades of disparlure and analogue synthesis. *New Journal of Chemistry*, 44(41), 17616–17636.
5. Regnier, F. E., & Law, J. H. (1968). Insect pheromones. *Journal of Lipid Research*, 9, 541–551.

6. Karlson, P., & Butenandt, A. (1959). Pheromones (ectohormones) in insects. *Annual Review of Entomology*, *4*, 39.
7. Abd El-Ghany, N. M. (2020). Pheromones and chemical communication in insects. In *Pests, weeds and diseases in agricultural crop and animal husbandry production* (pp. 16–30). IntechOpen.
8. Yadav, J. S., & Reddy, R. E. (1988). Synthesis of insect sex pheromones. *Current Science*, *57*(24), 1321–13330.
9. Bajikar, M. R., & Sarode, S. V. (1986). Future of pheromones in plant protection. *Proceedings of the National Academy of Sciences*, *52*(1), 129–133.
10. Millar, G. M., Hoddle, M., McElfresh, J. S., Zou, Y., & Hoggle, C. (2008). (9Z)-9,13-Tetradecadien-11-ynal, the sex pheromone of the avocado seed mot, *Stenoma catenifer*. *Tetrahedron Letters*, *49*(33), 4820–4823.
11. Sorensen, P. W., & Baker, C. (2014). Species-specific pheromones and their roles in shoaling, migration, and reproduction: A critical review and synthesis. *Fish Pheromones and Related Cues*, 11–32.
12. Witzgall, P., Kirsch, P., & Cork, A. (2010). Sex pheromones and their impact on pest management. *Journal of Chemical Ecology*, *36*(1), 80–100.
13. Howse, P., Stevens, J. M., & Jones, O. T. (2013). *Insect pheromones and their use in pest management*. Springer Science & Business Media.
14. Mori, K. (2011). *Chemical synthesis of hormones, pheromones and other bioregulators*. John Wiley & Sons.
15. Petkevicius, K., Löfstedt, C., & Borodina, I. (2020). Insect sex pheromone production in yeasts and plants. *Current Opinion in Biotechnology*, *65*, 259–267.
16. Holkenbrink, C., Ding, B. J., Wang, H. L., Dam, M. I., Petkevicius, K., Kildegaard, K. R., . . . Borodina, I. (2020). Production of moth sex pheromones for pest control by yeast fermentation. *Metabolic Engineering*, *62*, 312–321.
17. Beyond Discovery: The path from research to human benefit, Electronic distribution by the National Academy of sciences: January 2003. http://www.nasonline.org/publications/beyond-discovery/#:~:text=Beyond%20Discovery%E2%84%A2%3A%20The%20Path,important%20technological%20and%20medical%20advances.
18. Mayer, M. S. (2019). *Handbook of insect pheromones and sex attractants*. CRC Press.
19. Nojima, S., Classen, A., Groot, A. T., & Schal, C. (2018). Qualitative and quantitative analysis of chemicals emitted from the pheromone gland of individual Heliothis subflexa females. *PLoS One*, *13*(8), e0202035.
20. Roelofs, W. L., Liu, W., Hao, G., Jiao, H., Rooney, A. P., & Linn Jr, C. E. (2002). Evolution of moth sex pheromones via ancestral genes. *Proceedings of the National Academy of Sciences*, *99*(21), 13621–13626.
21. Cork, A., Souza, K. D., Hall, D. R., Jones, O. T., Casagrande, E., Krishnaiah, K., & Syed, Z. (2008). Development of PVC-resin-controlled release formulation for pheromones and use in mating disruption of yellow rice stem borer. *Crop Protection*, *27*, 248–255.
22. Peteu, S. F., Onacea, F., Sicua, O. A., Constantinesu, F., & Dinu, S. (2010). Responsive polymers for crop protection. *Polymers*, *2*, 229–251.
23. Krupke, C. H., Roitberg, B. D., & Judd, G. J. R. (2002). Field and laboratory responses of male coding moth (Lepidoptera:Tortricidae) to a pheromone—based attract-and-kill strategy. *Environmental Entomology*, *31*(2), 189–197.
24. Hussain, B., Ahmad, F., Ahmad, E., Yousuf, W., & Mehdi, M. (2021). Role of pheromone application technology for the management of codling moth in high altitude and cold arid region of Ladakh. *Moths and Caterpillars*. doi: 10.5772/intechopen.96438
25. Rizvi, S. A. H., George, J., Reddy, G. V., Zeng, X., & Guerrero, A. (2021). Latest developments in insect sex pheromone research and its application in agricultural pest management. *Insects*, *12*(6), 484.

26. Wang, Z., Ma, T., Mao, T., Guo, H., Zhou, X., Wen, X., & Xiao, Q. (2018). Application technology of the sex pheromone of the tea geometrid Ectropis grisescens (Lepidoptera: Geometridae). *International Journal of Pest Management*, *64*(4), 372–378.

27. Benelli, G., Lucchi, A., Thomson, D., & Ioriatti, C. (2019). Sex pheromone aerosol devices for mating disruption: Challenges for a brighter future. *Insects*, *10*(10), 308.

28. Yan, J. J., Mei, X. D., Feng, J. W., Lin, Z. X., Reitz, S., Meng, R. X., & Gao, Y. L. (2021). Optimization of the sex pheromone-based method for trapping field populations of Phthorimaea operculella (Zeller) in South China. *Journal of Integrative Agriculture*, *20*(10), 2727–2733.

29. Luo, Z., Su, L., Li, Z., Liu, Y., Cai, X., Bian, L., . . . Chen, Z. (2018). Field application technology of sex pheromone on Ectropis grisescens. *Journal of Tea Science*, *38*(2), 140–145.

30. Arn, H., Toth, M. and Priesner. *List of sex pheromones of female Lepidoptera and related male attractants*. www.pherolist.slu.se/pherolist.php.

31. Walker, K. R., & Walter S.C. (2001). Potential for outbreaks of leaf- rollers (Lepidoptera:Tortricidae) in California apple orchards using mating disruption for codling moth suppression. *Journal of Economic Entomology*, *94*, 373–380.

32. Islam, M. A. (2012). *Knowledge and practice of pheromone technologies: A case study of a representative district in Bangladesh*. Academic Research International. www.savap. org.pk

33. Rizvi, S. A. H., George, J., Reddy, G. V., Zeng, X., & Guerrero, A. (2021). Latest developments in insect sex pheromone research and its application in agricultural pest management. *Insects*, *12*(6), 484.

34. Reddy, G. V. P., Sharma, A., & Guerrero, A. (2020). Advances in the use of semiochemicals in integrated pest management: Pheromones. In N. Birch & T. Glare (Eds.), *Biopesticides for sustainable agriculture*. Cambridge: Burleigh Dodds Science Publishing.

13 Oil Pesticides

Leo M.L. Nollet

CONTENTS

DOI: 10.1201/9781003265139-16

13.1 OIL PESTICIDES

Oils have been used as pesticides for centuries. Some of them are very effective and safe alternatives to synthetic insecticides and fungicides. Most oil-based products sold as pesticides are regulated by the Environmental Protection Agency (EPA) under the Federal Insecticide, Fungicide and Rodenticide Act [1].

Safe and effective use of any oil as a pesticide, however, requires a basic understanding of its chemical nature, modes of action, and limitations of use.

Types of oils and oil products that are commercially available for use as pesticides are provided in Table 13.1 [2].

These products include oils distilled from petroleum (also known as horticultural or mineral oils) and oils extracted from plants and animals. Most oil-based pesticides are used for insect control, but in many cases, oil products also have fungicidal properties.

TABLE 13.1
Examples of Common Oil-Based Products and Sources

Oil Type/Source	Insecticide/Fungicide	Brand Name	Manufacturer
Petroleum/Paraffin	I, F	Orchex	Columet Lubricants Co.
Petroleum/Paraffin	I, F	Ultra-Fine Oil	Whitemite Micro-Gen
Petroleum/Paraffin	I	All Seasons Horticultural and Dormant Spray Oil	Bonide
Petroleum/Paraffin	I, F	Saf-T-Side™	Monterrey
Petroleum/Paraffin	I	Horticultural Oil Spray	Green Light
Petroleum/Paraffin	I, F	JMS Stylet-Oil and Organic JMS Stylet-Oil	JMS Flower Farms, Inc.
Plant/Canola	I	Vegol Growing Season Spray Oil	Lilly Miller Brands
Plant/Clove, Cotton Seed, Garlic	I, F	GC-Mite	JH Biotech, Inc.
Plant/Corn, Cotton Seed, Garlic	F	GC-3	JH Biotech, Inc.
Plant/Cotton Seed Oil	I, F	SeaCide	Omega Protein Corp.
Plant/Neem	I, F	Trilogy	Certis USA
Plant/Neem	I, F	Triact	OHP, Inc.
Plant/Neem	I, F	70% Neem oil	Monterrey
Plant/Neem	I, F	Neem Concentrate	Green Light
Plant/Neem	I, F	Rose Defense	Green Light
Plant/Rosemary & Clove	I, F	Phyta-Guard EC	Monterrey
Plant/Sesame Oil	I, F	Organocide	Organic Laboratories
Plant/Soybean	I	Golden Pest Spray Oil	Stoller Enterprises Inc.

13.1.1 Tips for Oil Pesticide Use

1. Always read and follow label instructions.
2. Apply a thin, even coating of oil over the plant or fruit surface. Cover all plant surfaces, especially the undersides of leaves and crevices of branches, fruit, twigs, and stems where pest can hide.
3. To minimize the risk of plant injury, avoid treating when temperatures are below 40 °F or above 85 °F or when the relative humidity is above 90%. Plant injury is due the residual activity of oils (how long they last). The activity is less under hotter conditions than at cooler temperatures. This is largely because the oil molecules evaporate and move within the plant more quickly in a warmer climate.
4. It is better to apply a higher volume of spray mix than to use a higher concentration of oil in the mix.
5. Apply sprays just to the point of runoff—i.e., when the spray on the leaves just starts to drip—in tropical, subtropical, and moderate temperature regions where multiple low concentration (< 0.5%) oil sprays are used annually for control of pests such as the citrus leaf miner.
6. Apply sprays at volumes which exceed the point of runoff—i.e., when the spray on the leaves is continuously dripping—in situations in subtropical and temperature regions, where only one or two applications of higher-concentration (1%) oil spray are used annually for control of pests such as scales.
7. Timings of oil sprays are critical. Apply when the target pest numbers exceed acceptable levels, and they are at their most susceptible stage or when susceptible new growth on trees starts to appear. For example, new summer and autumn growth is attacked annually by the citrus leaf miner in the Sydney region.
8. Not all pests are susceptible to oils, so proper pest identification is required when using any pesticide.
9. Do not mix oil sprays with sulphur- or copper-based pesticides and avoid application of oil before or after a sulphur- or copper-based insecticide treatment.
10. Apply dormant oil sprays only after winter hardening has occurred.

13.1.2 Modes of Action of Oil Pesticides

Almost all oil pesticides have similar modes of action. They are contact pesticides, and their modes of action are based on the disruption of respiration by controlling gas exchange, cell membrane function, or structure. They also can kill pests by disrupting their feeding on oil-covered surfaces. The toxic action of oils is more physical than chemical, and it is short lived. When oils are used to control plant pathogens, they may smother fungal growth and reduce spore germination on treated surfaces. Oils are mostly fungistatic, stopping fungal growth rather than killing the pathogens. Stylet oils are highly refined oils, and they may control insect-vectored plant viruses in addition to insects, mites, and fungal pathogens. These oils reduce the ability of aphids to acquire the virus from an infected plant and transmit it to healthy plants. Stylet oils may interfere with the virus's ability to remain in aphid mouth parts. The sulphur-containing oils, such as neem oil, may possess additional fungicidal activity compared to petroleum oils. Oil-based pesticides have low residual activity and must be sprayed directly on the insect or mite. The oils are generally applied prophylactically prior to infection in order to combat plant fungal pathogens. Repeated applications of oils are needed to achieve satisfactory insecticidal control.

13.1.3 TARGET PESTS AND DISEASES FOR OIL PESTICIDES

Oil pesticides are most commonly used against mites, aphids, whiteflies, thrips, mealy bugs, and scale insects: i.e., they are most effective against soft-bodied arthropods. Dormant oil sprays are also used against overwintering eggs and scales. Petroleum (horticultural) oils as well as plant oils are commonly used to suppress certain fungal diseases, like powdery mildew and black spot on roses. Stylet oils may be used to manage insect-vectored plant viruses.

13.1.4 APPLICATIONS OF OIL PESTICIDES

Oil pesticides have been historically utilized to protect fruit trees and preserve woody ornamentals. In Indian villages, mustard oil (seldom) and mobile oil (generally) are used to preserve woody furniture. Mustard oil is commonly used to preserve foodstuffs. Several types of oil pesticides including emulsions have recently been marketed for ornamental plants, flowers, and vegetables. The emulsifiers are considered inert, but they may have some insecticidal properties. Oil formulations are generally designed to be mixed with water at concentrations of 0.5% to 2.0% percent (volume/volume). All oil pesticides should be mixed properly before spraying and the tank sprayer agitators allowed to run for continuous mixing. This avoids the breakdown of the emulsion that could result in a clogged sprayer, uneven plant coverage, and possible plant injury. The application efficiency of oil pesticides can be enhanced by mixing them with other pesticides. It is well known the oils are surfactants, so they are capable of improving plant coverage and penetration of pesticides into leaf surfaces. Oil pesticides may be incompatible with copper applications on some crops.

13.1.5 PHYTOTOXICITY OF OIL PESTICIDES

Generally, oil pesticides are considered safe to use. It is also known that they can injure susceptible plants species. Effects of plant injury (phytotoxicity) may be acute or chronic, such as leaf scorching and browning, defoliation, reduced flowering, and stunted growth. The phototoxic effects may be associated with plant stress, ambient temperature, and rate of application. It can vary from plant species to plant species and cultivation. The risk of phytotoxicity can be reduced by avoiding the treatment of stressed plants and spraying when the temperature is below 85 °F, and the humidity is 90%. In the summer, applications may be made in the morning and late evening. The longer wet oil sprays remain on foliage, the greater the chance of phytotoxicity. In the winter, applications may be made only when temperature is above 40 °F, and dormant oils or higher rates of summer oils should be applied only after stems and buds have become winter hardened and before buds begin to swell in the spring. Evergreen trees should generally be treated only by summer rates of all-season oils. Some evergreens, especially those with a glaucous (waxy) coat, may become discolored following an oil application. It usually does not harm the tree or shrub. Before treating a new kind of plant, it is advisable to apply horticultural oils to a part of the plant or to a few small specimens before treating large quantities of foliage. It is also important to read and follow the instructions and recommendations on the label. Manufacturers generally provide useful information about sensitive plant species based on extensive testing. Some plants which are commonly listed as being oil sensitive include azalea, carnation, fuchsia, hibiscus, impatiens, photinia, rose, cryptomeria, juniper, Japanese holly, and spruce.

13.1.6 SAFETY OF OIL PESTICIDES

Oil pesticides have many characteristics that make them desirable to growers and home users. For example, they are low in toxicity to humans, wildlife, and pets. Oil pesticides are only active for a short period of time, so they do not scale but do control insects. Oils will separate out from the carrier, so agitation is necessary to keep them in solution. Oil pesticides do not affect insect predators or parasitoids unless they are exposed to the direct spray. Oils evaporate quickly and do not generally contaminate the soil or groundwater sources. Oils obtained from plants and animals are broken down rapidly by sun, air, water, and microorganism on plants or soil, so they pose minimum risk to non-target organisms. Oil pesticides are considered one of the few classes of pesticides to which insects and mites have not developed resistance.

13.2 PETROLEUM OIL PESTICIDES

Petroleum oils may be referred to by many names, including mineral oil/white mineral oil, horticultural oil, spray oil, dormant oil, summer oil, supreme oil, superior oil, or Volck oil. Petroleum products comprise kerosene, petrol, lubricating oils, asphalt, tar, etc. Petroleum oil pesticides are highly refined paraffinic oils which are used to manage pests and diseases of plants. Petroleum oils are composed of saturated and unsaturated hydrocarbons. The unsaturated hydrocarbons are volatile and readily react to form compounds with plants. Therefore, the lower the content of unsaturated hydrocarbons, the safer the oil is for plant treatment. Besides purity or degree of refinement as measured by the unsulphonated residues, some other important characteristics of spray oils are their viscosity and low volatility or distillation range.

Viscosity varies with molecular weight as well as composition. Penetration of the tracheae is inversely proportional to viscosity. Oils of low viscosity are more active because of their greater mobility. On the other hand, oils of high viscosity are less likely to be expelled from the tracheae. If an oil is highly volatile, it gives poor exposure to the treated plant. On the other hand, undue persistence or more stability (i.e., the least volatility of oil on plant surfaces) provides high absorption that may injure plant surfaces. The flash point is the maximum temperature at which an oil gives sufficient vapors to catch fire, so manufacturers provide guidelines for the safe use (spray), storage, and transport of unemulsified oil. Saturated hydrocarbons may oxidize in the presence of air and sunlight. The effects of oxidation on petroleum toxicity are little known. It has been claimed that the oxidation of kerosene increases both toxic action and spreading power. Unrefined oils possess a greater ability than refined oils to spread on the surface of water and the moist surfaces of plant. Petroleum oils generally contain sulphur as an impurity (0–4%). Sulphur often causes severe foliage injury. The addition of sulphur to hydrogen sulphide, organic sulphides, disulphides, and mercaptans in oil does not increase the oxidation number of the oil used. The addition of sulphur in any of the forms does not, however, result in severe leaf injury.

As stated earlier petroleum products are available in several forms and all most all have been tried in pest control.

Kerosene is subjected to a number of chemical treatments, depending upon the purity required. Widely used products are pure kerosene, bland (a carrier of other pesticides), or water white products which have been used in sprays containing pyrethrum, synthetic organic pesticides such as DDT, or a combination of these.

Light oils are the fraction of petroleum with viscosity between 40–65 seconds Saybolt. They are composed of almost entirely saturated hydrocarbons of 14–18 carbon atoms per molecule. The medium oils have characteristics similar to those of light oils, but their viscosity varies from 65 to 85 seconds Saybolt, so they are more viscous. Both these oils are highly refined and contain a small quantity of unsaturated hydrocarbons. These oils are emulsified by the users and applied to control pests.

Paraffinic oils are more toxic to pests than corresponding naphthenic and aromatic oils. Low molecular weight fractions of all these oils have low toxicity to pests. Paraffinic oils with a boiling point below 670 °F and of viscosity less than 55 seconds Saybolt are practically non-toxic to eggs of the oriental fruit moth. Naphthenic oils boil at 690 °F, and their viscosity is 110 seconds Saybolt. They have appreciable toxicity.

Oils having a high content of paraffinic hydrocarbons and a low content of aromatic and naphthenic hydrocarbons are called superior type oils. They are very popular because of their high effectiveness as insecticides and lower phytotoxicity than that of aromatic and naphthenic oils. Superior oils are preferred for orchard spray. Oils containing a high percentage of paraffinic hydrocarbons are more efficient pesticides than those containing a high percentage of aromatic and naphthenic oils.

Heavy oils have viscosity greater than 55 seconds Saybolt, and they are used as dormant sprays in pesticide control. The emulsions of these oils are also used as pesticides. These emulsions usually have 85% oil. Materials such as dinitro-o-cyclohexylphenol are added to dormant spray in order to enhance its efficiency in terms of toxicity and reduction in oil deposits on the plant surfaces. Mixtures of petroleum and tar oils are also used as dormant sprays.

Highly refined whale oils with a viscosity of 150–250 seconds Saybolt, known for medicinal purposes, have been used for protection of sweet corn ears at 100 °F. Heavy oils are mainly used for the protection of dormant vegetation, so their high refinement is not required.

Petroleum oils play a different role in the formation of emulsions of water-insoluble and water-soluble pesticides. Generally a small quantity (1–2%) of petroleum oil is added to the dust formulations in order to increase the adhesion and killing efficiency of the emulsion. It also reduces the cost of emulsion as a minimum quantity of oil is used. On the other hand, the emulsion provides safety to plant life. When emulsions of water-soluble pesticides such as nicotine (almost equally soluble in water and petroleum oil) are applied on plants, the water content drains off during and immediately after the application, and a large percentage of the oil remain. This process tends to extend the period of effectiveness of the nicotine.

It has been observed that most vegetable oils retain more nicotine than petroleum oils in combination with water. This may be due the higher polarity of vegetable oil. The presence of some additives such as cryolite may create a multi-purpose emulsion: e.g., the cryolite controls orange worms, and the oil controls scale insect and mites. Nicotine plays its own role. In this mixed emulsion, oil also improves the action of cryolite by increasing its spread and adhesion.

13.2.1 PESTICIDAL ACTION OF PETROLEUM OILS

Diverse observations have been recorded and published in the literature. A few salient features of petroleum oils (naphtha, gasoline, ether, benzene, kerosene, gas oil, fuel oil, lubricating oil, petroleum, etc.) for use as pesticides are summarized in the following paragraphs. Paraffinic oils are more toxic than corresponding naphthenic or aromatic oils. Some insects are more effectively controlled by oils of high paraffinicity while others are more susceptible to oils of high aromatic content. For example, the eggs of fruit tree leaf rollers and aphids are resistant to petroleum oils while they are easily affected by low concentrations of tar oil. The fumes of low-boiling oils such as gasoline kill insects readily but higher-boiling oils such as lubricating oils have so little volatility that they possess no apparent fumigant action.

Petroleum oils are being used extensively as scalecides and mosquito larvicides. De Ong [3] observed that when a drop of kerosene was placed on a scale insect, it penetrated through the tracheal system but was frequently expelled rapidly, apparently by the efforts of the insect. Penetration to the tracheae is inverse to viscosity, so low-viscosity oils are more active, and high-viscosity oils are less likely to be expelled from the tracheae. When plant surfaces are covered with tiny droplets of oil pesticides of high viscosity and boiling point, the pest may be inactive and finally may die of suffocation because their oxygen is totally cut off. However, suffocation is a very slow process, probably because an inactive pest carries sufficient air within its tracheal system to provide oxygen to its tissues and organs for a considerable time. It has been observed that it takes a few hours to kill larvae by suffocation while highly refined oil kills the larvae much more quickly. Furthermore, potato tuber moth larvae become motionless more rapidly in either nitrogen or water than they do in petroleum oils.

Similarly, several theories have been proposed for the ovicidal mode of action of oils: (1) oil may coat the egg in such a manner that the young insect cannot get energy, (2) the oil may soften the shell (corium) of the egg to interfere with its embryonic development, (3) the oil may act by contact with delicate integument of the young insect when it emerges from the egg, (4) oil may prevent respiration of the developing embryo by interfering with passage of oxygen or carbon dioxide through the corium, and (5) the oil may actually penetrate the egg and coagulate the proto-plasm within.

Oil deposits on the surface may prevent the development of newly hatched larvae. Sublethal contact of a scale insect with an oil deposit may result in a reduction in the emergence of the young scales, or an appreciable percentage of young may be born dead. Aphid and psyllid eggs are more effectively killed by the aromatic hydro-carbons which occur in tar oils, but geometrid and capsid eggs are destroyed more readily by what appears to be the physical environment and stiffing of the eggs.

13.2.2 HORTICULTURAL SPRAYS OF PETROLEUM OIL PESTICIDES

Generally, oil sprays are known as winter or dormant sprays and summer sprays. Winter or dormant sprays are applied largely to kill aphids, the overwintering eggs of any insect, and mite pests. Winter sprays contain heavy oils. Oils may not be highly refined and may contain tar oils. Some additives such as lime sulphur and dinitro-phenol are incorporated to increase the ovicidal effectiveness of the spray. Winter sprays are applied to deciduous trees in the dormant stage, so they are also known as dormant sprays.

Oils used for summer spray are highly refined and lighter. They are toxic and non-phytotoxic. They are widely used to control sap-feeding insects such as scale insects and mites on fruit trees. They may be applied on deciduous trees in full foli-age and on green house plants. Some synthetic insecticides, nicotine, lead arsenate, rotenone, or various fungicides may be added to provide complete pest control.

13.2.3 EFFECT OF PETROLEUM OIL PESTICIDES ON PLANTS

Petroleum-based pesticides adversely affect plants, so their use as insecticides and acaricides is restricted. Therefore, studies are focused on the nature of oil, the con-centration of the oil in the spray, the concentration of additives in the spray, the amount of residual oil on the plant surface, etc. It has been found that the oil emul-sions are injurious to plants even though the soap, the oil, and the additives present in the emulsion are not injurious individually. Oils purified by sulphur extraction are more injurious that those purified by sulphuric acid. This may be due to the impuri-ties left after purification.

Petroleum oils as such are not injurious to leaves, but the acids from their oxi-dation are injurious. The oxidation depends on the nature of the oil (saturated, unsaturated, naphthenic, or aromatic), the presence of foreign substances (impu-rities, additives), sunlight (temperature), and oxygen. Thus, plants have a greater tolerance to paraffinic oils which are free from impurities. Unsaturated hydrocar-bons have been found to be responsible for practically all leaf dropping. Con-tact between these oils and the roots may cause the death of a tree within a short period. Chronic injury may be controlled by avoiding repeated application in short intervals. Fully matured leaves, particularly those which are senile, are more sus-ceptible to injury than younger ones. However, young leaves are more susceptible to injury from unrefined oils than are mature but not senile leaves. Appreciable effects of oils have not been noted on the respiration rate of twigs when treated with light or heavy saturated oils. The rate of respiration depends on the stage or seasonal development of the plant, while the rate of respiration of dormant apple twigs was reported to be accelerated more by light unsaturated oils than by heavy unsaturated oils. The effects of oils on leaves are either physical or chemical. The physical effects tend to decrease respiration by plugging the stomata and intercellular spaces, and the chemical effects increase respiration due to toxicity. The transpiration of old leaves is reduced more than that of young ones after treatment with petroleum oils.

Oil sprays have a considerable effect on the photosynthesis of leaves. The starch contents of the foliage drop off markedly. Oil sprays intensify the green color of foliage. Treatment with summer oils causes foliage to drop later than normal. The chlorophyll content of apple leaves is increased up to 47% by the spray.

The absorption of oil by the leaves varies from species to species. The disappearance of oil sprayed on the leaves' surface appears to be due more to absorption than to evaporation. The evaporation of oil from the interior of living orchard foliage is negligible while much of the oil is retained by leaves.

13.2.4 PRODUCTS

SACOA [4] is a company with a market leadership position in spray oils and adjuvant technology in Australia. They supply and support innovative products for sustainable farming.

SACOA is a trusted and recognised brand with a diverse product portfolio and geographic diversity across all key crop markets, including broadacre cereal, cotton, bananas, almonds, citrus, olives, grapes, pome fruits, stone fruits, and vegetable crops.

SACOA has a proven, differentiated, and comprehensive high-quality product range, supported by an extensive field and laboratory research programme; established proprietary and trademarked brand names including BioPest, Enhance, Cropshield, and Antievap; and a strong pipeline of new products and technologies for future market launch.

All Seasons White Oil Insecticide [5] is for the control of scale insects and certain other insects on citrus, shrubs, roses, and ornamentals and for use as a spreading agent.

White Oil Insecticide is used to control scales, aphids, mites, mealybugs, citrus leaf miners, red and white wax mites, and other scale and spider mites in certain situations, including on deciduous fruit trees, roses, citrus trees, indoor plants, ornamental shrubs, and daphne. The active ingredient is 825 g petroleum oil per L.

13.2.5 PLANT DAMAGE AND CORRECT APPLICATION UNDER
THE RIGHT CONDITIONS

The phytotoxicity of oils has already been discussed in this text. The rate of penetration, migration, oxidation, and evaporation of spray oils is generally rapid but slows down with increasing the molecular weight of the oil and lowering atmospheric temperature. The presence of oil on plant surfaces inhibits photosynthesis, respiration, and transpiration which are responsible for decolourization of leaves, leaf burning, leaf drop, reduced yield and injury to fruit or wood. The risk of phytotoxicity can be minimized by application of correct concentration of oil spray under appropriate conditions. Beattie [6] has recommended the following spray volumes for 4 m–high orange trees with dense canopies by using an air blast sprayer:

1. Red scale and other hard scales at young stages by 10,000–14000 L/ha of 1% oil
2. Mealybugs at young stages by 10,000–14,000 L/ha of 1% oil
3. Greenhouse thrips at young stages before fruits touch in mid-to-late January by 5,000–8,000 L/ha of 0.25–0.5% oil
4. Soft scales at young stages by 8,000–10,000 L/ha of 1% oil
5. Mites at young stages by 5,000–8,000 L/ha of 0.75% oil, single spray, and 0.5% oil, multiple sprays
6. Citrus leaf miners at young growth in summer/autumn until most leaves are >4 cm long, 0.25–0.5% oil, multiple sprays

13.2.6 EFFECT OF PETROLEUM OIL PESTICIDES ON MAN AND ANIMALS

Petroleum oils are inflammable, so they can catch fire during application, transport, and storage. They also give off an unpleasant smell, so contamination of foodstuffs is unwanted and undesirable. Their contact with skin causes irritation because they are good solvents for lipids. The lipids-free skin becomes dry, rough, and prone to catch environmental pollutants. Similarly, they affect farm animals by inflaming the skin, causing an excessive rise in body temperature and burning the hide severely, producing visible lesions. Skin damage is proportional to the sulphonation number and viscosity index of the oil. The chemical reaction of petroleum oils with skin causes inflammation of the peripheral capillaries, thereby reducing water dissipation from the skin and impairing the cooling power of animals to critical limits at high atmospheric temperature. These hazards cause skin irritation and reduction in milk yield. The application of petroleum oil is commonly made to control the common fly, the stable fly, etc.

13.2.7 USES OF PETROLEUM OIL PESTICIDES

These are used (1) as insecticide carriers and (2) as insecticides as is. A genuine interest has been taken in researching the use of the oils as carriers in different formulations. Petroleum oil–based formulations containing micro levels of synthetic pesticides have been found to be more effective and selective as well as less hazardous due to the combined effects of the oil and the synthetic pesticide.

13.2.8 ADVANTAGES OF PETROLEUM OIL PESTICIDES

These pesticides possess the following advantages: (1) they are affordable: i.e., they are of relatively low cost compared to other pesticides; (2) they have good spreading power; (3) they are easy to mix and handle; (4) petroleum pesticides are not toxic to humans or other animals; (5) they are more effective than broad-spectrum synthetic pesticides for a wide range of pests and diseases: i.e., many pests can be controlled simultaneously; (6) they are less harmful to natural enemies of citrus pests; (7) they do not stimulate other pest outbreaks; (8) pests are known to develop resistance to petroleum pesticides; (9) the oil deposits are broken down within weeks to simple and harmless molecules; (10) when using oils, only minimal and low-cost clothing needs to be worn; and (11) they are suitable, depending on the emulsifiers and additives used to formulate the pesticide, for use in organic farming.

13.2.9 DISADVANTAGES OF PETROLEUM OIL PESTICIDES

The disadvantages are: (1) they have low insect toxicity; (2) they require safeguards for storage; (3) they are phytotoxic; (4) they require very attentive application: for example, the risk of phytotoxicity is higher if oil sprays are applied at temperatures above 35 °C; in hot, dry wind or excessive or prolonged cold temperature; on trees in poor health; on low-moisture soil or water-stressed plants; on waterlogged trees; or on trees sprayed with sulphur spray within one month; and (5) they cause damage to pesticide equipment, including rubber hoses, pistons, and other related parts of the sprayer; containers; etc.

13.2.10 CONCLUSION

In most cases, the concentration of oil should not need to exceed 1%. Focus on increasing spray volume rather than increasing the concentration of oil. Oils need to be applied in order to get even coverage of all plant surfaces on which the target pests are present. All the precautions and recommendations should be followed to avoid the phytotoxic effects. The appropriate spray volume should apply so that the oil goes off within one to two hours.

See further references [7–9].

13.3 COAL TAR DERIVATIVES AND THEIR CLASSIFICATION

Coal tar is obtained from the destructive distillation of coal. It is a byproduct of production of coke and illuminating gas from coal. The composition of coal tar depends on the types and composition of the coal.

Usually, the crude tar is obtained from the coking process. The fractional distillation during the process gives the following: <210 °C: light oil, benzene, toluene, xylene; 210–241 °C: middle or carbolic acid, phenols, naphthalene; 240–270 °C: heavy or creosote oil; and >270 °C: anthracene oil. Hence, it is clear that these products differ from petroleum oil because aromatic series predominate rather than paraffinic and naphthenic compounds. The fraction of high boiling point oil free from tar acids is most useful for spray purpose. Tar acids generally present in tar products are phenols (hydroxyl benzene), cresols (methyl phenols), xylenols (dimethyl phenols) and higher hydroxyl compounds. These compounds cause plant injury and are irritating to human skin.

The presence of acidic impurities reduces the pesticidal nature of the oil, but their presence in traces helps in the emulsification of the tar distillates. The acidic compounds can be removed as water-soluble sodium salts by treating the mixture with caustic soda. The tar bases are mainly nitrogen-containing heterocyclic compounds related to pyridine and quinolenes. They are soluble in acids.

Natural oil used as an insecticide is derived from the creosote and anthracene fractions. Naphthalenes and anthracene do not add much to the effectiveness of the oil, and they can be removed as crystalline solids. Coal tar creosote is a standard wood preservative, but the freshly treated wood can catch fire.

Wood tar can be obtained in the process of the destructive distillation of pure wood. It is high in organic acids, so it can injure plant foliage. However, pine tar oil has been claimed to be safe for foliage applications. It may be used on animal life. It gives an unpleasant odor if it is derived from hard wood.

The use of coal tars (also known as carbolineums, tar cresols, tar distillate, and tar oils) as insecticides is of comparatively recent origin.

13.4 ANIMAL AND VEGETABLE OILS

A number of animal and vegetable oils and fats have been used as insecticides and acaricides under limited and specialized conditions for a long time. Their soaps have most often been used in several insecticides formulations. A mixture of mustard oil and powdered rock salt is still being used as fungicide to protect the human body during rainy season in Indian villages. Several vegetable and animal oils are also widely used in Africa, China, and Europe to control a variety of insects and mites, usually as spray emulsions.

Animal and vegetable oils and fats are quite different from mineral oils. As stated earlier, mineral oils such as kerosene oil and lubricating oils are mixtures of hydrocarbons. Animal and vegetable oils also differ from essential oils. Animal and vegetable oils and fats are collectively known as lipids. Lipids are insoluble in water but soluble in organic solvents like chloroform and carbon tetrachloride. Most of them are glycerol esters of unbranched long chain carboxylic acids with an even number of carbon atoms. They can be hydrolyzed to yield glycerol and acids. Because of their occurrence in fats, these straight-chain acids have been called fatty acids. The important vegetable oils are castor oil, cotton oil, coconut oil, soybean oil, hempseed oil, palm oil, peanut oil, and tung oil. Pertinent examples of animal fats and oils are neatsfoot oil, cod liver oil, shark liver oil, fish oil, whale oil, lard, and lard oil.

13.4.1 ANIMAL OILS

Whale oil and some other varieties of fish oil—namely, menhaden and herring oils—have been reported to possess insecticidal characteristics. Bone oil is derived from the destructive distillation of bones for manufacturing charcoal. It is source of pyridine and other nitrogen-containing bases. However, the use of animal oils as insecticides is limited because of their cost and scarcity. Fish and whale oil soaps are frequently used as emulsifying and wetting agents. A wide variety of fatty acids are obtained by the hydrolysis of animal oils and fats. Some fatty acids have been found to possess insecticidal properties. Siegler and Popenoe [10] investigated a number of fatty acids as contact insecticidal sprays, particularly against aphids. The toxicity of fatty acids increases with increasing molecular weight up to C10–C12 and then decrease markedly with further increase in molecular weight. Free acids are more toxic than corresponding soluble neutral or alkaline salts.

13.4.2 VEGETABLE OILS

Vegetable oils and fats are classified as both edible and inedible. Inedible oils like tallow and palm oil are used for making soaps. Vegetable oils are also classified as non-drying oils (saturated carbon chain oils), semi-drying oils (some degree of unsaturation in the carbon chain), and drying oils (unsaturated carbon chain). Petroleum and tar oils are cheap and available in abundance as well as more effective insecticides, so vegetable oils are rarely used as pesticides. Many oils like castor, sesame, cotton seed, linseed, olive, coconut, corn, hemp, palm, groundnut, soybean, and rapeseed oils were tested as sprays. It was found that 0.5% rapeseed oil kills certain insects it may be used in oil emulsions at the pink stage. Its emulsions are used as summer sprays on apple. The presence of unsaturated fatty acids is responsible for consumption of oxygen while the presence of sulphur as an impurity may be responsible for its insecticidal activity. Some vegetable oils have been found to be efficient ovicides, but their cost is many times more than that of petroleum and tar oils.

13.4.3 GRAIN PROTECTION BY VEGETABLE OILS

Several vegetable oil pesticides have been developed and used for grain protection. Plant extracts of some species—namely, *Lantana camara*, *Illicium verum*, and *Tithonia diversifolia*—are known to control different insects on grains during storage. Plant-derived products such as azadirachtin from neem (*Azadirachta indica*), pyrethrin from pyrethrum, carvone from caraway (*Carum carvi*), and allyl isothiocyanate from mustard (*Brassica nigra*), and horseradish (*Armoracia rusticana*) oil have received active attention due to their pesticidal properties and potential to protect several foodstuffs [11]. Numerous vegetable oils including peanut, coconut, safflower, mustard, castor bean, cotton seed, soybean, neem, cucurbits, and maize have been used as protective additives successfully.

13.4.4 MECHANISM OF SEED PROTECTION BY VEGETABLE OILS

The mechanism of pesticidal action of oils is not completely clear. However, it has been reported that vegetable oils affect egg laying as well as embryo and larvae development on the surface of the seed. They cause the egg and larvae to die before they can bore the seed. In some cases, female insects are able to lay eggs, but the larvae are prevented from hatching by the oil coating.

Oil may also kill insect eggs by curative action. If the egg is already present on the surface of the seed or inside the seed, the oil coating interferes with gaseous exchange. As a result, the larvae inside the egg or the kernel may die for want of oxygen. A small amount of oil derived from neem, karanja, undi, or Kusum may kill

TABLE 13.2
Protection of Legume Pulses by Vegetable Oils

Oils	Application and Purpose
1. Sunflower, mustard, groundnut, sesame, soybean, olive, and palm oils	At 5–7.5 ml/kg of grain (0.5–1% v/w concentration) to protect legumes from pulse beetle
2. Sandbox seed oil	To protect from cowpea seed beetle
3. Domba, batu, cinnamon, mustard, neem, mee, castor, citronella, and sesame oils	To protect from pulse beetle
4. Indian tree neem oil	To protect from insect pests
Protection of Cereal Grains by Vegetable Oils	
1. Groundnut, rapeseed, and sunflower oils	At 10 mg/kg grain to protect from infestation
2. Neem, lemongrass, lantana, basil, and African marigold plants	To protect from maize weevils
3. Cotton seed, corn, groundnut, and palm oils	To suppress the population of *Cryptolestes pusillus* and *Rhyzopertha dominica* in maize and sorghum
4. Citrus oil	To protect from microbes and insecticides
5. Cotton seed oil	To protect from maize weevils
6. Pyrethrum oil	To protect from maize weevils

about 90% of cowpea weevils. These oils are not harmful to human beings, and their protective action lasts for three months. As stated earlier, the oils must be applied carefully in order to form an impermeable layer. Table 13.2 summarizes some pertinent examples of legume pulses and cereal grain protection by vegetable oils.

See also references [12] and [13].

13.4.5 ESSENTIAL OILS

These are pleasant-smelling, highly volatile liquids, widely distributed in various types of plants [12]. They are also known as volatile liquids. Commonly known essential oils are turpentine oil, oil of clove, and oil of eucalyptus. These oils are obtained from common natural sources like buds, flowers, petals, barks, leaves, roots, etc. They consist of different functional groups; for example, linalool (sandal oil, lavender oil) contains an alcoholic group; citral (lemongrass oil) contains an aldehyde group; eucalyptus oil of wintergreen contains an ester group; cymene contains hydrocarbons; and eugenol (bay oil) contains a phenolic group.

Essential oils are colorless in a pure state but light yellow in a crude state. They are soluble in organic solvents in all proportions. They are steam volatile, and, in some cases, they may decompose.

Certain essential oils have been found to be of practical value as insect repellents and attractants, namely: oil of citronella (citronella: Gran oil: 17–34%: 16–45%), lemongrass oil (citral: citronella: 65–85%: 35–15%), oil of cassia (cinnamic aldehyde: 85–95%), oil of pennyroyal (isomenthan: Pulegone: 50%: 30%), oil of thyme (thymol or carvacrol), clove oil (eugenol: 85–95%), pine needle oil (bornyl acetate: 49%), turpentine or pine oil, and camphor.

Sources of essential oils include:

1. Berries such as allspice and juniper
2. Bark such as cassia, cinnamon, and sassafras
3. Flowers such as cannabis, chamomile, clary sage, clove, echinacea, scanted geranium, hops, hyssop, jasmine, lavender, manuka, marjoram, orange, rose, and ylang-ylang
4. Leaves such as basil, bay leaf, buchu, cinnamon, common sage, eucalyptus, guava, lemongrass, melaleuca, myrrh, oregano, patchouli, peppermint, pine, rosemary, spearmint, tea tree, thyme, tsuga, and wintergreen
5. Peel such as bergamot, grapefruit, lemon, lime, orange, and tangerine
6. Resin such as benzoin, copaiba, frankincense, and myrrh
7. Rhizome such as galangal and ginger
8. Root such as valerian
9. Seed such as almond, anise, buchu, celery, cumin, nutmeg oil, and rhatany
10. Wood such as camphor, cedar, rosewood, sandalwood, and agarwood

The median lethal doses (oral LD50, dermal LD50 in g/kg) for some common oils are as follows:

1. Neem (14, >2)
2. Lemon myrtle (2.43, 2.25)
3. Frankincense (*Boswellia carterii*) (>5, >5)
4. Frankincense (*Boswellia sacra*) (>2, >2)
5. Indian frankincense (*Boswellia serrata*) (>2, >2)
6. Ylang-ylang (>5, >5)
7. Cedarwood (>5, >5)
8. Roman chamomile (>5, >5)
9. White camphor (*Cinnamomum camphora*, extracted from leaves) (>5, >5)
10. Yellow camphor (*Cinnamomum camphora*, extracted from bark) (3.73, >5)
11. Ho oil (*Cinnamomum camphora*, oil extracted from leaves) (3.80, >5)
12. Cassia (2.80, 0.32)

These data are intended as a guideline only as the reported values may vary widely due to difference in tested species and laboratory conditions [14]. It is also important to note that these data are far from relevant in everyday practice due to localized exposures. For example, a dose of any one of these essential oils may do no harm if swallowed in a diluted solution or emulsion, but it could do serious damage to eyes or lungs at a higher concentration [15].

13.4.6 Pesticidal Action of Essential Oils

A sequence of acaricidal activity of 35 commercial essential oils against Tetranychus urticae adult females at 0.1% concentration after 24 hours' treatment (mean±SD) is provided next [16]:

1. Sandalwood (89.2 ± 8.5)
2. Common thyme (62.2 ± 42.0)

3. Scotch pine (50.4 ± 18.5)
4. Sweet orange (45.6 ± 37.4)
5. Juniper (42.6 ± 21.8)
6. Clove bud (41.3 ± 36.3)
7. Lemon peel (34.9 ± 37.9)
8. Grapefruit (30.6 ± 35.2)
9. Geranium (30.0 ± 15.7)
10. Cypress (28.9 ± 18.4)
11. Tea tree (28.6 ± 33.5)
12. Hyssop (28.1 ± 22.3)
13. Eucalyptus (27.9 ± 21.7)
14. Citronella (27.6 ± 20.7)
15. Niaouli (26.8 ± 30.7)
16. True lavender (26.1 ± 21.2)
17. Frankincense (24.8 ± 27.7)
18. Ylang-Ylang (24.2 ± 1'4.9)
19. Peppermint (23.7 ± 18.2)
20. Cinnamon (23.6 ± 20.6)
21. Cajeput tree (23.5 ± 11.4)
22. Black pepper (22.8 ± 20.0)
23. Myrrh (22.8 ± 17.8)
24. Bitter orange (21.4 ± 21.5)
25. Sweet basil (21.0 ± 17.9)
26. Patchouli (20.3 ± 20.4)
27. Blue gum (19.7 ± 22.5)
28. Lemongrass (17.8 ± 9.5)
29. Cedar wood (12.4 ± 6.7)
30. Ginger (11.9 ± 13.0)
31. Rosemary (11.7 ± 22.0)
32. Bergamot (11.0 ± 6.0)
33. EtOH+Triton-X+water (10.9 ± 4.4)
34. Clary sage (7.1 ± 0.2)
35. Sweet marjoram (7.1 ± 0.2)

A lot of information on pesticidal activities of essential oils may be found in references [17] and [18].

Many plant essential oils show a broad spectrum of activity against insects and plant pathogenic fungi, ranging from insecticidal, antifeedant, repellent, and oviposition to deterrence, including growth regulatory and antivector activities. Some pertinent examples are given in Table 13.3. They are just examples. There are many more references.

The literature survey shows that considerable research has been published in the form of papers, reviews, and books on the pesticidal properties of essential oil and their constituents. Many national and international seminars, symposia, and conferences have been organized on the subject throughout the world. In spite of this,

TABLE 13.3
Pesticidal Action of Some Essential Oils

Essential Oil	Mode of Action	Reference
Insecticides and Growth Regulators		
1. Eugenol	Termiticide, fumigant, feeding deterrent, and toxic to many species	14
2. Orange oil extracts from citrus peel (containing 92% d-Limonene)	96% and 68% mortality to Formosan subterranean termite, *Coptotermes formusanus*, shiraki within five days	15
3. Citronellal, eugenol, menthol, pulegone, and thymol	Moderately active against various mites	16
4. Diterpene, 3-epicaryotin	Reduces growth of European corn borer larvae	17
5. Menthol	Reduces growth of European corn borer larvae	18
6. 1, 8-Cineole from *Artemisia annua*	A potential insecticide	19
7. Turmeric plant oil	Very useful in pest control	20, 21
Fumigants		
1. Pulegone, linalool, and limonene	Effective fumigants against rice weevils	20
2. Mentha citrata oil containing linalool and linalylacetate	Significantly toxic to rice weevils	21
3. Trans-anethol, thymol, 1, 8-cineole, carvacrol, terpineol, and linalool	Fumigants against *Tribolium castaneum*	22
4. Anethol combined with 1, 8-cineole (1:1). by 100%.	Reduces the population of *T. castaneum*	23
Antifeedants		
1. Thymol, citronellal, and a-terpineol	Effective as feeding deterrent against tobacco cutworm	24
2. 1, 8- Cineol	Antifeedant against *T. castaneum*	22
3. Terpenoid lactone	Antifeedant against granary weevil	25
4. Essential oil of marjoram and rosemary oil	Antifeedant against onion thrips	26
5. Essential oils from *Elsholtzia densa* and *E. piulosa*	Antifeedant against third instars of *S. litura*	27
Repellents		
1. Monoterpenes, eremophilane sesquiterpenes, eremophilane sesquiterpenes derivatives from Alaska, and *Aedes aegypti* adults	Active against *Ixodes scapularis* nymphs, yellow cedar	28
2. Carvacrol	Active against ticks, fleas, and mosquitoes	29
3. Citronellal	More effective than eugenol and cineole against mosquitoes	30

(Continued)

TABLE 13.3 (Continued)

Essential Oil	Mode of Action	Reference
4. Essential oils from *Cinnamomum* species	Effective mosquito larvicides	31
5. Fruit oil of *Piper retrofractum*	High repellency (52–90%) against *T. Castaneum* at 0.5–2% concentration	32
Oviposition Inhibitors and Ovicides		
1. 1, 8-Cineole and marjoram at concentrations of 1.0%	Reduce oviposition rate by 30–50%	33
2. Calamus oil	Prevents oviposition of *Callosobruchus maculatus*	34
3. Garlic oil	Oviposition deterrent, toxic to eggs	35
4. Essential oil of *Aegle marmelos*	Reduces 99.5% egg hatching in *Spilosoma obliqua* at 250 mg 250 egg	36
5. 1-carvone	Completely suppresses the egg hatching of *T. castaneum* at 7.22 mg/cm2 surface treatment	37
6. Carvacrol, carveol, geraniol, carvones, linalool, menthol, terpinol, thymol, fenchone, menthone, pulegone, thujone, verbenone, cinnamaldehyde, citral, citronellal, and cinnamic acid	Ovicides against *Musca domestica* egg	38
Attractants		
1. Geraniol and eugenol	Effective attractants used in traps to lure Japanese beetles	39
2. Methyl-eugenol	Used to trap oriental fruit flies	40
3. Cinnamyl alcohol, 4-methoxy-cinnamaldehyde, geranylacetone, and a-terpineol	Attractants to adult corn rootworm beetles	41
4. 1, 8-Cineole	Attractant to western flower thrips.	42
5. Terpenes and geraniol from lemon essential oil	Attractants to thrips, fungus, gnats, mealybugs, scale, and Japanese beetles	
6. Cis-jasmone	Effective attractant to adult Lepidoptera	43
7. Sandalwood oil, basil oil, grapefruit oil, and other aromatics	Attractants to greenhouse whitefly	44
Antifungal Agents		
1. Thymol and carvacrol	Active against most fungal species such as *Alternaria padwickii*, *Bipolaris oryzae*, and peanut fungi	45
2. Plant essential oils at 400 mg and 700 mg per liter of soil.	Effective fungicide against *R. solanacearum*	46
3. Thymol, palmarosa oil, and lemongrass oil at a concentration of 700 ml/liter	Effective against bacterial wilt and *R. sonalacearum*	47

(*Continued*)

TABLE 13.3 (Continued)

Essential Oil	Mode of Action	Reference
Antiviral Agents		
1. Essential oil of *Melaleuca alterifolia* at a concentration of 100–500 ppm	Effective in decreasing local lesions of TMV on host plant *Nicotiana glutinosa*	48
2. Essential oils of *Ageratum conyzoides*, *Callistemon lanceolatus*, *Carum copticum*, *Ocimum sanctum*, and *Peperomia pellucida*	Inhibitory active against cowpea mosaic virus, mung bean mosaic virus, bean common mosaic virus, and southern mosaic virus	49
3. *Tagetes minutaoil*	Active against carnation ring spot and carnation vein mottle viruses	50
4. Essential oils and kaolin	Effective to control viruses and to reduce insecticide use on tomatoes	51

Source: Commercialization of Essential Oil-Based Pesticides

surprisingly few pest control products are available in the market. It may be because of regulatory barriers to commercialization (e.g., the cost of toxicological and environmental evaluations) or the fact that the efficacy of essential oils toward pests and disease vectors is not as apparent or obvious as that of currently available chemical pesticides.

13.4.7 ADVANTAGES OF ESSENTIAL OIL–BASED PESTICIDES

1. Many plant essential oils perform a broad spectrum of activities against pest insects and plant pathogenic fungi, including insecticidal, antifeedant, repellent, oviposition deterrent, growth regulatory, and antivector activities.
2. These oils are also being used in the protection of stored foodstuffs and products.
3. Most essential oils are relatively non-toxic to mammals and fish, so they are reduced-risk pesticides. This is because they interfere with the octopaminergic to the nervous system in insects.
4. Some essential oils are widely used as flavoring agents in foodstuffs and beverages and are even exempt from pesticide registration.
5. Since some essential oils are available abundantly, the commercialization of essential oil– based pesticides is feasible.
6. Essential oil–based pesticides are known as green pesticides because of their use against home and garden pests and in organic farming.
7. Essential oils are complex mixtures, so pests may develop resistance to them more slowly. This is an issue for many synthetic chemical pesticides.
8. These pesticides may have their greatest impact in future integrated pest management (IPM) programmes, especially in developing countries which are rich in endemic plant biodiversity.

9. Essential oils are volatile, so they have limited persistence under field conditions, and therefore, although natural enemies are susceptible via direct contact, they are unlikely to be poisoned by residue contact as often occurs with conventional insecticides.
10. The predator, parasitoid, and pollinator insect populations will be less impacted because of the minimal residual activity.

13.4.8 LIMITATIONS OF ESSENTIAL OIL–BASED PESTICIDES

1. In spite of the various advantages of essential oil–based pesticides (EOBPs), few pest control products using them have appeared in the market. This may be due to the cost of toxicological and environmental evaluations.
2. In terms of specific constraints, the efficacy of essential oil–based pesticides falls short when compared to synthetic chemical pesticides.
3. Essential oil–based pesticides require somewhat greater application rates (as high as 1% active ingredient) and may require frequent reapplication when used out of doors. Therefore, EOBP crop protection is more costly, time consuming, and laborious than chemical crop protection.
4. Essential oil is a complex mixture of several compounds, so it acts as a broad-spectrum pesticide. The characterization and detection of specificity towards pests of each compound present in an essential oil are unaffordable for use on agricultural farms.
5. Other challenges to the commercial applications of EOBPs include the availability of sufficient quantities, protection of technology (patents), and regulatory approval [57].
6. The chemical profiles of plant species depend on geographic, genetic, climatic, and seasonal factors. Pesticide manufacturers must take additional steps to ensure the consistency of the product.
7. It is obvious that small-budget companies cannot be involve in this project due to these complexities.

REFERENCES

1. www.epa.gov/laws-regulations/summary-federal-insecticide-fungicide-and-rodenticide-act#:~:text=The%20Federal%20Insecticide%2C%20Fungicide%2C%20and%20Rodenticide%20Act%20(FIFRA),registered%20(licensed)%20by%20EPA
2. Bogran, C. E., Ludwig, S., & Metz, B. (2006). *Using oils as pesticides.* Texas FARMER Collection.
3. Ong, D. Q., Sitaram, N., Rajakulendran, M., Koh, G. C., Seow, A. L., Ong, E. S., & Pang, F. Y. (2010). Knowledge and practice of household mosquito breeding control measures between a dengue hotspot and non-hotspot in Singapore. *Annals Academy of Medicine Singapore, 39*(2), 146.
4. www.sacoa.com.au/about/
5. https://specialistsales.com.au/shop/farm-chemicals/pest-control-farm-chemicals/farm-insect-control/white-oil-insecticide-petroleum-oil/
6. https://ipmguidelinesforgrains.com.au/ipm-information/cultural-and-physical-control/petroleum-or-mineral-oils/#:~:text=Petroleum%20oils%20are%20highly%20refined,soft%2Dbodied%20insects%20and%20mites.

7. Jungers, G., Portet-Koltalo, F., Cosme, J., & Seralini, G. E. (2022). Petroleum in pesticides: A need to change regulatory toxicology. *Toxics, 10*(11), 670.
8. https://ipm.ucanr.edu/agriculture/citrus/precautions-for-using-petroleum-oil-sprays/
9. http://npic.orst.edu/ingred/petroleum-distillates.html
10. Siegler, E. H., & Popenoe, C. H. (1925). The fatty acids as contact insecticides. *Journal of economic entomology, 18*(2), 292–299.
11. Singh, A., Khare, A., & Singh, A. P. (2012). Use of vegetable oils as biopesticide in grain protection-a review. *J Biofertil Biopestic, 3*, 1–114.
12. Johnson, W. T. (2020). Oils as pesticides for ornamental plants. In *Handbook of integrated pest management for turf and ornamentals* (pp. 557–581). CRC Press.
13. Isman, M. B. (2020). Bioinsecticides based on plant essential oils: A short overview. *Zeitschrift für Naturforschung C, 75*(7–8), 179–182.
14. https://www.iso.org/ics/71.100.60/x/
15. Sapeika, N. (1972). *Actions and uses of drugs.* AA Balkema.
16. Roh, H. S., Lim, E. G., & Kim, J. (2011). Acaricidal and oviposition deterring effects of santalol identified in sandalwood oil against two-spotted spider mite, Tetranychus urticae Koch (Acari: Tetranychidae). *Journal of Pest Science, 84*, 495–501.
17. https://www3.epa.gov/pesticides/chem_search/reg_actions/registration/fs_G-114_01-Jul-01.pdf
18. Nollet, L. M. L., & Rathore, H. S. (2019). *Green pesticides handbook: Essential oils for pest control* (pp. 1–572). CRC Press.
19. Nisar, M. S., Nazir, T., Zaman, S., Hussain, S. I., Khan, N. A., Aslam, H. M. U., . . . Akhtar, M. (2022). Toxicity and repellency of plant extract and termiticide against fungus growing subterranean termites (Blattodea: Termitidae). *Journal of Bioresource Management, 9*(2), 13.
20. Raina, A., Bland, J., Doolittle, M., Lax, A., Boopathy, R., & Folkins, M. (2007). Effect of orange oil extract on the Formosan subterranean termite (Isoptera: Rhinotermitidae). *Journal of Economic Entomology, 100*(3), 880–885.
21. Vigad, N., Pelyuntha, W., Tarachai, P., Chansakaow, S., & Chukiatsiri, K. (2021). Physical characteristics, chemical compositions, and insecticidal activity of plant essential oils against chicken lice (*Menopon gallinae*) and mites (*Ornithonyssus bursa*). *Veterinary Integrative Sciences, 19*(3), 449–466. https://doi. org/10.12982/VIS. 2021.037.
22. Koul, O., Walia, S., & Dhaliwal, G. S. (2008). Essential oils as green pesticides: Potential and constraints. *Biopesticides International, 4*(1), 63–84.
23. Durden, K., Sellars, S., Cowell, B., Brown, J. J., & Pszczolkowski, M. A. (2011). Artemisia annua extracts, artemisinin and 1, 8-cineole, prevent fruit infestation by a major, cosmopolitan pest of apples. *Pharmaceutical Biology, 49*(6), 563–568.
24. Damalas, C. A. (2011). Potential uses of turmeric ('Curcuma longa') products as alternative means of pest management in crop production. *Plant Omics, 4*(3), 136–141.
25. Golden, G., Quinn, E., Shaaya, E., Kostyukovsky, M., & Poverenov, E. (2018). Coarse and nano emulsions for effective delivery of the natural pest control agent pulegone for stored grain protection. *Pest Management Science, 74*(4), 820–827.
26. Aarthi, K., Shanthi, M., Srinivasan, G., Vellaikumar, S., & Hemalatha, G. (2022). Repellent toxicity of mint essential oils against rice weevil, Sitophilus oryzae L. The *Pharma Innovation, 11.*
27. Mondal, M., & Khalequzzaman, M. (2010). Toxicity of naturally occurring compounds of plant essential oil against Tribolium castaneum (Herbst). *Journal of Biological Sciences, 10*(1), 10–17.
28. Koul, O, Singh, G., Singh, R., & Singh, J. (2007). Mortality and reproductive performance of Tribolium castaneum exposed to anethole vapours at high temperature. *Biopesticides International, 3*, 126–137.
29. Isman, M. B., Wan, A. J., & Passreiter, C. M. (2001). Insecticidal activity of essential oils to the tobacco cutworm, Spodoptera litura. *Fitoterapia, 72*(1), 65–68.

30. Paruch, E., Ciunik, Z., Nawrot, J., & Wawrzeńczyk, C. (2000). Lactones. 9. Synthesis of terpenoid lactones active insect antifeedants. *Journal of Agricultural and Food Chemistry*, *48*(10), 4973–4977.
31. Van Tol, R. W., James, D. E., De Kogel, W. J., & Teulon, D. A. (2007). Plant odours with potential for a push—pull strategy to control the onion thrips, Thrips tabaci. *Entomologia Experimentalis et Applicata*, *122*(1), 69–76.
32. Tandon, S., Mittal, A. K., Kasana, V. K., & Pant, A. K. (2004). Antifeedant activity of Elsholtzia essential oils against Spodoptera litura. *Annals of Plant Protection Sciences*, *12*(1), 197–199.
33. Jordan, R. A., Schulze, T. L., & Dolan, M. C. (2012). Efficacy of plant-derived and synthetic compounds on clothing as repellents against Ixodes scapularis and Amblyomma americanum (Acari: Ixodidae). *Journal of Medical Entomology*, *49*(1), 101–106.
34. de Souza, J. R. L., de Oliveira, P. R., Anholeto, L. A., Arnosti, A., Daemon, E., Remedio, R. N., & Camargo-Mathias, M. I. (2019). Effects of carvacrol on oocyte development in semi-engorged Rhipicephalus sanguineus sensu lato females ticks (Acari: Ixodidae). *Micron*, *116*, 66–72.
35. Kim, J. K., Kang, C. S., Lee, J. K., Kim, Y. R., Han, H. Y., & Yun, H. K. (2005). Evaluation of repellency effect of two natural aroma mosquito repellent compounds, citronella and citronellal. *Entomological Research*, *35*(2), 117–120.
36. Cheng, S. S., Liu, J. Y., Tsai, K. H., Chen, W. J., & Chang, S. T. (2004). Chemical composition and mosquito larvicidal activity of essential oils from leaves of different Cinnamomum osmophloeum provenances. *Journal of Agricultural and Food Chemistry*, *52*(14), 4395–4400.
37. Abdellaoui, K., Acheuk, F., Miladi, M., Boughattas, I., & Omri, G. (2018). Phytochemistry, biochemical and insecticidal activities of Ruta chalepensis essential oils on Tribolium confusum. *Poljoprivreda i Sumarstvo*, *64*(3), 31–45.
38. Arshad, Z., Hanif, M. A., Qadri, R. W. K., Khan, M. M., Babarinde, A., Omisore, G. O., ... Latif, S. (2014). Role of essential oils in plant diseases protection: A review. *International Journal of Chemical and Biochemical Sciences*, *6*, 11–17.
39. Shukla, R., Kumar, A., Prasad, C. S., Srivastava, B., & Dubey, N. K. (2009). Efficacy of Acorus calamus L. leaves and rhizome on mortality and reproduction of Callosobruchus chinensis L.(Coleoptera: Bruchidae). *Applied Entomology and Zoology*, *44*(2), 241–247.
40. Shah, S., Hafeez, M., Wu, M. Y., Zhang, S. S., Ilyas, M., Wu, G., & Yang, F. L. (2020). Downregulation of chitin synthase A gene by diallyl trisulfide, an active substance from garlic essential oil, inhibits oviposition and alters the morphology of adult Sitotroga cerealella. *Journal of Pest Science*, *93*, 1097–1106.
41. Thodsare, N. H., Bhatt, P., & Srivastava, R. P. (2014). Bioefficacy of Murraya koenigii oil against Spilosoma obliqua and Spodoptera litura. *Journal of Entomology and Zoology Studies*, *2*(4), 201–105.
42. Abdelgaleil, S. A., Mohamed, M. I., Badawy, M. E., & El-arami, S. A. (2009). Fumigant and contact toxicities of monoterpenes to *Sitophilus oryzae* (L.) and Tribolium castaneum (Herbst) and their inhibitory effects on acetylcholinesterase activity. *Journal of Chemical Ecology*, *35*, 518–525.
43. Xie, Y., Huang, Q., Rao, Y., Hong, L., & Zhang, D. (2019). Efficacy of Origanum vulgare essential oil and carvacrol against the housefly, *Musca domestica* L. (Diptera: Muscidae). *Environmental Science and Pollution Research*, *26*, 23824–23831.
44. Klein, M. G., Tumlinson, J. H., Ladd, T. L., & Doolittle, R. E. (1981). Japanese beetle (Coleoptera: Scarabaeidae) response to synthetic sex attractant plus phenethyl propionate: Eugenol. *Journal of Chemical Ecology*, *7*, 1–7.
45. Hammack, L. (1996). Corn volatiles as attractants for northern and western corn rootworm beetles (Coleoptera: Chrysomelidae: Diabrotica spp.). *Journal of Chemical Ecology*, *22*, 1237–1253.

46. Chermenskaya, T. D., Burov, V. N., Maniar, S. P., Pow, E. M., Roditakis, N., Selytskaya, O. G., . . . Woodcock, C. M. (2001). Behavioural responses of western flower thrips, Frankliniella occidentalis (Pergande), to volatiles from three aromatic plants. *International Journal of Tropical Insect Science*, *21*(1), 67–72.

47. Reitz, S. R., Maiorino, G., Olson, S., Sprenkel, R., Crescenzi, A., & Momol, M. T. (2008). Integrating plant essential oils and kaolin for the sustainable management of thrips and tomato spotted wilt on tomato. *Plant Disease*, *92*(6), 878–886.

48. Landolt, P. J., Adams, T., Reed, H. C., & Zack, R. S. (2001). Trapping alfalfa looper moths (Lepidoptera: Noctuidae) with single and double component floral chemical lures. *Environmental Entomology*, *30*(4), 667–672.

49. Górski, R. (2004). Effectiveness of natural essential oils in the monitoring of greenhouse whitefly (Trialeurodes vaporariorum Westwood). *Folia Hort*, *16*, 183–187.

50. Abbaszadeh, S., Sharifzadeh, A., Shokri, H., Khosravi, A. R., & Abbaszadeh, A. (2014). Antifungal efficacy of thymol, carvacrol, eugenol and menthol as alternative agents to control the growth of food-relevant fungi. *Journal de Mycologie Medicale*, *24*(2), e51–e56.

51. Pradhanang, P. M., Momol, M. T., Olson, S. M., & Jones, J. B. (2003). Effects of plant essential oils on Ralstonia solanacearum population density and bacterial wilt incidence in tomato. *Plant Disease*, *87*(4), 423–427.

52. Isman, M. B. (2005). Problems and opportunities for the commercialization of botanical insecticides. *Biopesticides of Plant Origin*, EDs BJR Philogene, C Regnault-Roger, C. Vincent Lavoisier Cachan France, 283–291.

14 Does RNAi-Based Technology Fit within EU Sustainability Goals?[*]

Clauvis Nji Tizi Taning, Bruno Mezzetti, Gijs Kleter, Guy Smagghe, and Elena Baraldi

CONTENTS

14.1 FUTURE AGRICULTURE: ENSURING SUSTAINABLE FOOD PRODUCTION IN THE EU

The Farm-to-Fork (F2F) strategy,[I] one of the pillars underneath the European Commission's New Green Deal, aims to ensure a more sustainable and food-secure society. Its aims include a reduction in agrochemical inputs, such as pesticides, fertilizers, and antimicrobials, to achieve greater sustainability and health and reduce loss of biodiversity while ensuring continued crop protection. It envisages various practices that promote less pesticide usage, such as integrated pest management (IPM), and the use of precision agriculture and artificial intelligence. It also recommends imposing maximum levels (tolerances) for pesticide residues in imported commodities to enforce sustainable production and pesticide use in countries exporting to the EU. The F2F strategy's pesticide reduction measures are also cited by the Commission's concurrent Biodiversity Strategy 2030[II] as a means of reversing the alarming decline in farmland birds and insects (especially pollinating ones). It also proposes IPM and the establishment of variable landscapes hospitable to natural pest regulators.

DOI: 10.1201/9781003265139-17

Furthermore, it envisages organic farming covering 25% of agricultural land in 2030, and it suggests establishing forests that are resilient to pests and the banning of chemical pesticides from use in urban green spaces.

At the same time, the Commission also published a report on the experience gained with its policy towards more sustainable use of pesticides under the so-called Sustainable Use Directive (SUD) of 2009.[III] Under the SUD, EU member states had to develop and implement national action plans for the reduction of pesticide volumes and risk. The report concluded that one prevalent shortcoming across the board was the lack of assessment of the actual implementation of IPM practices, while there was an upward trend in nonchemical, low-risk, and basic active substances, and it also concluded that the R&D basis for such alternatives should be broadened. Notably, the European pesticide and biopesticide producers' organization, the European Crop Protection Association (ECPA), recently pledged to support these with various commitments focused on innovation, sustainability, and health, including a €14 billion investment in the development of precision agricultural techniques for the more targeted (hence reduced) application of pesticides and of natural biopesticides with favorable IPM characteristics, complementing other pesticides.[IV] Concurrent with the EU's policy towards less risky pesticides, the United Nations Environmental Program (UNEP), other international and national organizations, and companies are proactively pursuing risk reduction for "highly hazardous pesticides."

As already noted in the review of the SUD implementation, R&D for alternatives to chemical pesticides has been lagging behind and needs to be broadened. The new EU policies make the case for innovative, enabling technologies using versatile platforms for the development of agents that are widely applicable to a host of different crop pests and diseases, yielding products with high specificity for the targeted pest or disease and a benign environmental and health profile, requiring a short development time and affordable in cost.

14.2 RNAI-BASED TECHNOLOGY ENABLES PESTICIDE RISK-REDUCTION GOALS

RNAi is a well-known natural biological process in most eukaryotes, in which double-stranded RNA (dsRNA) molecules regulate gene expression by targeting specific endogenous mRNA molecules in a sequence-specific manner (Figure 14.1). By exploiting this sequence-dependent mode of action, RNAi-based products with higher selectivity and better safety profiles (less mobile through the soil, less persistent, and less toxic) than contentious chemical pesticides are being developed [1–4]. RNAi-based control has several unique features that offer additional opportunities compared with contentious chemical pesticides. The dsRNA active molecules can be designed to target the expression of different genes without the need to change the sequence-dependent mode of action, and, depending on the gene targeted in the pest, various outcomes ranging from sublethal to lethal effects can be achieved. Although selecting effective RNAi targets can be a challenging step, a combination of the availability of in-silico tools and an increase in the availability of genome databases for various species have made it possible to design species-selective and efficient dsRNA molecules with zero to negligible off-target effects in non-target

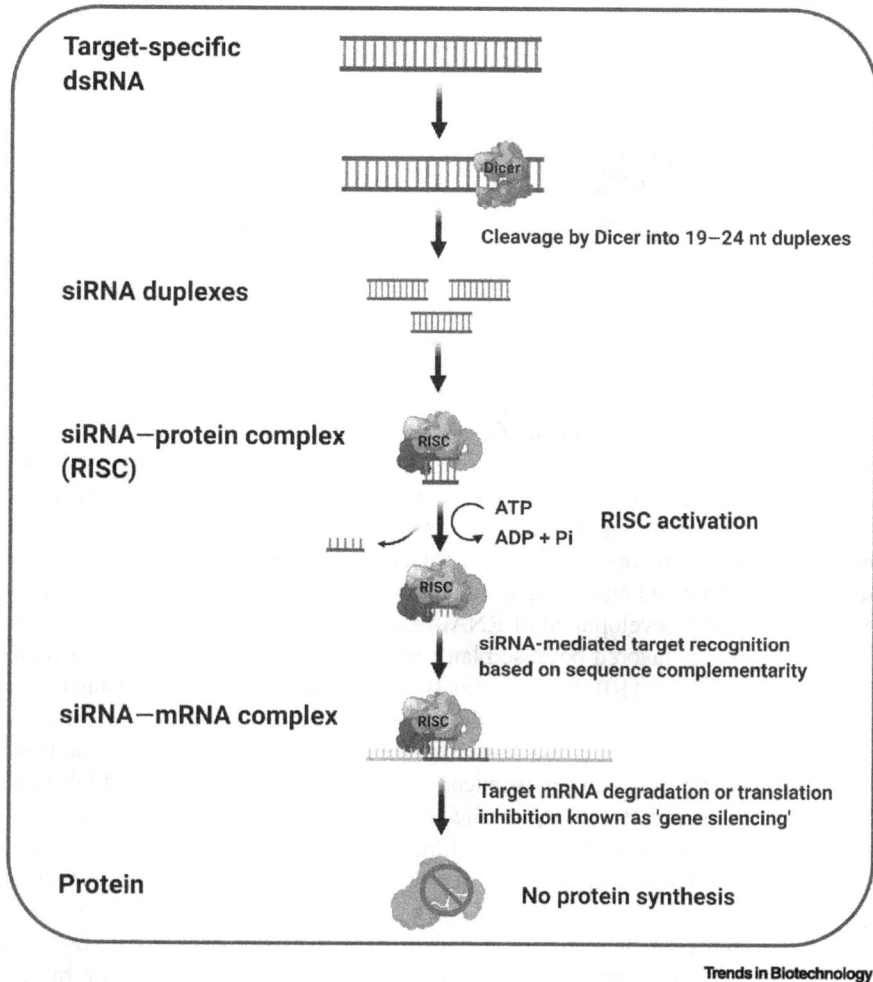

Target-specific dsRNA

Cleavage by Dicer into 19–24 nt duplexes

siRNA duplexes

siRNA–protein complex (RISC)

ATP
ADP + Pi   **RISC activation**

siRNA-mediated target recognition based on sequence complementarity

siRNA–mRNA complex

Target mRNA degradation or translation inhibition known as 'gene silencing'

Protein   **No protein synthesis**

Trends in Biotechnology

FIGURE 14.1   Illustration of RNAi-mediated gene silencing.

species [5, 6]. This presents an advantage over current contentious chemical pesticides with broad action spectra, usually affecting non-target species. Additionally, dsRNA is a natural molecule that is rapidly degraded by nucleases and UV radiation [7, 8], in contrast to some chemical pesticides with longer persistence in the environment.

RNAi-based biocontrol can be applied using two main approaches: in planta delivery through genetically modified (GM) crops or exogenous application of formulated RNAi-based products (Figure 14.2). RNAi-based GM plants differ from most other GM plants expressing new proteins because the introduced gene sequences give rise to the stable expression of small dsRNA molecules for agricultural purposes. This would further promote the cost-effective manufacture of the sizable volumes of dsRNA needed for large-scale pesticide applications. Furthermore, significant improvements

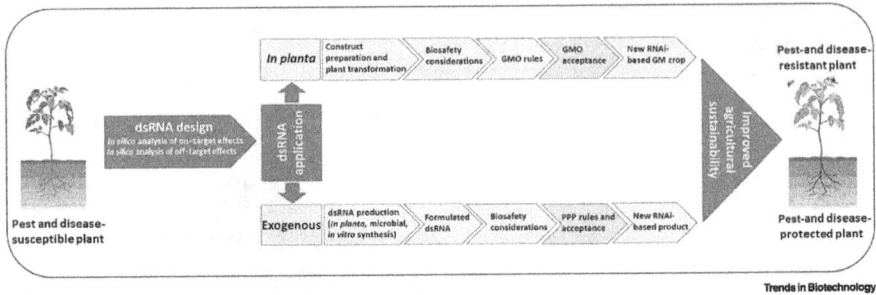

Trends in Biotechnology

FIGURE 14.2   Two main delivery approaches for RNAi-based control of pests and diseases.

are being made in the development of appropriate formulation technologies with very high target specificity as active compounds. Moreover, the expressed short (s)RNAs are highly mobile in the plant vascular system, allowing horticultural crops to acquire resistance when grafted onto transgenic rootstocks [9]. This type of dsRNA application is of particular interest to seed and nursery industries, which are mainly interested in propagating and marketing new resistant high-quality varieties. Nevertheless, the trend toward the development of RNAi-based products for exogenous application against crop pests is favored because plants treated with dsRNAs are not considered GM organisms (GMOs) [10]. These products can also be directly applied using current agricultural practices, such as spray application, trunk injection for tree species, seed soaking, and root drenching through hydroponic systems in greenhouses. The possibility of developing exogenously applicable RNAi-based products with high target pest selectivity and a better safety profile has also stimulated the creation of various start-up companies (albeit, at the moment, mostly non-European) that exploit available biotechnology tools to support the development of these products for field-scale applications.[v]

As a result of the severe acute respiratory syndrome coronavirus 2 (SARS-CoV2) pandemic, the existing production capacity for dsRNA molecules may be ramped up exponentially in the near future to provide the global community with RNA-based vaccines, owing to substantive private investments running into the hundreds of millions of dollars. The same platforms could be converted to dsRNA production (with less stringent quality requirements than for biomedicines) for agricultural purposes. This would further promote the cost-effective manufacture of the sizable volumes of dsRNA needed for large-scale pesticide applications. Furthermore, significant improvements are being made in the development of appropriate formulation technologies for the delivery of dsRNA molecules to target crop pests or pathogens [3]. Formulations are generally designed to either improve dsRNA stability and/or ensure effective delivery of dsRNA to pests or pathogens, thereby improving RNAi efficiency. Developing the right formulation can be challenging and depends on the interaction between plants and pests or pathogens. Consequently, risk assessment for formulated dsRNA should be performed on a case-by-case basis. At a time when the EU strategy is directed towards sustainability with significant reduction in contentious pesticides, the interest of agribusinesses in investing in the development of RNAi-based products with better biosafety profiles [11], coupled with the stimulation of start-up companies to support product development, indicate that RNAi-based pest control can contribute to the EU pesticide reduction goal in the F2F strategy.

14.3 SOCIETAL PERCEPTION AND ACCEPTANCE OF RNAI-BASED BIOCONTROL

Alternatives to pesticides address policy needs towards increased sustainability, an acknowledged field of science [12] and a term widely used in media but still a complex concept for the general public. Consumer perceptions of pesticides have often been studied as risk perception linked with adverse effects [13]. It is important to explicitly include the perception of both risks and benefits to capture the trade-offs that consumers are willing to make to accept new generations of pesticides. Perceived benefits alone might not be the decisive factor in societal acceptance of RNAi-based control; other relevant factors, such as the extent to which the application is perceived to be important or necessary, or unimportant, could shape societal acceptance [14]. Therefore, it is of fundamental importance to develop dialog with stakeholders to ensure that ethical and social concerns are addressed early on in the development process of RNAi-based control strategies. If the attitude of stakeholders towards RNAi-based control remains uncrystallized, a constant reevaluation of what the stakeholders think is advised [15] because such an attitude is unlikely to be fixed but rather influenced by external events, including the order of entry of RNAi-based products into the market. Findings from such evaluations would further shape communication strategies that could drive societal acceptance of RNAi-based control strategies.

14.4 CONCLUDING REMARKS AND FUTURE PERSPECTIVES

The sequence-specific mode of action of RNAi-based products makes them unique in selectivity and efficiency compared with conventional agrochemicals and suggests them as a promising solution to substitute contentious pesticides or reduce reliance in general. However, to enable society to accept RNAi-based products, several key tasks will have to be accomplished. The co-creation of new, effective, and safe RNAi-based products in collaboration with stakeholders under the responsible research and innovation paradigm promoted by the EU will foster greater knowledge and acceptance of technology. Furthermore, an advancement in under-standing consumers' perception will facilitate the successful market introduction of RNAi-based, sustainable products for crop protection.

RNAi is a natural cellular process that is typically initiated by the endogenous production or exogenous introduction of long double-stranded RNAs (dsRNAs) into the eukaryotic cell. These long dsRNA molecules can originate from different sources, such as direct introduction of exogenously produced dsRNA; dsRNA-viral intermediates; or hybridization of complementary RNA transcripts present in the same cell, from single-stranded RNAs (ssRNA) that contain near-complementary or complementary inverted repeats, or that are separated by a short spacer sequence capable of folding back onto itself to form a hairpin structure (hpRNA). Once present in the cell, the dsRNA molecule triggers the RNAi mechanism by recruiting an RNase-III-like enzyme known as Dicer and its cofactors, which leads to the cleavage of the dsRNA molecule into siRNA duplexes (19–24 bp). These siRNAs are then loaded into a protein complex, forming an RNA-induced silencing complex (RISC). The activation of RISC involves the retention of one strand of the siRNA (guide) while the other strand (passenger) is released and degraded in the cell. The RISC is then guided in a sequence-specific manner by the loaded guide stand of the siRNA to target and bind to target mRNAs that are nearly perfectly complementary. The formation of the siRNA-mRNA complex ultimately leads to degradation or translational inhibition of the target mRNA, preventing protein synthesis and leading to post-transcriptional gene silencing.

RNAi-based control typically starts with designing the double-stranded (ds) RNA active molecule to be specific to an essential gene in the pest or pathogen while remaining safe for non-target organisms. The increase in availability of genome/transcriptome databases for more organisms and the development of sophisticated high-throughput approaches, such as RNA-seq and digital gene expression tag profile (DGE-tag) technologies, has significantly improved the selection of potential RNAi targets. Although in-silico tools can help in the design of pest- or pathogen-specific dsRNA molecules, empirical evidence from bioassays is required to support effectiveness against the pest/pathogen and zero to negligible adverse effects in non-target organisms. There are two main dsRNA application approaches for RNAi-based control; one approach involves the in planta expression of RNAi constructs to target genes of pests/pathogens while the second approach involves the exogenous application of dsRNA. The choice of the application method depends on the interaction between the crop and the pest/pathogen and on the feasibility of developing an efficient application method against the pest/pathogen. These delivery approaches follow two separate paths during the product development pipeline, with the in planta approach falling under genetically modified organism (GMO) regulation while the exogenous approach is expected to fall under the plant protection product (PPP) regulation. The genetically modified (GM) approach can be challenged by a lack of technological tools to modify some plant species, high cost of production to registration of the GMO, and some public opposition to GMOs. The exogenous approach is challenged by the high cost of producing dsRNA for large-scale field application and the lack of appropriate formulation technologies to improve RNAi efficiency against recalcitrant pests/pathogens. Despite these challenges, both delivery approaches have successfully led to the development of safe and effective RNAi-based products, which can significantly contribute to agricultural sustainability by reducing the use of contentious pesticides.

14.4.1 Acknowledgments

This article is based on work from COST Action CA 15223 iPLANTA, supported by COST (European Cooperation in Science and Technology), www.cost.eu.

B.M. received funding from the MIUR-PRIN2017 national program via Grant No. 20173LBZM2-Micromolecule; C.N.T.T. and G.S. from the Special Research Fund of Ghent University (BOF) and the Research Foundation-Flanders (FWO); and G.K. from the Netherlands Ministry of Agriculture, Nature and Food Quality and from the International Union for Pure and Applied Chemistry.

14.4.2 Resources

I   https://food.ec.europa.eu/system/files/2020-05/f2f_action-plan_2020_strategy-info_en.pdf

II   https://eur-lex.europa.eu/legal-content/EN/TXT/?uri=CELEX:52011DC0244

III   https://ec.europa.eu/food/plant/pesticides/sustainable_use_pesticides_en

IV   www.ecpa.eu/commitments/2030-commitments/innovation-investment

V   www.agrorna.com/sub_05.html

REFERENCES

1. Wang, M., Weieberg, A., Lin, F.M., Thomma, B.P., Huang, H.D., Jin, H. (2016) Bidirectional cross-kingdom RNAi and fungal uptake of external RNAs confer plant protection. *Nat. Plants* 2(10), 1–10.

2. Koch, A., Biedenkopf, D., Fuch, A., Weber, L., Rossbach, O., Abdellatef, E., et al. (2016) An RNAi-based control of *Fusarium graminearum* infections through spraying of long dsRNAs involves a plant passage and is controlled by the fungal silencing machinery. *PLoS Pathog.* 12(10), e1005901.

3. Mitter, N., Worrall, E.A., Robinson, K.E., Li, P., Jain, R.G., Taochy, C., et al. (2017) Clay nanosheets for topical delivery of RNAi for sustained protection against plant viruses. *Nat. Plants*. 3(2), 1–10.

4. Jin, S., Singh, N.D., Li, L., Zhang, X., Daniell H. (2015) Engineered chloroplast dsRNA silences cytochrome p450 monooxygenase, V-ATPase and chitin synthase genes in the insect gut and disrupts *Helicoverpa armigera* larval development and pupation. *Plant Biotechnol. J.* 13(3), 435–446.

5. Bachman, P.M., Huizinga, K.M., Jensen, P.D., Mueller, G., Tan, J., Uffman, J.P., Levine, S.L. (2016) Ecological risk assessment for DvSnf7 RNA: A plant-incorporated protectant with targeted activity against western corn rootworm. *Regul. Toxicol. Pharmacol.* 81, 77–88.

6. Taning, C.N.T., Gui, S., De Schutter, K., Jahani, M., Castellanos, N.L., Christiaens, O., Smagghe, G. (2021) A sequence complementarity-based approach for evaluating off-target transcript knock-down in *Bombus terrestris*, following ingestion of pest-specific dsRNA. *J. Pest. Sci.* 94(2), 487–503.

7. Bachman, P., Fischer, J., Song, Z., Urbanczyk-Wochniak, E., Watson, G. (2020) Environmental fate and dissipation of applied dsRNA in soil, aquatic systems, and plants. *Front. Plant Sci.* 11, 21.

8. Parker, K.M., Barragán Borrero, V., van Leeuwen, D.M., Lever, M.A., Mateescu, B., Sander, M. (2019) Environmental fate of RNA interference pesticides: Adsorption and degradation of double-stranded RNA molecules in agricultural soils. *Environ. Sci. Technol.* 53(6), 3027–3036.

9. Zhao, D., Zhong, G.Y., Song, G.Q. (2020) Transfer of endogenous small RNAs between branches of scions and rootstocks in grafted sweet cherry trees. *PLoS ONE* 15(7), e0236376.

10. Shew, A.M., Danforth, D.M., Nalley, L.L., Nayga Jr, R.M., Tsiboe, F., Dixon, B.L. (2017) New innovations in agricultural biotech: Consumer acceptance of topical RNAi in rice production. *Food Cont.* 81, 189–195.
11. Mat Jalaluddin, N.S., Othman, R.Y., Harikrishna, J.A. (2019) Global trends in research and commercialization of exogenous and endogenous RNAi technologies for crops. *Crit. Rev. Biotechnol.* 39(1), 67–78.
12. Kates, R.W., Clark, W.C., Corell, R., Hall, J.M., Jaeger, C.C., Lowe, I., Svedin, U., et al. (2001) Sustainability science. *Science* 292(5517), 641–642.
13. Remoundou, K., Brennan, M., Hart, A., Frewer, L.J. (2014) Pesticide risk perceptions, knowledge, and attitudes of operators, workers, and residents: A review of the literature. *Hum. Ecol. Risk. Assess.* 20(4), 1113–1138.
14. Gupta, N., Fischer, A.R.H., Frewer, L.J. (2015) Ethics, risk and benefits associated with different applications of nanotechnology: A comparison of expert and consumer perceptions of drivers of societal acceptance. *NanoEthics* 9(2), 93–108.
15. Fischer, A.R., Van Dijk, H., de Jonge, J., Rowe, G., Frewer, L.J. (2013) Attitudes and attitudinal ambivalence change towards nanotechnology applied to food production. *Public Underst. Sci.* 22(7), 817–831.

15 Insect Growth Regulators

Kavita Munjal, Vinod Kumar Gauttam, Showkat Rasool Mir, Nikita Nain, and Sumeet Gupta

CONTENTS

15.1 INTRODUCTION

Farmers' main concern is their significant loss due to pests and illnesses, which occurs regardless of the production strategy used. Plant infections, insects, and weed pests destroy an estimated 40% of all potential food production annually. Despite using nearly three million tons of pesticide every year, as well as a range of nonchemical controls such as biological controls and crop rotation, this loss occurs. If some of this food could be preserved from pests, it could be used to feed the world's more than three billion starving people (Meenu & Ayushi, 2020). So concerns about pesticides' effects on human health and the environment have led to the hunt for alternative pest control methods. Environmental toxicity issues are largely caused by the use of chemicals for insect population control, which has as one of its goals the enhancement of economic investment (Munjal et al., 2022; Apolinário & Feder, 2021). Neonicotinoids (NEOs), a group of insecticides that are relatively highly selective, harm the environment and have negative effects on human health. The majority of NEOs consumed by humans are eliminated in urine, but certain NEOs, and particularly their metabolites, may circulate in the body and cause potential harm (D. Zhang & Lu, 2022). Metabolomic studies reveal changes in metabolite profiles due to growth and metamorphosis (Tyagi et al., 2010b). One of the contributing factors is undoubtedly the overuse of neonicotinoids and other broad-spectrum insecticides. As a result,

DOI: 10.1201/9781003265139-18

demand for ecofriendly insecticides, ideally ones that are safe for non-target species, has increased (Jindra, 2021).

Carbamates, organophosphates, pyrethroids, and organochlorines are only a few of the traditional insecticides that have been created (Mdeni et al., 2022). In recent decades, they have been utilized to manage insect pests, leading to a reduction in agricultural productivity losses. However, as resistance to pesticides reached crisis proportions, the extreme negative effects of pesticides on the environment and the public outcry grew, and stronger protocols and regulations aimed at reducing their use were enacted. The pest control business is always looking for new technologies and products to improve how pests are managed and prevented (Fang et al., 2019; Nicolopoulou-Stamati et al., 2016). So concerns about pesticides' effects on human health and the environment have led to the hunt for alternative pest control methods.

Insect growth regulators (IGRs) are insecticides that disturb the life cycle of insects by interfering with normal growth and development and result in the insect's mortality before it reaches adulthood. IGRs are a form of "birth control" for pests that helps keep the populations of undesired pests in check by preventing both present and future infestations (Masih & Bhat, 2019). The term *IGR* was introduced by Schneiderman (1972) for substances that are analogues or antagonists of these hormones and interfere with insect development. Since then, the compounds that adversely interfere with the growth and development of insects have been collectively referred to as insect growth regulators, or IGRs (Poppenga & Oehme, 2010). They are a class of biorational chemicals that inhibit insect-specific or mite-specific, physiological processes that do not exist in vertebrates, including humans. They basically work on hormone pathways in insects; they do not have effects on other organisms (Mondal & Parween, 2000). When IGRs first hit the market in the 1980s and 1990s, they gained popularity by being marketed as being safe for humans, animals, and pets (Hodgson, 2010). The LD50 values in rodents, which range from 2 to 10 g/kg, support their safety. IGRs do not harm adult insects or arthropods; only the developmental stages are impacted. Therefore, it takes a few weeks following treatment for them to start working effectively against pests (Hovda & Hooser, 2002). IGRs do not quickly knock down insects or cause mortality like conventional pesticides do, but long-term exposure to these chemicals limits population expansion because of the effects indicated in both the parents and progeny. As a result, many IGRs are combined with adulticidal pesticides such as pyrethrins and pyrethroids or fipronil (Mondal & Parween, 2000).

IGRs typically prevent insect reproduction or regulate metamorphosis in order to manage insects. IGRs work differently from other synthetic pesticides like organophosphates and carbamates in that these chemicals interfere with different physiological processes but do not regulate the development of typical insects. Therefore, an IGR need not necessarily be toxic to its target but may instead cause a variety of defects that reduce insect survival (Tunaz & Uygun, 2004).

IGRs substances seem to meet the criteria for third-generation pesticides. They are environmental friendly and safer grain protectants due to their selectivity of action. IGRs include non-hormonal substances such as precocenes (anti JH) and chitin synthesis inhibitors as well as synthetic analogues of insect hormones like ecdysoids and juvenoids. Commercially produced IGRs are used to manage insect pests in forestry, agriculture, public health, and stored goods. They have an impact on the biology of treated insects, including behaviour, reproduction, and mortality as well as embryonic and post-embryonic development. IGR activity in insects is seen to cause abnormal morphogenesis. Even against the eggs, several of them are more effective than pesticides used at present (Sun & Zhou, 2015).

In this chapter, we discuss types of IGRs with respect to their natural and synthetic analogues, particularly their role in pest management, along with a brief prospect of biological insecticides.

15.2 TYPES OF INSECT GROWTH REGULATORS (IGRS)

Natural hormones of insects which play a role in growth and development are:

Brain hormone: Activation hormone (AH) is another name for brain hormone. AH is secreted by neurons of the central nervous system called neurosecretory cells (NSC). Its function is to stimulate the production of juvenile hormone (JH) by the corpora allata (Oberlander & Silhacek, 2000).

Juvenile hormone (JH): Neotinin is another name for JH, secreted by corpora allata, which are paired glands located beneath the insect brains. They are responsible for maintaining the larva's juvenile state by inhibiting the genes that promote development of adult characteristics, causing the insect to remain a nymph or larva. The amount of JH is less concentrated as the larva matures and enters the pupal stage. JH has a unique terpenoid structure and is the methyl ester of epoxy farnesoic acid. There are at least six distinct variants of this sesquiterpene. JH I, JH II, JH III, and JH IV have been found in several insect species. The development of the ovary in adult female insects depends on JH I, II, and IV, which are found in larva, whereas JH III is detected in adults. Most insects have JH III, which is the most prevalent form, while JH I and II are the most prevalent in Lepidoptera (STAAL, 1982).

Ecdysone: Moulting hormone is another name for ecdysone. Insects wear their skeletons, known as exoskeletons, on the outside. A new exoskeleton must develop inside the old one as the insect grows, and the old one must be discarded. The new one then grows in size and becomes harder; the process is known as moulting. This moulting also involves the transition from a larval to an adult form, a process known as metamorphosis. Moulting happens in insects only in the presence of ecdysone. The level of ecdysone is usually lower and eventually absent in adult insects. A pair of endocrine glands that are found in the prothorax of Lepidoptera and other insects (prothoracic glands) or in the ventroposterior region of the head release the insect moulting hormone 20-hydroxyecdysone (20E) as ecdysone (ventral glands). Ecdysone is transformed into 20E, which interacts with amino acid residues in the ligand-binding domain of the ecdysone receptor (EcR) protein to produce moult-inducing actions (Rajendran & Singh, 2016; Tunaz & Uygun, 2004).

15.3 SYNTHETIC ANALOGUES OF INSECT GROWTH REGULATORS

IGRs include synthetic analogues of insect hormones such as ecdysoids and juvenoids and non-hormonal compounds such as precocenes (anti JH) and chitin synthesis inhibitors.

15.3.1 ECDYSOIDS

These substances are synthetic versions of ecdysone. When given to insects, they cause a damaged cuticle to develop, killing the insects. Scales or a wax layer are absent from the integument as a result of the growth processes being expedited and bypassing many typical events (Song et al., 2017).

15.3.2 JUVENOIDS OR JH MIMICS

These are juvenile hormone analogues made in a lab. As hormonal insecticides, they are most promising. In 1966, Williams and Slama made the first discovery of JH mimics. They discovered that the *Pyrrhocoris* bug died before reaching adulthood because of the paper towel that was stored in the glass jar used for its raising. The element from the paper was referred to as "paper factor" or "juvabione." They discovered that the paper was made from balsam tree (*Abies balsamea*) wood pulp, which contained the JH mimic (Azmi, 2021).

Juvenoids have an anti-metamorphic effect on insect larval stages. They maintain the situation as it is in insects (larva remains larva), and additional (supernumerary) moultings occur, generating super larva, larval-pupal, and pupal-adult intermediates that result in the death of the insects. Juvenoids interrupt insect diapause, are larvicidal and ovicidal in action, and prevent insect development (Sláma & Williams, 1966; Yang et al., 2022).

15.3.3 ANTI JH OR PRECOCENES

Precocenes are also considered an anti JH. They exert their influence by obstructing JH synthesis, which results in killing the corpora allata. When insects are treated at an early stage, they skip one or two larval instars and develop into tiny, immature adults. They will quickly expire because they are unable to reproduce or lay eggs.

15.3.4 CHITIN SYNTHESIS INHIBITORS

Chitin is an important component of the cuticle, an insect's outermost covering and an exoskeleton that shields it from physical harm, microbial infection, and dehydration. Insects continually synthesize and destroy chitin in a highly regulated manner so as to enable ecdysis and the renewal of the peritrophic matrices (Doucet & Retnakaran, 2012). It has been discovered that benzoyl phenyl ureas can prevent chitin synthesis from occurring in living organisms by suppressing the action of the chitin synthetase enzyme. Diflubenzuron (Dimilin) and penfluron are two significant substances in this group. Moulting is disrupted, mandibles and labrums are moved, the adult is unable to break free of the skin of the pupa and dies, and there is an ovicidal effect. Chitin synthesis inhibitors have been successfully utilized against pests of soybeans, cotton, apples, fruits, vegetables, forest trees, mosquitoes, and stored grain in many countries where they have been approved for use. The structures of JHAs, ecdysone agonists, and CSIs are given in Figures 15.1A–15.1D.

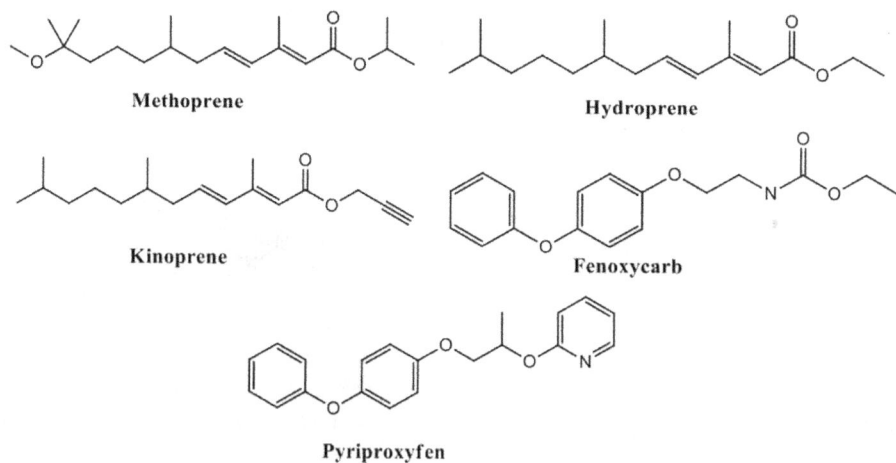

FIGURE 15.1A   Structures of juvenile hormone analogues.

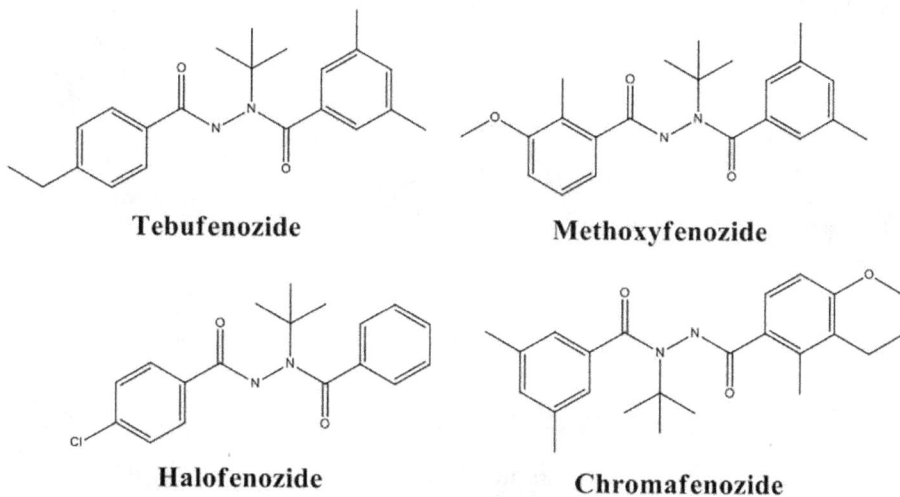

FIGURE 15.1B   Structures of ecdysone agonists.

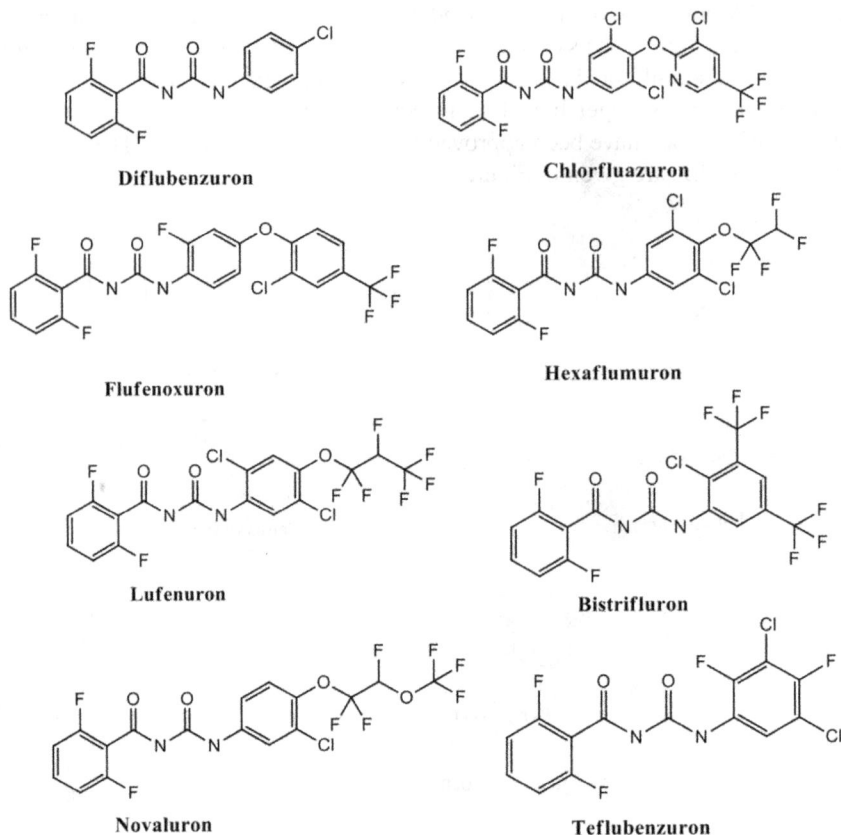

FIGURE 15.1C   Structures of chitin synthesis inhibitors (BPU derivatives).

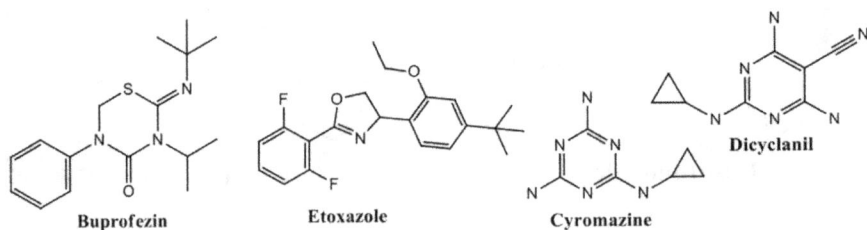

FIGURE 15.1D   Structures of chitin synthesis inhibitors (Non-BPU derivatives).

15.4 EFFECTS OF INSECT GROWTH REGULATORS (IGRS)

IGRs act through control of both hormonal and non-hormonal activities. IGRs control the stages of moulting by acting on the hormones required by the epidermis, which is a portion of the exoskeleton. Moulting occurs in insects to accommodate their growth under the influence of the steroidal insect moulting hormone 20-hydroxyecdysone

(20-E) and regulated by the sesquiterpenoid JH. These hormones change their roles to regulating reproductive processes in the adult stages of the insects. The exoskeleton becomes too small for the growing body tissues of an immature insect, and it must moult its cuticle to accommodate the growth of the internal organs and tissues. Stretch receptors in the insect body are stimulated by the increasing growth of body tissues, and when it attains a critical size, the brain secretes PTTH into the circulating hemolymph. PTTH binds with specific receptor proteins on prothoracic glands (PGLs). Multiple biochemical reactions are initiated that result in synthesis of ecdysone by PGL (Zitnanová et al., 2001). Ecdysone is converted to 20-E by epidermal cells. Epidermal cells enlarge and change shape to become columnar. They begin to divide by mitosis, and with more numerous cells per unit area, there is an increase in surface tension, leading to the separation of epidermal cells from the old cuticle (Ewer, 2005). Apolysis, the separation of epidermal cells from old cuticle, marks the beginning of a moult and of a new instar. The ecdysial space, a minute space created by the separation of epidermal cells from the old cuticle, is filled with moulting fluid containing inactive chitinolytic enzymes, proteinases for digesting the cross-linking proteins and chitin of the old endocuticle. The fall in the 20E titer triggers the activation of enzymes in the moulting fluid for digestion of the procuticle underlying the old cuticle. The preparation for ecdysis starts with the resorption of the moulting fluid. Finally, the clearance of 20E titer completely from the system triggers the release of eclosion hormone, leading to the ecdysis of the larva leaving behind the remnants of the old cuticle (J. Zhang et al., 2014).

15.5 USE OF INSECT GROWTH REGULATORS (IGRS) IN PEST MANAGEMENT

As explained earlier, when compared to first- and second-generation insecticides, which have significant non-target effects on human and animal health as well as the health of other elements of the ecosystem, IGR were deemed much safer for humans, animals, and the environment.

Juvenile hormone analogues (JHAs) can be roughly divided into two groups: terpenoid chemicals like methoprene, hydroprene, and kinoprene and phenoxy JHAs like fenoxycarb and pyriproxyfen. Fenoxycarb and pyriproxyfen have been found to be more photostable with broad-spectrum action compared to methoprene, hydroprene, and kinoprene. The structures of these synthetic IGRs are analyzed using various spectroscopic techniques such as infrared spectroscopy, mass spectroscopy, and nuclear magnetic resonance (Alkefai et al., 2022). JHAs have been proven to be particularly efficient against adult stages of dipteran and homopteran insect pests, but not against lepidopteran insect pests (Kotaki et al., 2009). Methoprene treatment of water near dwellings is recommended by the World Health Organization for mosquito control (Parthasarathy & Palli, 2021). Hydroprene has also been shown to effectively control a variety of pests found in stored products. Hydroprene and methoprene have both been used successfully to control a variety of urban pests. Both have been shown to cause larval mortality and to interfere with egg production and hatching in fleas, as well as genitalia and ootheca deformities and sterility in German and oriental cockroaches (Sierras & Schal, 2020). Hydroprene (Gentrol) has larvicidal activity (Aboelhadid et al., 2018). Kinoprene is very effective against *Aphis gossypii* nymphs (Homoptera: Aphididae), but it does not provide complete control due to its lack of persistence (Hamaidia & Soltani, 2014). Kinoprene (Enstar) acts on aphids, caterpillars, leaf miners, mealybugs, mites, and thripes (Berwal et al., 2021; Hickin et al., 2022; Rebek & Cloyd, 2015). Fenoxycarb was found to impair both growth and lipid content in the mud crab *Rhithropanopeus harrisii*, as well as to reduce fertility in the hymenoptera *Aphytis melinus* (Arambourou et al., 2018). Finoxycarb (Insegar) can interfere with reproduction. Pyriproxyfen is effective for mosquito larval control and adult sterilization, which may lead to increased use of this chemical for vector control (Parthasarathy & Palli, 2021). Pyriproxyfen (Admiral 10EC, NyGuard 10%) is available to control mosquitoes and other pests in urban areas, on pets, and in greenhouses (Su et al., 2019).

The next synthetic analogues, chitin synthesis inhibitors (CSIs), prevent the polymerization of chitin. CSIs lead to aberrant endocuticle deposits that build up during moulting, notably uridine diphospho-N-acetylglucosamine monomers, which stop the formation of chitin. When the pro-cuticle is subjected to the pressures of ecdysis and cuticular expansion, this result in a weaker cuticle and causes mortality. It has been discovered that benzoyl phenyl ureas (BPU) can prevent chitin synthesis from occurring in living organisms by suppressing the action of the chitin synthetase enzyme. Since benzoyl phenyl urea has a distinct mode of action, is very active against the pests it is intended to control, and has low toxicity against non-target organisms, it has garnered significant interest and was developed into a new tool for integrated pest management. Diflubenzuron (Dimilin) and penfluron are two significant substances in this group. They have also been shown to reduce egg fertility and hatching (Harðardóttir et al., 2019). In Norway, diflubenzuron is used in salmon farming. Despite many years of use, no drug resistance has been reported in salmon. Diflubenzuron (Dimilin) is toxic to zooplankton and other aquatic invertebrates. Because the mosquito *Aedes aegyptis*, the vector that causes dengue fever, has developed resistance to other pesticides, Dimilin has been used to control it (Marcon et al., 2016). Genomic and proteomic approaches help in understanding the mechanisms of such insecticidal effects of benzyl urea compounds (Merzendorfer et al., 2012; S. Tyagi et al., 2010a, 2010b). Another example, Chlorfluazuron (Atabron) is extremely effective against a variety of insects that attack stored products as well as other pests that have developed resistance to organic insecticides (El-Monairy et al., 2012). The activity of flufenoxuron (Cascade) is the result of its inhibition of chitin synthesis against *Helicoverpa armigera* at various stages (Woo & Lim, 2015). In addition, sub-lethal and lethal doses of hexaflumuron caused mortality and abnormality in larvae, prepupae, and pupae, as well as abnormality in adults (Taleh et al., 2015). Lufenuron (Match 5% EC) inhibits chitin synthesis in immature ticks (larvae and nymphs) It reduces eggs and larvae of wild *Drosophila suzukii* infesting rabbit eye blueberries, *Vaccinium virgatum*. Furthermore, lufenuron inhibits embryonic development in *Spodoptera frugiperda* and *Anthonomus grandis* eggs (Aboelhadid et al., 2018). Match (lufenuron) affects the synthesis of the peritrophic matrix, leading to histopathological and histochemical alterations in the midgut epithelium (Costa et al., 2017). Bistrifluron (Xterm), a benzoylphenyl urea compound, was evaluated for effectiveness against *Coptotermes lacteus* (frogatt), a mound-building species found in southern Australia (Webb, 2017). Novaluron precisely controls the most destructive red palm weevil larval instars (Hussain et al., 2019). Novaluron-containing products include Pedestal (10%) for controlling insect pests on ornamentals in greenhouses, shadehouses, and nurseries and Mosquiron (0.12%) for mosquito control (Su et al., 2019). In the United States, noviflumuron, 0.5% AI on a laminated cellulose matrix (Recruit III) was approved for termite control in 2004 (Eger et al., 2012). Teflubenzuron is available under trade name Nomolt SC (Yoon et al., 2012). Triflumuron (Alsystin) is the least toxic pesticide to earthworms, causing the greatest reduction in worm growth rate. This could be attributed to the pesticide's high persistence in soil or to the slow degradation in the worms, resulting in less elimination of the metabolites (Badawy et al., 2013). Fluazuron (Acatak) has been widely used to control fleas in dogs and cats, but its effectiveness as an acaricide is limited to a few species (Calligaris et al., 2013). Flucycloxuron is available under the trade name Andalin (Seal et al., 2013).

A number of chitin ysnthesis inhibitors (CSIs) which are not related to the BPU category of compounds, such as buprofezin, etoxazole, and dicyclanil, have also been developed. Buprofezin is a thiadiazine insecticide with very low environmental and human risks. It is particularly effective against sucking pests such as planthoppers (Ali et al., 2017). Buprofezin (commercialized with the trade name Applaud) is attributed to pesticide persistence in soil or to slow degradation in worms, resulting in less elimination of metabolites. It could also indicate that the earthworm is regulating pesticide intake by reducing consumption, thereby affecting growth. This latter strategy has been shown to be commonly used to avoid heavy metal and pesticide poisoning (Badawy et al., 2013). Etoxazole (commercialized with the trade name TetraSan 5 WDG) is a new acaricide that belongs to the diphenyloxazole class of miticides/ovicides and acts by causing adults to lay sterile eggs. It is recommended for controlling tea and brinjal mites (Karmakar & Patra, 2013; Li et al., 2014). Etoxazole is available under trade name TetraSan 5 WDG (Bruinsma et al., 2021). Cyromazine (Larvadex, Neporex) is commonly used to control immature houseflies on poultry farms. Cyromazine is formulated as a pre-mix (1%), which is added to poultry food; it is also formulated as a water-soluble granule and a soluble powder (50%) for topical application to manure containing fly larvae (Crespo et al., 2002). Dicyclanil (CLiKZiN) is highly effective at protecting ewes against early season strike challenge (Walters & Wall, 2012).

Ecdysone regulates the growth of developing adult organs and has an impact on developmental time (Nogueira Alves et al., 2022). Tebufenozide is thought to have limited direct non-target effects because the active ingredient mimics the insect moulting hormone 20-hydroxyecdysone, causing lepodopteran larvae to moult prematurely and incompletely (Edge et al., 2022). Tebufenozide is a bisacylhydrazine ecdysteroid agonist that mimics the target insect's natural ecdysone (20-hydroxyecdysone). Tebufenozide (Mimic) binds to the ecdysone receptor in the gut of larvae, triggering the moulting process. Unlike the natural ecdysone-induced initiation, this moulting process is not completed and is ultimately fatal to the affected individual. The high specificity against lepidopteran targets, while exhibiting low toxicity to non-targets from other orders, distinguishes these agonists from traditional insecticides such as organophosphates. Tebufenozide is toxic to a variety of lepidopteran pests (Roscoe et al., 2020). The most recent commercially developed compound in this class is methoxyfenozide. Methoxyfenozide (Runner 24 SC) is the most effective analogue against lepidopteran larvae, including *S. littoralis* and *S. exigua* (Hübner) (Lepidoptera: Noctuidae), as well as dipteran pests like *Culex pipiens* and *Musca domestica*. Furthermore, methoxyfenozide is a non-toxic compound that is safe for mammals, birds, and fish (Ahmed et al., 2022). Halofenozide (Mach2) is used selectively in turf to control the Japanese beetle, *Popillia japonica* Newman, and the oriental beetle, *Exomala orientalis* (Waterhouse), with no adverse effects on ground beetles (Joseph & Braman, 2016). Chromafenozide (MATRIC FL, MATRIC DL) has toxic effects on various lepidopteran pest larvae, primarily through digestion, and thus has excellent efficacy in crop protection from serious lepidopteran pest damage to vegetables, fruit trees, tea, rice, ornamental plants, and other crops (Yanagi et al., 2006). Table 15.1 lists trade names for different types of IGRs available on the market.

TABLE 15.1
Commercially Available Insect Growth Regulators (IGRs)

Sr. No.	IGR Compound	Trade Name(s)	References
		JH Analogues	
1.	Methoprene	Altosid	(Harbison et al., 2018)
2.	Hydroprene	Gentrol	(Aboelhadid et al., 2018)
3.	Kinoprene	Enstar AQ	(Rebek & Cloyd, 2015)
4.	Fenoxycarb	Insegar 25 WG	(Mahmoudvand & Moharramipour, 2015)
5.	Pyriproxyfen	NyGuard, Admiral 10 EC	(Su et al., 2019)
		Chitin Synthesis Inhibitors	
		BPU Derivatives	
6.	Diflubenzuron	Dimilin	(Marcon et al., 2016)
7.	Chlorfluazuron	Atabron	(El-Monairy et al., 2012)
8.	Flufenoxuron	Cascade	(Woo & Lim, 2015)
9.	Hexaflumuron	Consult 10 EC	(Taleh et al., 2015)
10.	Lufenuron	Match®	(Costa et al., 2017)
11.	Bistrifluron	Xterm	(Webb, 2017)
12.	Novaluron	Pedestal, Mosquiron	(Su et al., 2019)
13.	Noviflumuron	Recruit III	(Eger et al., 2012)
14.	Teflubenzuron	Nomolt SC	(Yoon et al., 2012)
15.	Triflumuron	Alsystin	(Badawy et al., 2013)
16.	Fluazuron	Acatak	(Calligaris et al., 2013)
17.	Flucycloxuron	Andalin	(Seal et al., 2013)
		Non-BP Derivatives	
18.	Buprofezin	Applaud	(Badawy et al., 2013)
19.	Etoxazole	TetraSan 5 WDG	(Bruinsma et al., 2021)

Sr. No.	IGR Compound	Trade Name(s)	References
20.	Cyromazine	Larvadex, Neporex	(Crespo et al., 2002)
21.	Dicyclanil	CLiKZin	(Walters & Wall, 2012)
		Ecdysone Agonists	
22.	Tebufenozide	Mimic	(Roscoe et al., 2020)
23.	Methoxyfenozide	Runner 24% SC	(Ahmed et al., 2022)
24.	Halofenozide	Mach 2	(Joseph & Braman, 2016)
25.	Chromafenozide	Matric Fl, Matric Dl	(Yanagi et al., 2006)

There are many products on the market as a combination of two or three IGRs. Table 15.2 lists such commercially available IGRs. The efficacy of formulation (Certifect) is comparable to that of *Ctenocephalides canis* strains treated with fipronil alone and the fipronil/(S)-methoprene combination (Bouhsira et al., 2011). Another product tested, Sergeant's Gold, has a therapeutic efficacy of 83.1% against *Haemaphysalis elliptica*. This pattern of efficacy makes it a suitable parasiticide for use at the start of the tick season to prevent a rapid seasonal increase in *H. elliptica* numbers (Fourie et al., 2010). Another combination, Inesfly IGR FITO, is an insecticide paint that contains 3.0% chlorpyrifos and 0.063% pyriproxyfen, a formulation with microencapsulation (Llácer et al., 2010). Insecticide paint, Inesfly 5A IGR, produces longer and higher mortality rates than other conventional products (Mosqueira et al., 2010). Further, in the bioassay with *Rhipicephalus sanguineus*, white and dark fecal spots, products of metabolic function, indicated that novaluron and pyriproxyfen (Tekko Pro) were affecting larval and nymphal metabolism regardless of the concentrations (Showler et al., 2019). Another important commercially available IGR is Lufox EC, comprising 7.5% fenoxycarb and 3% lufenuron (Kavallieratos et al., 2012). Plethora (Novaluron 5.25% + Indoxacarb 4.5% SC) and other insecticides are used to control *Helicoverpa armigera* and *Spodoptera litura* infesting tomatoes. Plethora at 875 ml/ha showed only 3.75% fruit damage, whereas the control plot showed 45.6% (Ghosal et al., 2016).

15.6 BOTANICAL INSECTICIDES

Currently, botanical insecticides are gradually replacing synthetic pesticides due to environmental and health concerns. Essential oils (EOs) and their primary chemical components are substitutes for pest management (Haye et al., 2022). Secondary metabolites from essential oil plants have been studied as sources of potentially valuable bioactive compounds because of their high organoleptic qualities and low molecular weight, and it is well known that they play a significant part in the interactions between insects and plants (Magierowicz et al., 2019). Monoterpenoids isolated from essential oils can induce several types of biological activities in pests. They are a class of natural products that are generally safe. Monoterpenoids are ten carbon hydrocarbons or closely related compounds with functional groups such as hydroxyl, carbonyl, and carboxylic that are relatively simple to modify into different

TABLE 15.2
IGR Combination Products

Sr. No	Combination Products	Mixture	Target Pest
	Juvenile Hormone Analogues		
1.	(S)-Methoprene+ Fipronil+Amitraz	Certifect (Merial Limited, GA, USA)	Ticks (Bouhsira et al., 2011)
2.	Pyriproxyfen+Cyphenothrin	Sergeant's Gold (Sergeant's Pet Care Products, Omaha, NE, USA)	African yellow dog tick, *Haemaphysalis elliptica* (Acari: Ixodidae), and cat flea, *Ctenocephalides felis* (Fourie et al., 2010)
3.	Pyriproxyfen+chlorpyriphos	Inesfly IGR FITO (Industrias Químicas Inesba S.L., Paiporta, Spain)	Rhynchophorus ferrugineus (Llácer et al., 2010)
4.	Pyriproxyfen+Chlorpyrifos and Diazinon	Inesfly 5A IGR	Pyrethroid resistant populations of *Anopheles gambiae* and *Culex quinquefasciatus* (Mosqueira et al., 2010)
5.	Pyriproxyfen 5% EC+Fenpropathrin 15% EC	Sumiprempt	*Spodoptera frugiperda*
6.	Novaluron+Pyriproxyfen	Tekko Pro	*Amblyomma americanum* (Acari: Ixodidae), *Rhipicephalus* (*Boophilus*) *annulatus*, *Rhipicephalus* (*Boophilus*) *microplus*, and *Rhipicephalus sanguineus* (Showler et al., 2019)
	Chitin Synthesis InhibitorsC		
1.	Lufenuron+Fenoxycarb	Lufox EC	*Prostephanus truncatus and Rhyzopertha dominica* (Kavallieratos et al., 2012)
2.	Novaluron+Indoxacarb	Plethora	Tomato: fruit borer and leaf- eating caterpillar (Ghosal et al., 2016)
3.	Buprofezin+Deltamethrin	Dadeci	Rice planthopper (Kumar et al., 2020)

derivatives. They are widely distributed in plants, marine algae, and insects and are naturally occurring. More than 1,000 monoterpenoids have been discovered so far, and many of them exhibit intriguing chemical and ecological characteristics (Munjal et al., 2020; Reis et al., 2016). They tend to be more lipophillic substances and, hence, could affect an insect's ability to perform its core metabolic, biochemical, physiological, and behavioural processes (Malik et al., 2015). It has been discovered that monoterpenes extracted from essential oils can bind to ionotropic GABA receptors in insects, rodents, and other mammals (García et al., 2006).

In this category, carvacrol (2-methyl-5-(1-methylethyl) phenol) is obtained from the essential oils of thyme (*Thymus vulgaris*), marjoram (*Origanum majorana*), and oregano (*Origanum vulgari*). It is a phenolic compound with antimicrobial, antifungal, and insecticidal activities (Natal et al., 2020; Vinod et al., 2018). A number of carvacrol ethers and esters have been created and tested for their ability to inhibit bacterial growth and development. Some of these compounds show promise for future development as antifungal and antibacterial drugs after modifications. However, there have been reports of synthetic monoterpenoids as potential novel insecticides. Recently, the naturally occurring carvacrol has undergone structural modifications to become thiadiazole and oxadiazole moieties with the purpose of creating novel IGRs: namely, two series of carvacrol analogues with 1,3,4-thiadiazole and 1,3,4-oxadiazole derivatives (Bagul et al., 2018).

Apart from monoterpenoids, azadirachtin, a tetranortriterpenoid derived from the seed of the Indian neem tree *Azadirachta indica* is one of the prominent biopesticides commercialized and remains the most successful botanical pesticide in agricultural use worldwide. Azadirachtin is a potent antifeedant and insect development disruptor with exceptionally low residual power and little toxicity to predators, parasitoids, and biocontrol agents (Subrahmanyam, 1990; Tulashie et al., 2021).

Azadirachtin has demonstrated exceptional selectivity and low mammalian toxicity. It has an LD_{50} value of more than 5,000 mg/kg, placing it in class U (unlikely to provide an acute hazard) of the WHO (https://www.who.int/publications/i/item/9789240005662) toxicity assessment. The Environmental Protection Agency (EPA) has classified azadirachtin as a general-use pesticide in toxicity class IV (relatively non-toxic). It has also been said to be less dangerous for beneficial and non-target organisms because it appears to be selective, non-mutagenic, and easily degradable (Dai et al., 2019).

Juvenile hormone (JH) and 20-hydroxyecdysone (20E), as mentioned earlier, are crucial for controlling growth and development in insects. Azadirachtin is well known for being an antagonist of these two key hormones. Its main action is its capacity to alter or suppress hemolymph ecdysteroid and JH titers by inhibiting the release of morphogenetic peptide hormone (PTTH) and allatotropins from the corpus cardiacum complex. This is the main cause of its well-documented IGR effects, which are primarily reduced pupation, malformation, and failure of adults (Bezzar-Bendjazia et al., 2017). Moreover, azidirachtin also modifies or stops the transcription and/or expression of a number of proteins at the molecular level. In one study, it suppressed the expression of ferritin and thioredoxin peroxidase genes in the sweet potato whitefly *Bemissia tabaci* related oxidative stress defence mechanisms (Asaduzzaman et al., 2016). In various insect orders, azadirachtin was found to have detrimental effects on reproduction. Many insects, including *Spodoptera littoralis*, *Galleria mellonella*, *Dysdercus cingulatus*, *Tuta absoluta*, and *Helicoverpa armigera*, have shown reduced fecundity and fertility. This reduction may be caused by azadirachtin's interference with the synthesis of yolk proteins or by the substance's uptake into oocytes, causing mating disruption (Ahmad et al., 2015; Oulhaci et al., 2018).

15.7 CONCLUSION

Pests are competing with vertebrae for their food, and that results in loss of production of food that could feed around three billion starving people all over the world. The conventional pesticides that are in use for pest management are not sufficient and also cause environmental toxicity and harm to natural flora that can result in compromised fertility of the soil. The use of IGRs is an effective and less harmful approach to pest management. Both natural and synthetic IGRs are available on the market that affect the developmental stages of pests and do not cause direct harm to the pest, which reduces the chances of the development of pest resistance. Furthermore, biopesticides are also gaining importance as a pest management tool; when combined with IGRs, the results can be more favourable. Though IGRs and natural biopesticides are safe, efficacious and, environmental friendly, conventional chemicals are not replicable as a popular tool for pest management.

REFERENCES

Aboelhadid, S. M., Arafa, W. M., Wahba, A. A., Mahrous, L. N., Ibrahium, S. M., & Holman, P. J. (2018). Effect of high concentrations of lufenuron, pyriproxyfen and hydroprene on Rhipicephalus (*Boophilus*) annulatus. *Veterinary Parasitology, 256*, 35–42. https://doi. org/10.1016/j.vetpar.2018.05.005

Ahmad, S., Ansari, M. S., & Muslim, M. (2015). Toxic effects of neem based insecticides on the fitness of Helicoverpa armigera (Hübner). *Crop Protection, 68*, 72–78. https://doi. org/10.1016/j.cropro.2014.11.003

Ahmed, F. S., Helmy, Y. S., & Helmy, W. S. (2022). Toxicity and biochemical impact of methoxyfenozide/spinetoram mixture on susceptible and methoxyfenozide-selected strains of Spodoptera littoralis (Lepidoptera: Noctuidae). *Scientific Reports, 12*(1), 6974. https://doi.org/10.1038/s41598-022-10812-w

Ali, E., Liao, X., Yang, P., Mao, K., Zhang, X., Shakeel, M., . . . Li, J. (2017). Sublethal effects of buprofezin on development and reproduction in the white-backed planthopper, Sogatella furcifera (Hemiptera: Delphacidae). *Scientific Reports, 7*(1), 16913. https://doi. org/10.1038/s41598-017-17190-8

Alkefai, N., Mir, S. R., Amin, S., Ahamad, J., Munjal, K., & Gauttam, V. (2022). *NMR in analysis of food toxins* (pp. 131–139). Boca Raton, FL: CRC Press.

Apolinário, R., & Feder, D. (2021). Existing potentials in Insect Growth Regulators (IGR) for crop pest control. *Research, Society and Development, 10*(1), e35910111726. https://doi. org/10.33448/rsd-v10i1.11726

Arambourou, H., Fuertes, I., Vulliet, E., Daniele, G., Noury, P., Delorme, N., . . . Barata, C. (2018). Fenoxycarb exposure disrupted the reproductive success of the amphipod Gammarus fossarum with limited effects on the lipid profile. *PLoS One, 13*(4), e0196461. doi:10.1371/journal.pone.0196461

Asaduzzaman, M., Shim, J.-K., Lee, S., & Lee, K.-Y. (2016). Azadirachtin ingestion is lethal and inhibits expression of ferritin and thioredoxin peroxidase genes of the sweet potato whitefly Bemisia tabaci. *Journal of Asia-Pacific Entomology, 19*(1), 1–4. doi:https://doi. org/10.1016/j.aspen.2015.10.011

Azmi, N. A. (2021). Juvenile hormone: Production, regulation, current application in vector control and its future applications. *Tropical Biomedicine, 38*, 254–264. doi:10.47665/tb.38.3.066

Badawy, M. E. I., Kenawy, A., & El-Aswad, A. F. (2013). Toxicity Assessment of Buprofezin, Lufenuron, and Triflumuron to the Earthworm *Aporrectodea caliginosa*. *International Journal of Zoology, 2013*, 174523. doi:10.1155/2013/174523

Bagul, S. D., Rajput, J. D., Srivastava, C., & Bendre, R. S. (2018). Insect growth regulatory activity of carvacrol-based 1,3,4-thiadiazoles and 1,3,4-oxadiazoles. *Molecular Diversity*, *22*(3), 647–655. doi:10.1007/s11030-018-9823-6

Berwal, R., Munjal, K., Sharma, S., Sharma, S., Deepak, K., Choudhary, A., & Kumar, A. (2021). Comparison of serum 25-hydroxyvitamin D levels after a single oral dose of vitamin D3 formulations in mild vitamin D3 deficiency. *Journal of Pharmacology and Pharmacotherapeutics*, *12*, 163–167.

Bezzar-Bendjazia, R., Kilani-Morakchi, S., Maroua, F., & Aribi, N. (2017). Azadirachtin induced larval avoidance and antifeeding by disruption of food intake and digestive enzymes in Drosophila melanogaster (Diptera: Drosophilidae). *Pesticide Biochemistry and Physiology*, *143*, 135–140. https://doi.org/10.1016/j.pestbp.2017.08.006

Bouhsira, E., Yoon, S. S., Roques, M., Manavella, C., Vermot, S., Cramer, L. G., . . . Franc, M. (2011). Efficacy of fipronil, amitraz and (S)-methoprene combination spot-on for dogs against adult dog fleas (*Ctenocephalides canis*, Curtis, 1826). *Veterinary Parasitology*, *179*(4), 351–353. https://doi.org/10.1016/j.vetpar.2011.03.048

Bruinsma, K., Salehipourshirazi, G., Zhurov, V., Dagher, F., Grbic, M., & Grbic, V. (2021). Effect of neo-boost pesticide on mortality and development of different life stages of two-spotted spider mite, tetranychus urticae. *Frontiers in Agronomy*, *3*. https://doi.org/10.3389/fagro.2021.701974

Calligaris, I. B., De Oliveira, P. R., Roma, G. C., Bechara, G. H., & Camargo-Mathias, M. I. (2013). Action of the insect growth regulator fluazuron, the active ingredient of the acaricide Acatak®, in Rhipicephalus sanguineus nymphs (Latreille, 1806) (Acari: Ixodidae). *Microscopy Research and Technique*, *76*(11), 1177–1185. https://doi.org/10.1002/jemt.22282

Costa, H. N., da Cunha, F. M., Cruz, G. S., D'Assunção, C. G., Rolim, G. G., Barros, M. E. G., . . . Teixeira, V. W. (2017). Lufenuron impact upon Anthonomus grandis Boheman (Coleoptera: Curculionidae) midgut and its reflection in gametogenesis. *Pesticide Biochemistry and Physiology*, *137*, 71–80. https://doi.org/10.1016/j.pestbp.2016.10.002

Crespo, D., Lecuona, R., & Hogsette, J. (2002). Strategies for controlling house fly populations resistant to cyromazine. *Neotropical Entomology*, *31*. https://doi.org/10.1590/S1519-566X2002000100019

Dai, W., Li, Y., Zhu, J., Ge, L.-Q., Yang, G.-Q., & Liu, F. (2019). Selectivity and sublethal effects of some frequently-used biopesticides on the predator Cyrtorhinus lividipennis Reuter (Hemiptera: Miridae). *Journal of Integrative Agriculture*, *18*(1), 124–133. https://doi.org/10.1016/S2095-3119(17)61845-8

Doucet, D., & Retnakaran, A. (2012). Chapter six—insect chitin: Metabolism, genomics and pest management. In T. S. Dhadialla (Ed.), *Advances in insect physiology* (Vol. 43, pp. 437–511). Berlin: Academic Press.

Edge, C., Baker, L., Smenderovac, E., Heartz, S., & Emilson, E. (2022). Tebufenozide has limited direct effects on simulated aquatic communities. *Ecotoxicology*, *31*(8), 1231–1240. https://doi.org/10.1007/s10646-022-02582-y

Eger, J. E., Jr., Lees, M. D., Neese, P. A., Atkinson, T. H., Thoms, E. M., Messenger, M. T., . . . Tolley, M. P. (2012). Elimination of subterranean termite (Isoptera: Rhinotermitidae) colonies using a refined cellulose bait matrix containing noviflumuron when monitored and replenished quarterly. *Journal of Economic Entomology*, *105*(2), 533–539. https://doi.org/10.1603/ec11027

El-Monairy, O. M., El-barky, N. M., Bakr, R. F. A., & El-shourbagy, N. M. B. (2012). Some factors affect the perception and production of pheromone and ultrastructure of antennae after treatment of Tribolium castaneum with Chlorfluazuron. *Egyptian Academic Journal of Biological Sciences. A, Entomology*, *5*(2), 39–47. https://doi.org/10.21608/eajbsa.2012.14789

Ewer, J. (2005). How the ecdysozoan changed its coat. *PLoS Biology*, *3*(10), e349. https://doi.org/10.1371/journal.pbio.0030349

Fang, Y., Shi, W. Q., Wu, J. T., Li, Y. Y., Xue, J. B., & Zhang, Y. (2019). Resistance to pyrethroid and organophosphate insecticides, and the geographical distribution and polymorphisms of target-site mutations in voltage-gated sodium channel and acetylcholinesterase 1 genes in Anopheles sinensis populations in Shanghai, China. *Parasit Vectors, 12*(1), 396. https://doi.org/10.1186/s13071-019-3657-7

Fourie, J. J., Fourie, L. J., Horak, I. G., & Snyman, M. G. (2010). The efficacy of a topically applied combination of cyphenothrin and pyriproxyfen against the southern African yellow dog tick, *Haemaphysalis elliptica*, and the cat flea, *Ctenocephalides felis*, on dogs. *Journal of the South African Veterinary Association, 81*(1), 33–36. https://doi.org/10.10520/EJC99860

García, D. A., Bujons, J., Vale, C., & Suñol, C. (2006). Allosteric positive interaction of thymol with the GABAA receptor in primary cultures of mouse cortical neurons. *Neuropharmacology, 50*(1), 25–35. https://doi.org/10.1016/j.neuropharm.2005.07.009

Ghosal, A., Dolai, A., & Chatterjee, M. (2016). Plethora (Novaluron + Indoxacarb) insecticide for the management of tomato fruit borer complex. *Journal of Applied and Natural Science, 8*, 919–922. https://doi.org/10.31018/jans.v8i2.897

Hamaidia, K., & Soltani, N. (2014). Laboratory evaluation of a biorational insecticide, kinoprene, against culex pipiens larvae: Effects on growth and development. *Annual Research & Review in Biology, 4*, 2263–2273. https://doi.org/10.9734/ARRB/2014/9729

Harbison, J. E., Runde, A. B., Henry, M., Hulsebosch, B., Meresh, A., Johnson, H., & Nasci, R. S. (2018). An operational evaluation of 3 methoprene larvicide formulations for use against mosquitoes in catch basins. *Environmental Health Insights, 12*. https://doi.org/10.1177/1178630218760539

Harðardóttir, H. M., Male, R., Nilsen, F., & Dalvin, S. (2019). Effects of chitin synthesis inhibitor treatment on Lepeophtheirus salmonis (Copepoda, Caligidae) larvae. *PLoS One, 14*(9), e0222520. https://doi.org/10.1371/journal.pone.0222520

Haye, A., Ansari, M., Saini, A., Ahmed, Z., Munjal, K., Shamsi, Y., & Sharma, M. (2022). Polyherbal formulation improves glucose-lipid metabolism and prevent hepatotoxicity in streptozotocin-induced diabetic rats: Plausible role of IRS-PI3K-Akt-GLUT2 signaling. *18*(77), 52–65. https://doi.org/10.4103/pm.pm_318_21

Hickin, M. L., Kakumanu, M. L., & Schal, C. (2022). Effects of Wolbachia elimination and B-vitamin supplementation on bed bug development and reproduction. *Scientific Reports, 12*(1), 10270. https://doi.org/10.1038/s41598-022-14505-2

Hodgson, E. (2010). Chapter 35—introduction to pesticide disposition. In R. Krieger (Ed.), *Hayes' handbook of pesticide toxicology* (3rd ed., pp. 863–864). New York: Academic Press.

Hovda, L. R., & Hooser, S. B. (2002). Toxicology of newer pesticides for use in dogs and cats. *Veterinary Clinics of North America: Small Animal Practice, 32*(2), 455–467. https://doi.org/10.1016/s0195-5616(01)00013-4

Hussain, A., AlJabr, A. M., & Al-Ayedh, H. (2019). Development-disrupting chitin synthesis inhibitor, novaluron, reprogramming the chitin degradation mechanism of red palm weevils. *Molecules, 24*(23). https://doi.org/10.3390/molecules24234304

Jindra, M. (2021). New ways and new hopes for IGR development. *Journal of Pesticide Science, 46*(1), 3–6. https://doi.org/10.1584/jpestics.M21-03

Joseph, S. V., & Braman, S. K. (2016). Influence of Turf Taxa and Insecticide Type on Survival of Geocoris spp. (Hemiptera: Geocoridae). *Journal of Entomological Science, 47*(3), 227–237. https://doi.org/10.18474/0749-8004-47.3.227

Karmakar, K., & Patra, S. (2013). Bio-Efficacy of New Acaricide Molecule, Etoxazole 10% Sc (W/W) Against Red Spider Mite, Tetranychus urticae Koch in Brinjal. *Vegetos-An International Journal of Plant Research, 26*, 396. https://doi.org/10.5958/j.2229-4473.26.2.104

Kavallieratos, N., Athanassiou, C., Vayias, B., & Tomanović, Ž. (2012). Efficacy of insect growth regulators as grain protectants against two stored-product pests in wheat and maize. *Journal of food protection*, *75*, 942–950. https://doi.org/10.4315/0362-028X.JFP-11-397

Kotaki, T., Shinada, T., Kaihara, K., Ohfune, Y., & Numata, H. (2009). Structure determination of a new juvenile hormone from a heteropteran insect. *Organic Letters*, *11*(22), 5234–5237. https://doi.org/10.1021/ol902161x

Kumar, A. A., Rao, R. V., & Rao, M. N. (2020). Insecticides and resurgence of rice brown plant hopper nilaparvata lugens (Stal). *Indian Journal of Entomology 2020*, *82*(4), 809–8012.

Li, Y., Yang, N., Wei, X., Ling, Y., Yang, X., & Wang, Q. (2014). Evaluation of etoxazole against insects and acari in vegetables in China. *Journal of Insect Science (Online)*, *14*, 1–14. https://doi.org/10.1673/031.014.104

Llácer, E., Dembilio, O., & Jacas, J. A. (2010). Evaluation of the efficacy of an insecticidal paint based on chlorpyrifos and pyriproxyfen in a microencapsulated formulation against Rhynchophorus ferrugineus (Coleoptera: Curculionidae). *Journal of Economic Entomology*, *103*(2), 402–408. https://doi.org/10.1603/ec09310

Magierowicz, K., Górska-Drabik, E., & Sempruch, C. (2019). The insecticidal activity of Satureja hortensis essential oil and its active ingredient-carvacrol against Acrobasis advenella (Zinck.) (Lepidoptera, Pyralidae). *Pesticide Biochemistry and Physiology*, *153*, 122–128. https://doi.org/10.1016/j.pestbp.2018.11.010

Mahmoudvand, M., & Moharramipour, S. (2015). Sublethal Effects of Fenoxycarb on the Plutella xylostella (Lepidoptera: Plutellidae). *Journal of Insect Science*, *15*, 1–6. https://doi.org/10.1093/jisesa/iev064

Malik, J., Munjal, K., & Deshmukh, R. (2015). Attenuating effect of standardized lyophilized Cinnamomum zeylanicum bark extract against streptozotocin-induced experimental dementia of Alzheimer's type. *Journal of Basic and Clinical Physiology and Pharmacology*, *26*(3), 275–285. https://doi.org/10.1515/jbcpp-2014-0012

Marcon, L., Lopes, D. S., Mounteer, A. H., Goulart, A. M., Leandro, M. V., & Dos Anjos Benjamin, L. (2016). Pathological and histometric analysis of the gills of female Hyphessobrycon eques (Teleostei:Characidae) exposed to different concentrations of the insecticide Dimilin(®). *Ecotoxicology and Environmental Safety*, *131*, 135–142. https://doi.org/10.1016/j.ecoenv.2016.05.016

Masih, S., & Bhat, R. (2019). Insect growth regulators for insect pest control. *International Journal of Current Microbiology and Applied Sciences*, *8*, 208–218. https://doi.org/10.20546/ijcmas.2019.812.030

Mdeni, N., Adeniji, A., Okoh, A., & Okoh, O. (2022). Analytical evaluation of carbamate and organophosphate pesticides in human and environmental matrices: A review. *Molecules*, *27*, 618. https://doi.org/10.3390/molecules27030618

Meenu, A., & Ayushi, V. (2020). Modern technologies for pest control: A review. In N. Mazen Khaled & Z. Hongbo (Eds.), *Heavy metals* (p. 15). Rijeka: IntechOpen.

Merzendorfer, H., Kim, H. S., Chaudhari, S. S., Kumari, M., Specht, C. A., Butcher, S., . . . Muthukrishnan, S. (2012). Genomic and proteomic studies on the effects of the insect growth regulator diflubenzuron in the model beetle species Tribolium castaneum. *Insect Biochemistry and Molecular Biology*, *42*(4), 264–276. https://doi.org/10.1016/j.ibmb.2011.12.008

Mondal, K., & Parween, S. (2000). Insect growth regulators and their potential in the management of stored-product insect pests. *Integrated Pest Management Reviews*, *5*(4), 255–295. https://doi.org/10.1023/A:1012901832162

Mosqueira, B., Duchon, S., Chandre, F., Hougard, J. M., Carnevale, P., & Mas-Coma, S. (2010). Efficacy of an insecticide paint against insecticide-susceptible and resistant mosquitoes—part 1: Laboratory evaluation. *Malar Journal*, *9*, 340. https://doi.org/10.1186/1475-2875-9-340

Munjal, K., Ahmad, S., Gupta, A., Haye, A., Amin, S., & Mir, S. (2020). Polyphenol-enriched fraction and the compounds isolated from *Garcinia indica* fruits ameliorate obesity through suppression of digestive enzymes and oxidative stress. *Health and Medicine*, *16*(70), 236–245. https://doi.org/10.4103/pm.pm_587_19

Munjal, K., Amin, S., Mir, S. R., Gauttam, V. K., & Gupta, S. (2022). Furanocoumarins and lectins as food toxins. In *Analysis of naturally occurring food toxins of plant origin* (pp. 67–80). Boca Raton, FL: CRC Press, USA.

Natal, C. M., Pereira, D. M., Pereira, R. B., Fernandes, M. J. G., Fortes, A. G., Castanheira, E. M., & Gonçalves, M. S. T. (2020). Carvacrol derivatives with potential insecticidal activity. *Chemistry Proceedings*, *3*(1), 37.

Nicolopoulou-Stamati, P., Maipas, S., Kotampasi, C., Stamatis, P., & Hens, L. (2016). Chemical pesticides and human health: The urgent need for a new concept in agriculture. *Frontiers in Public Health*, *4*. https://doi.org/10.3389/fpubh.2016.00148

Nogueira Alves, A., Oliveira, M. M., Koyama, T., Shingleton, A., & Mirth, C. K. (2022). Ecdysone coordinates plastic growth with robust pattern in the developing wing. *Elife*, *11*. https://doi.org/10.7554/eLife.72666

Oberlander, H., & Silhacek, D. L. (2000). Insect growth regulators. In B. Subramanyam & D. W. Hagstrum (Eds.), *Alternatives to pesticides in stored-product IPM* (pp. 147–163). Boston: Springer.

Oulhaci, C. M., Denis, B., Kilani-Morakchi, S., Sandoz, J.-C., Kaiser, L., Joly, D., & Aribi, N. (2018). Azadirachtin effects on mating success, gametic abnormalities and progeny survival in Drosophila melanogaster (Diptera). *Pest Management Science*, *74*(1), 174–180. https://doi.org/10.1002/ps.4678

Parthasarathy, R., & Palli, S. R. (2021). Stage-specific action of juvenile hormone analogs. *Journal of Pesticide Science*, *46*(1), 16–22. https://doi.org/10.1584/jpestics.D20-084

Poppenga, R. H., & Oehme, F. W. (2010). Chapter 7—pesticide use and associated morbidity and mortality in veterinary medicine. In R. Krieger (Ed.), *Hayes' handbook of pesticide toxicology* (3rd ed., pp. 285–301). New York: Academic Press.

Rajendran, T. P., & Singh, D. (2016). Chapter 1—insects and pests. In Omkar (Ed.), *Ecofriendly pest management for food security* (pp. 1–24). San Diego, CA: Academic Press.

Rebek, E. J., & Cloyd, R. A. (2015). *Management of insects and mites in greenhouse floral crops*. Oklahoma Cooperative Extension Service. ISO 690. https://extension.okstate.edu/fact-sheets/management-of-insects-and-mites-in-greenhouse-floral-crops.html#:~:text=thoroughly%20cleaning%20the%20greenhouse%20after,biological%20control%20agents%20when%20appropriate

Reis, S. L., Mantello, A. G., Macedo, J. M., Gelfuso, E. A., da Silva, C. P., Fachin, A. L., . . . Beleboni, R. O. (2016). Typical monoterpenes as insecticides and repellents against stored grain pests. *Molecules*, *21*(3), 258. https://doi.org/10.3390/molecules21030258

Roscoe, L. E., Forbes, G., Lamb, R., & Silk, P. J. (2020). Effects of topical tebufenozide application to choristoneura fumiferana pupae (Lepidoptera: Tortricidae). *Insects*, *11*(3). https://doi.org/10.3390/insects11030184

Schneiderman, H. A. (1972). Insect hormones and insect control. In J. J. Menn & M. Beroza, *Insect juvenile hormones* (pp. 3–27). New York: Academic Press.

Seal, D., Kumar, V., Kakkar, G., & Mello, S. (2013). Abundance of adventive thrips palmi (Thysanoptera: Thripidae) populations in Florida during the first sixteen years. *Florida Entomologist*, *96*, 789–796. https://doi.org/10.1653/024.096.0312

Showler, A. T., Donahue, W. A., Harlien, J. L., Donahue, M. W., Vinson, B. E., & Thomas, D. B. (2019). Efficacy of novaluron + pyriproxyfen (Tekko Pro) insect growth regulators against amblyomma americanum (Acari: Ixodidae), rhipicephalus (*Boophilus*) annulatus, rhipicephalus (*Boophilus*) microplus, and rhipicephalus sanguineus. *Journal of Medical Entomology*, *56*(5), 1338–1345. https://doi.org/10.1093/jme/tjz075

Sierras, A., & Schal, C. (2020). Lethal and sublethal effects of ingested hydroprene and metho-prene on development and fecundity of the common bed bug (Hemiptera: Cimicidae). *Journal of Medical Entomology, 57*(4), 1199–1206. https://doi.org/10.1093/jme/tjaa038

Sláma, K., & Williams, C. (1966). The juvenile hormone. V. The sensitivity of the bug, pyrrhocoris apterus, to a hormonally active factor in American paper-pulp. *Biological Bulletin, 130*. https://doi.org/10.2307/1539700

Song, Y., Villeneuve, D. L., Toyota, K., Iguchi, T., & Tollefsen, K. E. (2017). Ecdysone recep-tor agonism leading to lethal molting disruption in arthropods: Review and adverse out-come pathway development. *Environmental Science & Technology, 51*(8), 4142–4157. https://doi.org/10.1021/acs.est.7b00480

Staal, G. B. (1982). Insect control with growth regulators interfering with the endocrine system. *Entomologia experimentalis et Applicata, 31*(1), 15–23. https://doi.org/10.1111/j.1570-7458.1982.tb03115.x

Su, T., Thieme, J., Lura, T., Cheng, M.-L., & Brown, M. Q. (2019). Susceptibility profile of aedes aegypti L. (Diptera: Culicidae) from Montclair, California, to commonly used pes-ticides, with note on resistance to pyriproxyfen. *Journal of Medical Entomology, 56*(4), 1047–1054. https://doi.org/10.1093/jme/tjz019

Subrahmanyam, B. (1990). Azadirachtin—A naturally occurring insect growth regulator. *Proceedings: Animal Sciences, 99*(3), 277–288. https://doi.org/10.1007/BF03186397

Sun, J., & Zhou, Y. (2015). Design, synthesis, and insecticidal activity of some novel diacyl-hydrazine and acylhydrazone derivatives. *Molecules, 20*(4), 5625–5637.

Taleh, M., Farshbaf Pourabad, R., Geranmaye, J., & Ebadollahi, A. (2015). Toxicity of Hexaflumuron as an insect growth regulator (IGR) against Helicoverpa armigera Hubner (Lepidoptera: Noctuidae). *Journal of Entomology and Zoology Studies, 3*, 274–277.

Tulashie, S. K., Adjei, F., Abraham, J., & Addo, E. (2021). Potential of neem extracts as natu-ral insecticide against fall armyworm (Spodoptera frugiperda (J. E. Smith) (Lepidoptera: Noctuidae). *Case Studies in Chemical and Environmental Engineering, 4*, 100130. https://doi.org/https://doi.org/10.1016/j.cscee.2021.100130

Tunaz, H., & Uygun, N. (2004). Insect growth regulators for insect. *Turkish Journal of Agriculture and Forestry, 28*, 377–387.

Tyagi, S., Raghvendra, R., Singh, U., Kalra, T., Munjal, K., & Vikas. (2010a). Practical applications of proteomics-a technique for large- scale study of proteins: An overview. *International Journal of Pharmaceutical Sciences Review and Research, 3*, 87–90.

Tyagi, S., Singh, U., Kalra, T., & Munjal, K. (2010b). Applications of metabolomics—a sys-tematic study of the unique chemical fingerprints: An overview. *International Journal of Pharmaceutical Sciences Review and Research, 3*(1), 83–86.

Vinod, K. G., Kavita, M., & Nitisha, N. J. (2018). Bioactivity guided fractionation of potent antiacne plant extract against Propionibacterium acnes. *African Journal of Biotechnology, 17*(13), 458–465.

Walters, M., & Wall, R. (2012). Evaluation of dicyclanil (CLiKZiN®) treatment for the early-season protection of ewes against blowfly strike. *Veterinary Parasitology, 188*(1), 200–202. https://doi.org/https://doi.org/10.1016/j.vetpar.2012.03.016

Webb, G. A. (2017). Efficacy of bistrifluron termite bait on coptotermes lacteus (Isoptera: Rhinotermitidae) in Southern Australia. *Journal of Economic Entomology, 110*(4), 1705–1712. https://doi.org/10.1093/jee/tox133

Woo, J. H., & Lim, Y. S. (2015). Severe human poisoning with a flufenoxuron-containing insecticide: Report of a case with transient myocardial dysfunction and review of the lit-erature. *Clinical Toxicology (Phila), 53*(6), 569–572. https://doi.org/10.3109/15563650.2015.1040158

Yanagi, M., Tsukamoto, Y., Watanabe, T., & Kawagishi, A. (2006). Development of a novel Lepidopteran insect control agent, chromafenozide. *Journal of Pesticide Science, 31*, 163–164. https://doi.org/10.1584/jpestics.31.163

Yang, Z., Wu, Y., Yan, Y., Xu, G., Yu, N., & Liu, Z. (2022). Regulation of juvenile hormone and ecdysteroid analogues on the development of the predatory spider, Pardosa pseudoannulata, and its regulatory mechanisms. *Ecotoxicology and Environmental Safety, 242,* 113847. https://doi.org/https://doi.org/10.1016/j.ecoenv.2022.113847

Yoon, J.-Y., Park, J.-H., Han, Y.-h., & Lee, K.-S. (2012). Residue patterns of buprofezin and teflubenzuron in treated peaches. *Journal of Agricultural Chemistry and Environment, 1*(1), 5. https://doi.org/10.4236/jacen.2012.11002

Zhang, D., & Lu, S. (2022). Human exposure to neonicotinoids and the associated health risks: A review. *Environment International, 163,* 107201. https://doi.org/10.1016/j.envint.2022.107201

Zhang, J., Lu, A., Kong, L., Zhang, Q., & Ling, E. (2014). Functional analysis of insect molting fluid proteins on the protection and regulation of ecdysis. *Journal of Biological Chemistry, 289*(52), 35891–35906. https://doi.org/10.1074/jbc.M114.599597

Zitnanová, I., Adams, M. E., & Zitnan, D. (2001). Dual ecdysteroid action on the epitracheal glands and central nervous system preceding ecdysis of Manduca sexta. *The Journal of Experimental Biology, 204*(Pt 20), 3483–3495. https://doi.org/10.1242/jeb.204.20.3483

16 The Use of Plant Extracts and Wastes from Agroindustry as Pest Management Agents

Leo M.L. Nollet

CONTENT

The study of natural products is an old science that has been evolving and changing through the years (Torssell 1983). From the 18th century on, all the myths and magic of the use of natural products has been set apart, and its use has increased its rationality. This phenomenon is based in chemistry and started with the systematic development for separation, purification, and analysis of living organisms compounds (Nakanishi 1992). The natural products area has accompanied this development, and the basic concepts of this area have also evolved. When a new substance is isolated and analyzed, new data are accumulated. This led to identifying big metabolic origin pathways for the diverse compounds isolated.

This idea has been growing to explain the presence of secondary metabolites in living organisms and justify the high energetic cost that their biosynthesis involves (Baas 1989).

Chemical ecology is the study of the interactions between living organisms mediated by chemical compounds and is one of the newest concepts incorporated into natural products chemistry (Oldham & Boland 1996). Nowadays, natural products studies include the chemistry, structural elucidation, biosynthesis and bioactivity of the isolated compounds (Shapiro 1991).

Natural products are classified as primary or secondary metabolites. Lipids, carbohydrates, proteins, and nucleic acids are primary metabolites. They are related to the growth and reproduction of living organisms (Fuller & Nes 1985, Pickett et al. 1997), and they are universal constituents of all organisms' cells.

Secondary metabolites have particular distribution characteristics, from the taxonomic point of view to the ontogenic one (Hegnauer 1969, Waterman 1999). Secondary metabolites are compounds that participate in primary physiological processes. In plants, secondary metabolites such as alkaloids, coumarins, glucosinolates, and cyanogenic glucosides are derived from amino acids that are primary metabolites

DOI: 10.1201/9781003265139-19

(Berenbaum 1995). Some other secondary metabolites such as poliketides and terpenes have a common precursor: acetyl coenzyme A (Samuelson 1992).

Chlorophyll can be considered a primary metabolite because it is the main photosynthetic pigment present in angiosperms and gymnosperms, and it is present in all development stages and organs on the plants. Other vegetal pigments such as carotenes or flavonoids are secondary metabolites (Evans 1991a) and are present only in some kinds of angiosperms. Their distribution depends on the location, evolutional stage, and climate that surround them.

Even though glucose and some other sugars are primary metabolites, they play a fundamental role in secondary metabolism because their glycosylated forms have physiological importance for vegetables; mainly, they allow the lipophilic substances transport to the aqueous internal environment (Swain 1977).

Not long ago, secondary metabolites were considered useless or failure-derived metabolic substances. However, many different functions have been described for these compounds. For example (Pagare 2015, Tiwari & Rana 2015, Vijayakumar 2018):

- Regulating chemical composition of plants and/or fruits colors
- Starting or finishing seeds' and tubers' dormancy
- Promoting rooting and propagations
- Controlling the size of the plant or the organ
- Controlling the flowering process
- Controlling fruit maturation and development
- Selecting the minerals absorbed from the floor
- Changing the development time of plantations
- Increasing pest resistance
- Enduring climate factors such as temperature and water and air pollution (Baas 1989)

Within the last 50 years, it has been demonstrated that many secondary metabolites are important communication factors between individuals and their environment (Harborne 2001). It has been proposed that secondary metabolites play an important role as plant mediators and, as a consequence, in their adaptation (Swain 1977, Harborne 1986). The presence of secondary metabolism is directly related to the capacity of organisms to move. As plants do not have such ability, they have the most developed secondary metabolism among living beings. They have to survive and reproduce themselves in the place where they grow, either sunny or not, in drought or wetlands, at sea level or in the mountains, in the cold or in the tropics.

Plants have developed defense mechanisms against environmental aggressions; one of the most important is the defense of vegetables against parasites and predators. It is postulated that most of the defensive mechanisms of plants have a chemical character, and its existence refers to secondary metabolites (Harborne 1993, Schlee 1991, Torssell 1983).

The use of plants with insecticidal properties is an ancient practice; prehistoric men were the first who used vegetal extracts in an empiric or intuitive way, based on case-to-case observations, which are now studied in ethnobotany science (Evans 1991b). Before the organosynthetic insecticides were discovered in the first half of the last century, the substances extracted from vegetables were widely used to control insects (Gallo 2002).

Natural products found in plants have an important role in pest control. Many research studies have focused on plants' secondary metabolites that affect specific pest-insects processes, such as oviposition, reproduction, fertility, and feeding behavior. These abnormalities are related to physiological changes resulting from modifications of the endocrine system, which controls growth and ecdysis (Mordue & Blackwell 1993, Nisbet 2000).

The use of insecticides obtained from plants has many advantages in pest control. In general, they are less persistent and can be used without modifying the natural balance of the ecosystem, respecting the sustainability principles (Smith 1989). Insects and plants interact continuously, suffering the consequences of this interaction and adapting themselves to each other. As they live together in the same environment, herbivores develop eating strategies, and plants develop defense mechanisms in order to reduce or block the insects' attack. These mechanisms are developed through physical or chemical barriers.

Plants' chemical defenses against insects use basically the antixenosis (undesirability, avoidance by insect) and antibiosis (adverse effects or prevention of insect activity) mechanisms (Renwick 1983, Sulistyo & Inayati 2016, Kordan 2019). Antixenosis occurs because of stimuli caused by allelochemicals that cause repellency (orients the insect in the opposite direction of the plant), arrestancy (stops insect's movements), or suppression (stops insect's feeding or oviposition). Antibiosis occurs when an insect attacks a plant, and it responds, causing an adverse effect to the insect: for example, death; lifetime modification; and/or size, weight, or fecundity reduction (Gallo 2002).

Natural products with insecticidal activity are an important alternative for integrated pest management (IPM), and they can be applied as powders, oils, emulsions, etc. Particularly as subjects of scientific investigation, plant essential oils are attractive due to their relatively easy preparation, simple analytical methodologies, and the wide diversity of plants that produce them (Isman et al. 2011). Nevertheless, crude extracts obtained from different vegetal structures can also be used as pest management agents; for example, one of the most studied biopesticide is obtained from *Azadirachta indica* (Meliaceae), commonly known as the neem tree (Rodrigues 2008).

The growing accumulation of experience demonstrates that neem products work by intervening at several stages of an insect's life. The components of this tree approximate the shape and structure of some insects' vital hormones (ecdysones) that are absorbed by them, blocking their endocrine systems, resulting in behavioral and physiological aberrations. Neem extracts can influence more than 200 insect species; as a result, it can be said that neem products are medium- to broad-spectrum pesticides of phytophagous insects (Vietmeyer et al. 1992).

At least nine neem limonoids (triterpenes) have demonstrated an ability to block insect growth in some of the most deadly pests for agriculture and human health. However, the most studied is azadirachtin A. Azadirachtin was isolated from the seeds of *Azadirachta indica* by Butterworth and Morgan (1968), and its full structural determination was completed 17 years later, as shown in Figure 16.1. Its biosynthesis is thought to involve tirucallol, a tetracyclic triterpenoid, and a series of oxidation and rearrangement reactions.

Azadirachtin is a classic example of a natural plant defense chemical that affects feeding, primarily through chemoreception (primary antifeedancy) but also through a reduction in food intake due to its toxic effects if consumed (secondary antifeedancy).

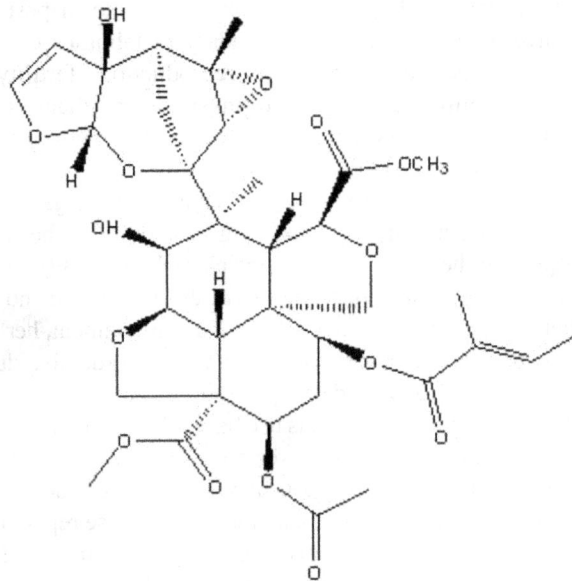

FIGURE 16.1   Azadirachtin A structure.

From initial observations of its antifeedant effects, it was a natural progression to test azadirachtin and azadirachtin-containing compounds for commercial use in pest control (Mordue & Blackwell 1993).

The regulatory growth effects of azadirachtin on larval stages of insects are well known. Treatment of insects or their food with azadirachtin causes growth inhibition, malformation, and mortality in a dose-dependent manner (Mordue & Blackwell 1993).

Further information can be found in recent articles: Fernandes et al. 2019, Kilani-Morakchi 2021, Khalil 2013, Sarah 2019, Nicoletti 2012 and Mafara & Bakura 2021.

Hence, increasing doses of azadirachtin in larval stages result in:

- Adults with reduced longevity and fecundity
- Wingless adults or adults with unplasticized wing lobes seen as crippled wings
- Nymphs or larvae that die during ecdysis, unable to complete the molting process
- Pupae with severe deformities to the head and thoracic appendages
- Nymphs or larvae that die immediately before the molt after a normal instar length
- Insects that remain "over-aged" larvae for a greatly extended period
- Insects that die within hours of treatment

(Karnavar 1987)

Some other Meliaceae present anti-insect activity: for example, the study of *Cabralea canjerana* extracts over *Anastrepha fraterculus* (fruit fly) highlights toxic and deterrent properties (antifeedant and anti-oviposition). These effects were observed for at least 72 hours. The different modes of action of the extracts of the fruits and seeds of *C. canjerana* on *A. fraterculus*, particularly the antifeedant activity and oviposition deterrence for at least 72 hours, indicate their possible use in the integrated pest management of this fruit fly (Magrini et al. 2014).

C. canjerana extracts were also tested over *Spodoptera frugiperda* (fall armyworm), showing adverse larvicidal effects at all assayed concentrations. Several physiological abnormalities were observed and could be attributed to alterations of the endocrine system, which controls the growth and ecdysis of the insect. The compounds found in these extracts were polar triterpenoids (Braga 2006; Sarria et al. 2011; Magrini et al. 2014, 2015).

These extracts can potentially be used for the alternative control of *S. frugiperda* and other pest insects after toxicity and environmental risk evaluations.

But not just the neem tree (or neem tree–related plants from the Meliaceae family) has been studied in reference to insect resistance; many other plants have demonstrated negative effects on different species (not only herbivores but also aggressive plants), and their extracts can be used as pest management agents.

Derris negrensis (Fabaceae) commonly, known as Timbó, is a liana found in firm land areas, from which venom is extracted for killing river fish (Almeida Neto 2004).

Phytochemical studies of a different number of *Derris* species performed at the Natural Products Laboratory of the Instituto Nacional de Pesquisas da Amazônia (INPA selected *D. negrensis* as the species with the highest concentration of rotenone (Maia 1976).

Rotenone is an insecticidal compound present in leguminous plants in Latin America, Asia, and the Caribbean (Caminha Filho 1940, Kathrina & Antonio 2004, Cloyd 2004). It is structurally an isoflavonoid, odorless and tasteless, that appears accompanied by other flavonoid compounds that also present insecticidal activity (Corbett 1940, Cravero et al. 1976, Lima 1987, Silva et al. 2001).

Rotenone is highly efficient in controlling beetles and leaf-chewing caterpillars, not only by ingestion but also by contact inhibiting insects' motor systems, which ends in death (Saito & Luchini 1998). Rotenone presents acaricide and insecticide action over caterpillars, lice, mosquitoes, aphids, fleas, flies, and ants (Maklouf 1986, Buss 2002, Kathrina & Antonio 2004, Cloyd 2004, Zubairi et al. 2016).

Castro Silva (2011) studied the ethyl acetate leaf extract of *Palicourea marcgravii* on *Rhipicephalus microplus*, which showed high acaricidal activity due to the monofluoroacetic acid detected in it. The high toxicity of this compound, as it targets one of the more important metabolic pathways of the tick, could be used as an acaricidal biorational after further studies on its environmental fate, species toxicity, and skin toxicity, among others.

The same author (Castro Silva 2009) showed that the hexane extract and essential oil from *Piper aduncum* were toxic for both larval and adult *Rhipicephalus microplus*. They could be alternatives to synthetic acaricides for tick control; the volatile oil is a particularly promising bioinsecticide, due to its toxicity to larvae at a micro molar range. The composition of *P. aduncum* essential oil showed the presence of nerodiol (0.74%), globulol (0.65%), spathulenol (0.64%), croweacin (1.91%), dillapiole (94.84%), and apiole (0.38%). Dillapiole is an interesting lead for the developing new compounds for tick control (Durofil 2021, Silva 2019).

Solanaceae plants are a wealth of sources of bioactive natural products. Nicotine, the tobacco alkaloid, has been used for centuries to protect crops, and it is widely used in organic agriculture against an array of pests, although its use is strictly controlled in Europe nowadays due to its toxicity. Nicotine acts as postsynaptic blocker of the neurotransmitter acetylcholine, hampering the repolarization of the neuronal membrane and thus interrupting the motor impulse, leading to the insect's death.

Other compounds from Solanaceae that have been suggested as crop protection agents are the steroidal alkaloids. Chemically, they are glycosidic steroidal alkaloids and showed some insecticidal activity, but their most interesting application is against some phytopathogenic fungi. Glycoalkaloids form relatively stable complexes with ergosterol, the fungal membrane steroid. As they can also interact with cholesterol, their toxicity to mammals is an open question.

Nevertheless, among the most interesting examples of promising bioactive natural compounds from Solanaceae plants are sugar esters, shown in Figure 16.2. They are esters of short, ramified fatty acids like 3-methyl valerianic acid and simple sugars like sucrose and glucose. Some *Nicotiana* species produce large amounts of sugar esters whose use as biorationals has been proposed. They have shown activity against a wide array of pests, based on the inability of the insects' lipolytic enzymatic activity to metabolize them. They have been reported to be stored in trichomes, not only in *Nicotiana* sp. but also in the *Datura* and *Solanum* genera (Hare 2005, Kennedy et al., 1995, Neal et al. 1990). Potato breeders took advantage of sugar esters producing the genetic character of some wild potato species, creating new potato varieties after crossing the Andean *Solanum berthaultii* with *Solanum tuberosum* strains to yield potato varieties resistant to aphids and other insect pests (Hackbar Medeiros & Tingley 2006).

Some other examples of plant extracts as pest management agents involve the antifeedant and repellent effects of common weed extracts from the *Solanum* species, such as *S. bonariense*, *S. bistellatum*, and *S. sisymbrifolium*. The genus *Solanum* is particularly heterogeneous and widely distributed on the American continent (Vázquez 1997). Most species of the genus have trichomes in their stems, leaves, and inflorescences (Mentz et al. 2000), which produce chemical substances with anti-insect activity (Silva 2003, 2005, Lovatto et al. 2004, Leite 2004, Srivastava & Gupta 2007, Szafranek et al. 2008).

In this case, the characteristics of the secondary metabolites of the *Solanum* species are completely different from the examples mentioned earlier; the anti-insect effect of sugar esters isolated from *S. sisymbrifolium* leaves was preliminary reported by Cesio et al. 2006 (Cesio 2006). As the relationship between chemical family and biological activity deserves more in-depth investigations, some other research has been done in order to evaluate the activity of this type of compound, searching in the Latin American biodiversity for more *Solanum* species that could, according to the leaf morphology and chemical composition of the obtained extracts, contain compounds with structural features similar to the ones described for *S. sisymbrifolim*, which could be the basis for a "combinatorial natural library" of bioactive acyl sugars with specific anti-insect activity.

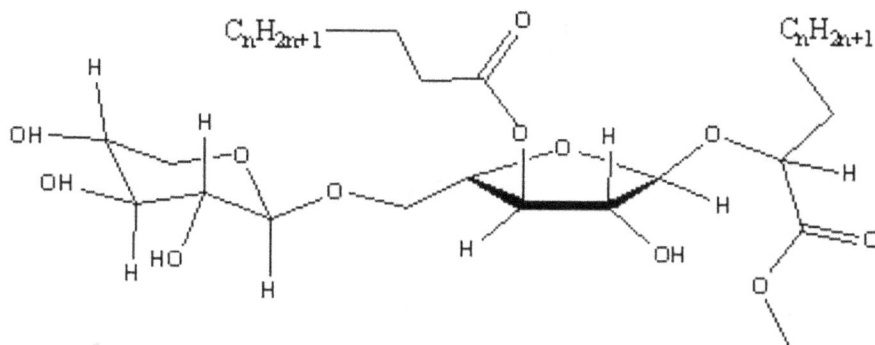

FIGURE 16.2   Sugar esters' general structure found in *S. sisymbrifolium* extracts.

The botanicals obtained from these leaf extracts were also tested over several monocotyledons and dicotyledons having different levels of inhibitory activity. Therefore, acyl sugars could be a useful alternative to synthetic herbicides, seeking their direct application in "biological" productive systems (Migues et al. 2011, 2012).

Recent articles on this topic are Rech-Cainelli 2015, Barik 2020, and Banumathi et al. 2016.

The same chemical ecology approach used to search for anti-insect compounds in plants can be applied to investigate the potential of plant extracts to control weeds. Weeds are plants growing in the wrong place at the wrong time, interfering with developing crops and competing with them for soil nutrients and light. The ability of plants to influence other plants' growth by secreting inhibitory chemicals is known as allelopathy. The described inhibition caused by plant metabolites in other plant development goes from hampering seed germination and seedling development to plant post-emergence development. Many compounds have been described with such properties (Vyvyan 2002). Of particular importance are some amino acids that mimic essential amin oacids like the m-hydroxyphenyl alanine (m-tyrosine) in the turf grass *Festuca* sp. (Bertin et al. 2007) that interrupts the metabolism of the normal p-hydroxyphenyl alanine, tyrosine in lignin production, causing a similar effect to that of the worldwide applied synthetic lignin biosynthesis inhibitor glyphosate. Other important natural products that are in the herbicide market nowadays are the tripeptide bialaphos and the mimic of glutamine, gluphosinate, which derive from *Streptomyces sp* secondary metabolism (Hoagland 1999).

Many other compounds have been investigated for their plant development inhibitory properties (Macias, 1995). Terpenoids compounds like monoterpenes (menthol, 1,8 cineol, carvone) have been shown to have important inhibitory properties, but the metabolism inhibition caused by monoterpenes is rather nonspecific, and therefore, their use is very restricted. Nevertheless, the menthol esters of fatty acids showed promising inhibitory properties. Particularly, the methyl ester caprylic acid has been the most active methyl ester against some monocotiledoneos weeds (Macías et al 2008). A post-emergence mode of action for triterpenes like lupeol was described by Macías et al. (1999). The same authors showed previously that sesquiterpene lactones have important allelopathic properties and a straightforward structure activity relationship (SAR). An explanation of their mechanism of action can be proposed (Macías 1995). Some of these lactones are present in sunflower debris and act as soil protectant covertures when they are left in the field between crop rotations. Sunflower stems and leaves also free to the environment many aromatic acids of known allelopathic activity, like ferulic, vainillinic, and other coumaric acids. The inhibition of weed development provided by these compounds is a useful tool used by farmers for centuries that finds application in modern production systems. The comprehension of allelopathic interactions allows the rational use of different farming practices, seeking minimization of herbicide use in an effort to protect the environment against the indiscriminate use of agrochemicals.

In some cases, the extraction of these secondary metabolites from natural origin is difficult and requires too much work, and the natural sources to obtain it are also restricted, hampering its wide use.

An alternative is to search industrial wastes or natural byproducts from agroindustry that could be a continuous source of compounds with anti-insect activity.

A very innovative example of this was research done by Sbeghen-Loss et al. (2011), in which they observed that lime wax influences the feeding behavior of *Cryptotermes brevis*, one of the most harmful termites of South America.

Termites are polymorphic social insects that build their nests in warm to hot zones, either natural or modified by man. They are considered an urban plague and can be classified as drywood termites or underground termites; their control and management varies, depending on type.

From the *Cryptotermes* genera, *C. brevis* is considered the most voracious in domestic environments, being widely distributed worldwide (Edwards & Mill 1986, Bacchus 1987). They infest dry wood and living tree branches, and they can be seen at night flying around light sources.

Citrus peels contain many commercial products: essential oils, dyes, pectins, sugars, acids, proteins, enzymes, vitamin C, and free heterosides. Many studies have shown that the essential oils obtained from various citrus peels and seeds have compounds with insecticidal and antifungal activity (Siskos 2009).

Citrus latifolia is a tropical fruit, and its economic value has been exploited recently. The volume of industrial wastes from juice extraction borders 45% of the fruit's weight.

At an industrial scale, the essential oil extraction occurs along with the juice extraction. First, mechanical pressure on the peel liberates the essential oil, which is removed from the fruits with water; then this emulsion is centrifuged, and the oil is separated from water. Then the fruit continues its process. The oil is finally purified using winterization, in which waxes and acids crystallize.

The reader will find further information on limonoids in recent articles: Shi et al. 2020, Galovičová et al. 2022, and Adusei-Mensah et al. 2014).

This nonvolatile portion also consists of steroids, carotenoids, coumarins, and furanocoumarins (e.g., psoralenes and poly methoxy flavones). Coumarins and psoralenes represent around 10–15% of acid lime nonvolatile residue (Dugo et al., 1999), shown in Figure 16.3.

Coumarins and furanocoumarins have a wide biological activity spectrum; some of the activities are insecticidal, antifeedant, and growth inhibition (Berenbaum & Feeny 1981, Oketch-Rabah et al. 2000, Sharma et al. 2007).

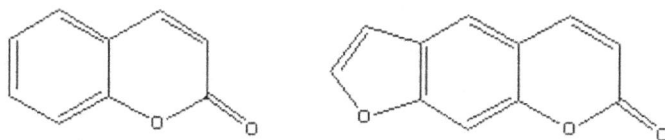

FIGURE 16.3 Coumarin and psoralene, found in citrus fruits.

Acid lime (*C. latifolia*) and orange (*C. sinensis*) waxes were tested against *C. Brevis* dry wood termite by Barros et al. (2008), who obtained better results with the acid lime waxes. The identified compounds present in the most active elements were limonene and furanocoumarins such as bergaptene, pimpineline, isopimpineline and xantotoxin. Based on this study, Sbeghen-Loss et al. (2011) studied *C. brevis* behavior, finding that the citrus wax waste (CWW), the essential oil, and the wax after hydrodistillation (CWFR) showed different levels of antifeedant activity when tested in a non-choice assay, using wood blocks that had been treated with the elements under study against pseudergates of *C. brevis*.

The results showed that CWW and CWFR had the greatest antifeedant effect.

The essential oil was less active, but at high concentrations, it had statistically significant antitermitic properties.

The fixed residue is a very interesting raw material for antitermitic formulations. It has a relatively high melting point (75–79 °C), suitable for industrial and household products.

The wax can be used as it is or may be included in commercial formulations that provide wood protection for six months, as was demonstrated when EC50 was evaluated.

This unused waste from the citrus industry could be developed into a sustainable natural agent for termite control.

The knowledge of the real potential of natural products derived from plants against plague insects showed here and the study of the activity of industrial acid lime wax and its compounds, which seem to be good candidates against *C. brevis*, promises a good commercial product that works as a wood preservative, especially because it is a natural residue, abundant and easy to obtain from the citrus industry.

But the citrus industry is not the only example of useful industrial wastes; Dhillon et al. (2013) recently reported an interesting issue about apple waste.

Apples and their products are among the most consumed fruits all over the world. Apple-processing industries generate millions of tons of agro-industrial waste worldwide with low nutritional value. The solid residues consist of a mixture of skin, pulp, and seeds known as "apple pomace" (Dhillon et al. 2011a). Because they are highly biodegradable, the disposal of these wastes represents a serious environmental problem and presents many challenges (Dhillon et al. 2011b, Shalini & Gupta 2010). Often only 20% is retrieved as animal feed, and the other 80% goes to landfill, is incinerated, or is sent to composting sites. The inefficient management of apple industry wastes results in significant greenhouse gas emissions (Matthews & Themelis 2007). However, advancement in technology has led to the alternative options of utilization of apple pomace. The solid-state fermentation (SSF), also known as koji fermentation, is gaining wide interest in bioproduction. Agro-industrial residues are generally considered the best substrates for koji fermentation processes, and several high-value products can be obtained from apple wastes, such as organic acids, cellulases, ligninolytic enzymes, natural antioxidants, dietary fibers, aroma compounds, biofuels, biopolymers, and animal/livestock feed, among others. But the most interesting uses of apple wastes is as insect diet or biocontrol formulation preparations. Some agriculturally important polyphagous pests affect a wide variety of fruits. Conventionally, the control of infestation has been carried out by the use of chemical pesticides with adverse impacts, such as the loss of natural enemies and pollinators, insecticide residues, and environmental consequences (Horton et al. 1997).

Biological control agents, such as *Bacillus thuringiensis* or baculovirus-based biopesticides, are effective using a special diet to attract the insects, but the cost is very high, limiting the growth of biopesticides on a commercial scale. However, some of these ingredients could be replaced by the ubiquitous apple pomace. The use of agro-industrial wastes such as apple pomace as diet supplements is a novel concept of sustainable use in larval diet formulations. The economical diet development will have a broader application as it can also be utilized for rearing different types of insects, such as cabbage loopers, gypsy moths, Bertha armyworms, spruce bud worms, cotton armyworms, hemlock loopers, and tussock moths, which have agricultural and economical importance. The main properties that may lead to the utilization of these wastes are its rheological characteristics and the high energy value (Gnepe 2011).

As can be seen, recent developments in fermentation and bioprocess technology provide promising alternatives for the biotransformation of these abundant wastes into high-value products that can help in reducing the environmental damage of synthetic pesticides.

Residues of almonds (Fernandez-Bayo 2020), coffee (Poopathi & Mani 2015), and olives (El-Abbassi 2017) are also used as sources of biopesticides.

Further information on biopesticide production with wastes and byproducts can be found in Sala et al. 2021, Balasubramanian 2017, Avishek et al. 2017, and Huang et al. 2020.

Stability and formulation methods are the keystones of field utilization of natural products. Compared to synthetic pesticides, botanical insecticides are relatively unstable and break down significantly faster when exposed to elements such as light, temperature, and air. Once plant chemicals have been removed from their protective compartments by destructive extraction methods, their constituents are prone to oxidative damage, chemical transformations, and polymerization reactions. Furthermore, as plant extracts age, their quality declines further. Over time, they might lose some of their attributes, such as odor, flavor, color, and consistency (Pokorny et al. 1998, Grassmann & Elstner 2003). The compositional diversity of the botanical extracts and the instability of their constituents can make botanical insecticides unsuitable for application where residual effects over long periods of time are desirable. To overcome the instability of botanical extracts and essential oils when used as pesticides, several formulation techniques and methods have been developed in recent years. Microencapsulation, for example, is a method used to protect sensitive materials that can easily suffer degradation (Cabral Marques 2010). Encapsulation techniques can be divided into three classes:

1. Chemical processes, such as molecular inclusion or interfacial polymerization (Chung et al. 2013)
2. Physicochemical techniques, such as coacervation and liposome encapsulation (Dong et al. 2011)
3. Physical processes, such as spray drying, spray chilling or cooling, co-crystallization, extrusion, and fluidized bed coating (Fang & Bhandari 2010, Laohasongkram et al. 2011)

Microencapsulation techniques are generally used to prepare pesticide nanoemulsions that provide some level of controlled release of the active botanical ingredient (Sakulku 2009). These microencapsulation techniques generally slow down the release or decay of the entire mixture obtained by the destructive extraction of plant tissues; however, no specific attention is paid to the behavior of individual constituents of the mixture. By contrast, plants rely on specific structural features, cellular compartments, and chemical pathways to control proactively the production, storage, and release of individual compounds within their defensive chemical arsenal (Niinemets & Reichstein 2003, Niinemets et al. 2004, Pichersky et al. 2006, Colquhoun et al. 2010). Novel technologies that consider the behavior and control level of the individual constituents of botanical insecticides are paving the way for a new generation of botanical insecticides that are applied in a manner closer to the natural defense methods used by plants against herbivores.

For recent references, see Miro Specos et al. 2017, Eski 2019, Sciortino et al. 2021, and Enascuta et al. 2020.

Another important issue that goes along with formulation is the application technology associated with the potential pest management agent. It is well known that application technology has an important influence on the efficacy of synthetic pesticides, but its effects on botanicals and bioherbicides has been poorly investigated (Doll et al. 2005, Lawrie et al. 2002a, 2002b). Retention of spray droplets can be affected by surface characteristics and the morphology of the weed and its biotypes, as well as the adjuvants, travel speed, and droplet sizes (Katovich et al. 1996, Ramsdale & Messersmith 2002, Singh 2002). All these factors are important in determining the effectiveness of the applied botanical compound.

Once the formulation and the application technology are defined and field trials, done it can be said that the compound found is a botanical pesticide, and it can be commercialized, obtaining the intellectual property of the formulated compound (Ash 2010).

In conclusion, we can say that lots of work has been done in order to find species with pest management agent potential. Many compounds have been isolated and showed interesting bioactivity in laboratory experiments; only a few have gone through field trials, but that does not mean that others could not reach the success that azadirachtin (as an example) found years ago. It is necessary to focus on the next step of the research; a better understanding of the behavior and bioactivity of the individual components of botanical insecticides, coupled with more advanced methods of compartmentalization and formulation, will allow greater degrees of control over the availability and activity of the individual components of complex botanical mixtures and, consequently, should enhance the efficacy of botanical insecticides, allowing them to get into the industry.

REFERENCES

Adusei-Mensah, F., Inkum, E., Mawuli Agbale, C., Eric, A. Comparative evaluation of the insecticidal and insect repellent properties of the volatile oils of Citrus aurantifolia (Lime), Citrus sinensis (Sweet orange) and Citrus limon (Lemon) on Camponotus nearcticus (Carpenter ants). International Journal of Novel Research in Interdisciplinary Studies, 1 (2), 19–25, 2014.

Almeida Neto, P. P. Os sistemas agroflorestais dos Katitaurlu "território dos saberes, geografia da biodiversidade" Revista de Geografia da UFC, ano 3, n.5, 2004.

Ash, G. J. The science, art and business of successful bioherbicides. Biological Control, 52, 230–240, 2010.

Avishek, D., Hayat, U., Ferdous, Z. Utilization of by-products from food processing as biofertilizers and biopesticides. Food Processing by-Products and Their Utilization, 175–193, 2017.

Baas, W. J. Secondary plant compounds, their ecological significance and consequences for the carbon budget. In *Causes and consequences of variation in growth rare and productivity of higher plants* (ed H. Lumbers). SPB Academic Publishing Inc., La Haya, Holanda, 1989.

Bacchus, S. A taxonomic and biometric study of the genus Cryptotermes (Isoptera: Kalotermitidae). Tropical Pest Bulletin, 7, 1–91, 1987.

Balasubramanian, S., Tyagi, R. D. Biopesticide production from solid wastes. In *Current developments in biotechnology and bioengineering* (pp. 43–58). Elsevier, 2017.

Banumathi, B., Malaikozhundan, B., Vaseeharan, B. Invitro acaricidal activity of ethnoveterinary plants and green synthesis of zinc oxide nanoparticles against Rhipicephalus *(Boophilus)* microplus. Veterinary Parasitology, 216, 93–100, 2016.

Barik, S., Dash, L., Das, M. P., Panigrahi, T., Naresh, P., Kumari, M., . . . Acharya, G. C. Screening of brinjal (*Solanum melongena L.*) genotypes for resistance to spotted beetle, Henosepilachna vigintioctopunctata (Coccinellidae, Coleoptera). Journal of Entomology and Zoology Studies, 8, 297–301, 2020.

Barros, N. M. Heinzen, H., Cesio, M. V., Acosta, S., Mato, M., Sbeghen-Loss, A. C., . . . Frizzo, C. The use of residues from the citrus industry to control dry-wood termites *Cryptotermes brevis* (Isoptera-Kalotermitidae). Sociobiology, 51 (1), 283–293, 2008.

Berenbaum M. R., Feeny P. Toxicity of angular furanocoumarins to swallowtails: escalation in the coevolutionary arms race. Science, 212, 927–929, 1981.

Berenbaum, M. R. The chemistry of defense: Theory and practice. In *Chemical ecology: The chemistry of biotic interaction* (eds T. Eisner & J. Meinwald), pp. 1–17. National Academy of Sciences, Washington, DC, 1995.

Bertin, C., Weston, L. A., Huang, T., Jander, G., Owens, T., Meinwald, J., Schroeder, F. C. Grass roots chemistry: meta-tyrosine, an herbicidal nonprotein amino acid. Proceedings of the National Academy of Sciences, 104 (43), 16964–16969, 2007.

Braga, P. A. C., Soares, M. S., da Silva, M. F. D. G., Vieira, P. C., Fernandes, J. B., Pinheiro, A. L. Dammarane triterpenes from *Cabralea canjerana* (Vell.) Mart. (Meliaceae): Their chemosystematic significance. Biochemical Systematics and Ecology, 34 (4), 282–290, 2006.

Buss, E. A. Natural products for insect pest management. Available (online) http://edis.ifas. ufl.edu/pdffiles/IN/IN19700.pdf, 2002.

Butterworth, J. H., Morgan, E. D. Isolation of a substance that suppresses feeding in locusts. Journal of the Chemical Society, 23–24, 1968.

Cabral Marques, H. M. A review on cyclodextrin encapsulation of essential oils and volatiles. Flavour and Fragrance Journal, 25, 313–326, 2010.

Caminha Filho, A., Timbós e rotenona: uma riqueza nacional inexplorada. 2. ed. Serviço de Informação Agrícola, Rio de Janeiro, p. 14, 1940.

Castro Silva W., de Souza Martins, J. R., de Souza, H. E. M., Heinzen, H., Cesio, M. V., Mato, M., . . . de Barros, N. M. Toxicity of *Piper aduncum* L. (Piperales: Piperaceae) from the Amazon forest for the cattle tick *Rhipicephalus* (*Boophilus*) *microplus* (Acari: Ixodidae). Veterinary Parasitology, 164 (2–4), 267–274, 2009.

Castro Silva W., de Souza Martins, J. R., Cesio, M. V., Azevedo, J. L., Heinzen, H., de Barros, N., Acaricidal activity of *Palicourea marcgravii*, a species from the Amazon forest, on cattle tick *Rhipicephalus* (*Boophilus*) *microplus*. Veterinary Parasitology, 179 (1–3), 189–194, 2011.

Cesio, V., Dutra, C., Moyna, P., Heinzen, H. Morphological and chemical diversity in the type IV glandular trichomes of *Solanaceae* (*S. sisymbrifolium* and *N. glauca*) as germplasm resources for agricultural and food uses. Electronic Journal of Biotechnology, 9 (3), 2006.

Chung, S. K., Seo, J. Y., Lim, J. H., Park, H. H., Yea, M. J., Park, H. J. Microencapsulation of essential oil for insect repellent in food packaging system. Journal of Food Science, 78 (5), 709–714, 2013.

Cloyd, R. A., Natural indeed: are natural insecticides safer and better than conventional insecticides? Illinois Pesticide Review, 17 (3), 1–8, 2004.

Colquhoun, T. A., Verdonk, J. C., Schimmel, B. C., Tieman, D. M., Underwood, B. A., Clark, D. G. Petunia floral volatile benzenoid/phenylpropanoid genes are regulated in a similar manner. Phytochemistry, 71 (2–3), 158–167, 2010.

Corbett, C. E., Plantas ictiotóxicas: farmacologia da rotenona. Faculdade de Medicina da Universidade de São Paulo, São Paulo, p. 157, 1940.

Cravero, E. S., Guerra, M. S., Silveira, C. P. D. Manual de inseticidas e acaricidas: aspectos toxicológicos. Aimara Ltda., Pelotas, p. 229, 1976.

Dhillon, G. S., Oberoi, H. S., Kaur, S., Bansal, S., & Brar, S. K. Value-addition of agricultural wastes for augmented cellulose and xylanase production through solid state tray fermentation employing mixed-culture of fungi. Industrial Crops and Products, 34 (1), 1160–1167, 2011a.

Dhillon, G. S., Brar, S. K., Verma, M., Tyagi, R. D. Apple pomace ultrafiltration sludge—a novel substrate for fungal bioproduction of citric acid: optimization studies. Food Chemistry, 128 (4), 864–871, 2011b.

Dhillon, G. S., Kaur, S., Brar, S. K. Perspective of apple processing wastes as low-cost substrates for bioproduction of high value products: A review. Renewable and Sustainable Energy Reviews, 27, 789–805, 2013.

Doll, D. A., Sojka, P. E., Hallett, S. G. Effect of nozzle type and pressure on the efficacy of spray applications of the bioherbicidal fungus *Microsphaeropsis amaranthi*. Weed Technology, 19, 918–923, 2005.

Dong, Z., Ma, Y., Hayat, K., Jia, C., Xia, S., Zhang, X. Morphology and release profile of microcapsules encapsulating peppermint oil by complex coacervation. Journal of Food Engineering, 104 (3), 455–460, 2011.

Dugo, G., Bartle, K. D., Stagno D'Alcontres, I., Trozzi, A., Verzera, A., Bonaccorsi, I. Advanced Analytical techniques for the analysis of citrus essential oils. Part 3. Oxygen heterocyclic compounds: HPLC, HPLC/MS, OPLC, SFC, fast HPLC analysis. Essence Derivati Agrumari, 69, 251–283, 1999.

Durofil, A., Radice, M., Blanco-Salas, J., Ruiz-Téllez, T. Piper aduncum essential oil: a promising insecticide, acaricide and antiparasitic. A review. Parasite, 28, 2021.

Edwards, R., Mill, A. E. Termites in buildings. Their biology and control. Rentokil Ltd., East Grinstead, p. 261, 1986.

El-Abbassi, A., Saadaoui, N., Kiai, H., Raiti, J., Hafidi, A. Potential applications of olive mill wastewater as biopesticide for crops protection. Science of the Total Environment, 576, 10–21, 2017.

Enascuta, C. E., Oprescu, E. E., Capră, L., Gidea, M., Niculescu, M., Marius, B., Lavric, V. Bioproducts based on microencapsulated oils and biostimulants used in agriculture crops. Multidisciplinary Digital Publishing Institute Proceedings, 57 (1), 40, 2020.

Eski, A., Demirbağ, Z., Demir, İ. Microencapsulation of an indigenous isolate of Bacillus thuringiensis by spray drying. Journal of Microencapsulation, 36 (1), 1–9, 2019.

Evans, W. G. Drogas de Origen biológico. In *Farmacognosia* (13th ed.). Amsterdam Holland: Interamericana & McGraw-Hill, 1991a.

Evans, W. G. Fuentes biológicas y geográficas de las drogas. In *Farmacognosia* (13th ed.). Interamericana & McGraw-Hill, p. 65, 1991b.

Fang, Z., Bhandari, B. Encapsulation of polyphenols—a review. Trends in Food Science and Technology, 21, 510–523, 2010.

Fernandes, S. R., Barreiros, L., Oliveira, R. F., Cruz, A., Prudêncio, C., Oliveira, A. I., . . . Morgado, J. Chemistry, bioactivities, extraction and analysis of azadirachtin: State-of-the-art. Fitoterapia, 134, 141–150, 2019.

Fernandez-Bayo, J. D., Shea, E. A., Parr, A. E., Achmon, Y., Stapleton, J. J., VanderGheynst, J. S., . . . Simmons, C. W. (2020). Almond processing residues as a source of organic acid biopesticides during biosolarization. Waste Management, 101, 74–82.

Fuller, G., Nes, W. D. Plant lipids and their Interaction. In Ecology and metabolism of plant lipids (eds G. Fuller, W. D. Nes), Vol. 1, pp. 2–8. American Chemical Society, Miami, FL, 1985.

Gallo, D. FEALQ. Entomologia Agrícola. Piracicaba, 920p, 2002.

Galovičová, L., Borotová, P., Vukovic, N. L., Vukic, M., Kunová, S., Hanus, P., . . . Kačániová, M. The potential use of Citrus aurantifolia L. essential oils for decay control, quality preservation of agricultural products, and anti-insect activity. Agronomy, 12 (3), 735, 2022.

Gnepe J.R., Tyagi, R. D., Brar, S. K., Valero, J. R. Rheological profile of diets produced using agroindustrial wastes for rearing codling moth larvae for baculovirus biopesticides. Journal of Environmental Science and Health B, 46 (3), 220–230, 2011

Grassmann, J., Elstner, E. F. Essential oils, properties and uses. In Encyclopedia of food science and nutrition (ed B. Caballero, et al.), pp. 2177–2184. Amsterdam Holland: Elsevier, 2003.

Hackbar Medeiros, A., Tingley, W. Glandular trichomes of *Solanum berthaultii* and its hybrids with Solanum tuberosum affect nymphal emergence, development, and survival of Empoasca fabae (Homoptera: Cicadellidae). Journal of Economic Entomology, 99 (4), 1483–1489, 2006.

Harborne, J. B., Recent advances in ecological chemistry. Natural Product Reports, 3, 323–344,1986.

Harborne, J. B. Advances in chemical ecology. Natural Product Reports, 6, 327–348, 1993.

Harborne, J. B. Twenty-five years of chemical ecology. Natural Product Reports, 18, 361–379, 2001.

Hare, J. D. Biological activity of acyl glucose esters from *Datura wrightii* glandular trichomes against three native insect herbivores. Journal of Chemical Ecology, 31 (7), 141475–141491, 2005.

Hegnauer, R. Chemical evidence for the classification of some Plant Taxa. In Perspectives in phytochemistry (eds J. B. Harborne, T. Swain), pp. 121–123. Academic Press, Londres, 1969.

Hoagland, R. E. Biochemical interactions of the microbial phytotoxin phosphinothricin and analogs with plants and microbes. In Biologically active natural products: agrochemicals (eds. H. G. Cutler& S. J. Cutler). CRC Press, Boca Raton, FL, 1999.

Horton D. L. et al. Summary of losses from insect damage and costs of control in Georgia 1997 II Apple Insects. The bugwood network. www.bugwood.org, 1997.

Huang, X., Li, T., Shan, X., Lu, R., Hao, M., Lv, M., . . . Xu, H. High value-added use of citrus industrial wastes in agriculture: semisynthesis and anti-tobacco mosaic virus/insecticidal activities of ester derivatives of limonin modified in the B ring. Journal of Agricultural and Food Chemistry, 68 (44), 12241–12251, 2020.

Isman, M. B., Miresmailli, S., Machial, C. Commercial opportunities for pesticides based on plant essential oils in agriculture, industry and consumer products. Phytochemistry Reviews, 10 (2), 197–204, 2011.

Karnavar, G. K. Influence of azadirachtin on insect nutrition and reproduction. Proceedings of the Indian Academy of Sciences Animal Sciences, 96, 341–347, 1987.

Kathrina G. A., Antonio, L. O. J., Controle biológico de insectos mediante extractos botánicos. In Control Biologico de Plagas Agrícolas, pp. 137–160. CATIE, Managua, Manual Técnico/CATIE, 53, 2004.

Katovich, E. J. S., Becker, R. L., Kinkaid, B. D. Influence of nontarget neighbors and spray volume on retention and efficacy of triclopyr in purple loosestrife (Lythrum salicaria). Weed Science, 44, 143–147, 1996.

Kennedy, B.S., Nielsen, M. T., Severson, R. F. Biorationals from Nicotiana protect cucumbers against Colletotrichum lagenarium (Pass.) disease development. Journal of Chemical Ecology, 21 (2), 221–231, 1995.

Khalil, M. S. Abamectin and azadirachtin as eco-friendly promising biorational tools in integrated nematodes management programs. Journal of Plant Pathology & Microbiology, 4 (174), 10–4172, 2013.

Kilani-Morakchi, S., Morakchi-Goudjil, H., Sifi, K. Azadirachtin-based insecticide: Overview, risk assessments, and future directions. Frontiers in Agronomy, 3, 2021.

Kordan, B., Stec, K., Słomiński, P., Laszczak-Dawid, A., Wróblewska-Kurdyk, A., Gabryś, B. Antixenosis potential in pulses against the pea aphid (Hemiptera: Aphididae). Journal of Economic Entomology, 112 (1), 465–474, 2019.

Laohasongkram, K., Mahamaktudsanee, T., Chaiwanichsiri, S. Microencapsulation of macadamia oil by spray drying. Procedia Food Science, 1, 1660–1665, 2011.

Lawrie, J., Greaves, M. P., Down, V. M., Western, N. M. Studies of spray application of microbial herbicides in relation to conidial propagule content of spray droplets and retention on target. Biocontrol Science and Technology, 12 (1), 107–119, 2002a.

Lawrie, J., Greaves, M. P., Down, V. M., Western, N. M., Jaques, S. J. Investigation of spray application of microbial herbicides using Alternaria alternata on Amaranthus retroflexus. Biocontrol Science and Technology, 12 (4), 469–479, 2002b.

Leite, G. L. D., Resistência de tomates a pragas. Unimontes Científica, 6, 129–140, 2004.

Lima, R. R. Informações sobre duas espécies de timbó: Derris urucu (Killip et Smith) Macbride e Derris nicou (Killip et Smith) Macbride, como plantas inseticidas. Embrapa-CPATU (Documentos, 42), Belém, p. 23, 1987.

Lovatto, P. B., Goetze, M., Thomé, G. C. H. Efeito de extratos de plantas da família Solanaceae sobre o controle de Brevicoryne brassicae em couve (Brassica oleraceae var. Acephala). Ciência Rural, 34 (4), 971–978, 2004.

Macías, F. A. Allelopathy in the search for natural herbicide models. In Allelopathy. organisms, processes, and applications, Inderjit (eds K. M. M. Dakshini, F. A. Einhellig). ACS Symposium Series 582). ACS, Washignton DC, 1995.

Macías, F. A., Molinillo, J. M. G., Galindo, J. C. G., Varela, R. M., Torres, A., Simonet, A. M. Terpenoids with potential use as natural herbicide templates. In Biologically active natural products: Agrochemicals (eds H. G. Cutler, S. J. Cutler). CRC Press, Boca Raton, FL, 1999.

Macías, F. A., Oliveros-Bastidas, A., Marín, D., Carrera, C., Chinchilla, N., Molinillo, J. M. G. Plant biocommunicators: their phytotoxicity, degradation studies and potential use as herbicide models. Phytochemistry Reviews, 7, 179–194, 2008.

Mafara, S. M., Bakura, T. L. Multiple potentials of neem tree (Azadirachta Indica); a review. Bakolori Journal of General Studies, 11(1), 1–2, 2021.

Magrini F. E., Specht, A., Gaio, J., Girelli, C. P., Migues, I., Heinzen, H., . . . Miller, T. Viability of *Cabralea canjerana* extracts to control the South American fruit fly, *Anastrepha fraterculus*. Journal of Insect Science, 14 (1), 2014.

Magrini, F. E., Specht, A., Gaio, J., Girelli, C. P., Migues, I., Heinzen, H., . . . Cesio, V. Antifeedant activity and effects of fruits and seeds extracts of Cabralea canjerana canjerana (Vell.) Mart.(Meliaceae) on the immature stages of the fall armyworm Spodoptera frugiperda (JE Smith) (Lepidoptera: Noctuidae). Industrial Crops and Products, 65, 150–158, 2015.

Maia, J. G. S. Estudo químico de *Derris negrensis*. Acta Amazonica, 1, 59–61, 1976.

Maklouf, L. A. A volta do timbó: o terror das prgas. Globo Rural, São Paulo, 11, 96–89, 1986.

Matthews, E., Themelis, N. J. Potential for reducing global methane emissions from landfills. In Proceedings Sardinia, 11th international waste management and landfill symposium, Cagliari, 1–5, pp. 2000–2030, 2007.

Mentz, L. A., De Oliveira, P. L., Da Silva, M. V. Tipologia dos tricomas das espécies do gênero *Solanum* (Solanaceae) na Região Sul do Brasil. Iheringia, 54, 75–106, 2000.

Migues, I., et al. Evaluación de la actividad fitotóxica de seis especies de *Solanum* (Solanaceae) para el desarrollo de herbicidas bioracionales Exposed at 2° ENAQUI Montevideo-Uruguay, p. 122 (Reviews book), 2011.

Migues I., et al. Effects on the development of *Spodoptera frugiperda* using *Solanum sisymbrifolium* extracts. Exposed at XXII Simpósio de Plantas Medicinais do Brasil, Bento Gonçalves-RS, Brasil, 2012.

Miro Specos, M. M., Garcia, J. J., Gutierrez, A. C., Hermida, L. G. Application of microencapsulated biopesticides to improve repellent finishing of cotton fabrics. The Journal of the Textile Institute, 108(8), 1454–1460, 2017.

Mordue, J. A., Blackwell, A. Azadirachtin: an update. Journal of Insect Physiology, 39 (11), 903–924, 1993.

Nakanishi, K. Bioactive compounds from nature. Acta Pharm Nord, 4, 319–328, 1992.

Neal, J. J., Tingey, W. M., Steffens, J. C. Sucrose esters of carboxylic acids in glandular trichomes of *Solanum berthaultii* deter settling and probing by green peach aphid. Journal of Chemical Ecology, 16 (2), 487–497, 1990.

Nicoletti, M., Maccioni, O., Coccioletti, T., Mariani, S., Vitali, F. Neem tree (*Azadirachta indica* A. Juss) as source of bioinsectides. Insecticides-Advances in Integrated Pest Management, 1–20, 2012.

Niinemets, U., Reichstein, M. Controls on the emission of plant volatiles through stomata: differential sensitivity of emission rates to stomatal closure explained. Journal of Geophysical Research, 108, 4208, 2003.

Niinemets, U., Loreto, F., Reichstein, M. Physiological and physicochemical controls on foliar volatile organic compound emissions. Trends in Plant Science, 9 (4), 180–186, 2004.

Nisbet, A. J. Azadirachtin from the neem tree *Azadirachta indica*: its action against insects. Anais da Sociedade Entomológica do Brasil, 29, 615–632, 2000.

Oketch-Rabah H. A., Mwangi, J. W., Lisgarten, J., Mberu, E. K. A new antiplasmodial coumarin from *Toddalia asiatica* roots. Fitoterapia, 71 (6), 636–640, 2000.

Oldham, N. J., Boland, W. Chemical ecology: Multifunctional compounds and multitrophic interactions. Naturwissenschaften, 83, 248–254, 1996.

Pagare, S., Bhatia, M., Tripathi, N., Pagare, S., Bansal, Y. K. Secondary metabolites of plants and their role: Overview. Current Trends in Biotechnology and Pharmacy, 9 (3), 293–304.2015.

Pichersky, E., Noel, J. P., Dudareva, N. Biosynthesis of plant volatiles: Nature's diversity and ingenuity. Science, 31, 808–811, 2006.

Pickett, J. A., Wadhams, L. J., Woodcock, C. M. Developing sustainable pest control from chemical ecology. Agriculture, Ecosystems & Environment, 64, 149–156, 1997.

Pokorny, J. Pudil, F., Volfová, J., Valentová, H. Changes in the flavour of monoterpenes during their autoxidation under storage conditions. Developments in Food Science, 40, 667–677, 1998.

Poopathi, S. and Mani, C. Use of coffee husk waste for production of biopesticides for mosquito control. In Coffee in health and disease prevention (pp. 293–300). Academic Press, 2015.

Ramsdale, B. K., Messersmith, C. G. Adjuvant and herbicide concentration in spray droplets influence phytotoxicity. Weed Technology, 16, 631–637, 2002.

Rech-Cainelli, V., de Barros, N. M., Gianni, S., Sbeghen-Loss, A. C., Heinzen, H., Díaz, A. R., . . . Cesio, M. V. Antifeedant and repellent effects of neotropical Solanum extracts on drywood termites (*Cryptotermes brevis*, Isoptera: Kalotermitidae). Sociobiology, 62 (1), 82–87, 2015.

Renwick, J. A. A. Plant resistance to insects edited by Hedin, P. A. ACS Symposium Series (USA), no. 208. American Chemical Society, Washington, DC, p. 200, 1983.

Rodrigues, S. R. Atividade inseticida de extratos etanólicos de plantas sobre (J. E. Smith) (Lepidoptera: Noctuidae). Agrarian, 1 (1), 133–144, 2008.

Saito, M. L., Luchini, F. Substâncias obtidas de plantas e a procura por praguicidas eficientes e seguros ao meio ambiente. Embrapa Meio Ambiente, Jaguariúna, 46, 1998.

Sakulku, U., Nuchuchua, O., Uawongyart, N., Puttipipatkhachorn, S., Soottitantawat, A., Ruktanonchai, U. Characterization and mosquito repellent activity of citronella oil nanoemulsion. International Journal of Pharmaceutics, 372 (1), 105–111, 2009.

Sala, A., Barrena, R., Sanchez, A., and Artola, A. Fungal biopesticide production: Process scale-up and sequential batch mode operation with Trichoderma harzianum using agroindustrial solid wastes of different biodegradability. Chemical Engineering Journal, 425, 131620, 2021.

Samuelson, G. Natural products biosynthetically derived from acetate. In Drugs of natural origin (3rd ed.). Stockholm Sweden: Swedish Pharmaceutical Press, 1992.

Sarah, R., Tabassum, B., Idrees, N., Hussain, M. K. Bio-active Compounds isolated from Neem tree and their applications. In Natural bio-active compounds (pp. 509–528). Springer, Singapore, 2019.

Sarria, L. F., Soares, M. S., Matos, A. P., Fernandes, J. B., Vieira, J. B., da Silva, M. G. Effect of Triterpenoids and Limonoids Isolated from Cabralea canjerana and Carapa guianensis (Meliaceae) against *Spodoptera frugiperda* (J. E. Smith). Zeitschrift der Naturforschung C, 66 (5–6), 245–250, 2011.

Sbeghen-Loss A. C., Mato, M., Cesio, M. V., Frizzo, C., de Barros, N. M., Heinzen, H. Antifeedant activity of citrus waste wax and its fractions against the dry wood termite, *Cryptotermes brevis*. Journal of Insect Science, 11, 1–7, 2011.

Schlee, D. Ecologic significance of secondary natural products. Pharmazie, 46, 19–23, 1991.

Sciortino, M., Scurria, A., Lino, C., Pagliaro, M., D'Agostino, F., Tortorici, S., . . . Ciriminna, R. Silica-microencapsulated orange oil for sustainable pest control. Advanced Sustainable Systems, 5(4), 2000280, 2021.

Shalini, R., Gupta, D. K. Utilization of pomace from apple processing industries: A review. Journal of Food Science and Technology, 47, 365–371, 2010.

Shapiro, J. P. Phytochemicals at the plant-insect interface. Archives of Insect Biochemistry and Physiology, 17, 191–200, 1991.

Sharma, V. K. Mohan, D., Sahare, P. D. Fluorescence quenching of 3-methyl 7-hydroxyl Coumarin in presence of acetone. Spectrochimica Acta Part A: Molecular and Biomolecular Spectroscopy, 66 (1), 111–113, 2007.

Shi, Y. S., Zhang, Y., Li, H. T., Wu, C. H., El-Seedi, H. R., Ye, W. K., . . . Kai, G. Y. Limonoids from Citrus: Chemistry, anti-tumor potential, and other bioactivities. Journal of Functional Foods, 75, 104213, 2020.

Silva, G., Lagunes, A., Rodrígues, J. C., Rodrígues, D., Insecticidas vegetales; una vieja nueva opción en el manejo de insectos. Manejo Integrado de Plagas y Agroecología, 66, 4–12, 2001.

Silva, T. M. S., Carvalho, M. G. D., Braz-Filho, R., Agra, M. D. F. Ocorrência de flavonas, flavonóis e seus glicosídeos em espécies do gênero *Solanum* (Solanaceae). Química Nova, 26 (4), 517–522, 2003.

Silva, T. M. S., Agra, M. D. F., Bhattacharyya, J. Studies on the alkaloids of Solanum of northeastern Brazil. Revista Brasileira de Farmacognosia, 15 (4), 292–293, 2005.

Silva, L. S., Mar, J. M., Azevedo, S. G., Rabelo, M. S., Bezerra, J. A., Campelo, P. H., . . . Sanches, E. A. Encapsulation of Piper aduncum and Piper hispidinervum essential oils in gelatin nanoparticles: a possible sustainable control tool of Aedes aegypti, Tetranychus urticae and Cerataphis lataniae. Journal of the Science of Food and Agriculture, 99(2), 685–695, 2019.

Singh, M., Tan, S. Y., Sharma, S. D. Adjuvants enhance weed control efficacy of foliar-applied diuron. Weed Technology, 16, 74–78, 2002.

Siskos, E. P., Konstantopoulou, M. A., Mazomenos, B. E. Insecticidal activity of *Citrus aurantium* peel extract against *Bactrocera oleae* and *Ceratitis capitata* adults (Diptera: Tephritidae). Journal of Applied Entomology, 133 (2), 108–116, 2009.

Smith, C. M. Use of plant resistance in insect pest management systems. In Plant resistance to insects. Wiley, 1989.

Srivastava, M., Gupta, L. Effect of formulations of *Solanum surratense* (Family: Solanaceae) an Indian desert plant on oviposition by pulse beetle *Callosobruchus chinensis* Linn. African Journal of Agricultural Research, 2 (10), 552–554, 2007.

Sulistyo, A., Inayati, A. Mechanisms of antixenosis, antibiosis, and tolerance of fourteen soybean genotypes in response to whiteflies (Bemisia tabaci). Biodiversitas Journal of Biological Diversity, 17(2), 2016.

Swain, T. Secondary compounds as protective agents. Annual Review of Plant Biology, 28 (1), 479–501, 1977.

Szafranek, B., Synak, E., Waligóra, D., Szafranek, J., Nawrot, J. Leaf surface compouds of potato (*Solanum tuberosum*) and their influence on Colorado potato beetle (*Leptinotarsa decemlineata*) feeding. Chemoecology, 18 (4), 205–216, 2008.

Tiwari, R., Rana, C. S. Plant secondary metabolites: a review. International Journal of Engineering Research and General Science, 3(5), 661–670, 2015.

Torssell, K. B. G. Chemical Ecology (2nd ed.). Swedish Pharmaceutical Press, Stockholm, 1983.

Vázquez, A. Química y Biologia de Solanaceas. Estructura y actividad biológica de los glicósidos del género *Solanum*. Universidad de la República. Facultad de Química. Montevideo, 1997.

Vietmeyer et al. Neem a tree for solving global problems. National Research Council. *Neem.* The National Academies Press, Washington, DC, p. 39, 1992.

Vijayakumar, R., Raja, S. S. (Eds.). Secondary Metabolites: Sources and Applications. BoD—Books on Demand, 2018.

Vyvyan, J. R. Allelochemicals as leads for new herbicides and agrochemicals. Tetrahedron, 58 (9), 1631–1646, 2002.

Waterman, P. G., The chemical systematics of alkaloids: A review emphasising the contribution of Robert Hegnauer. Biochemical Systematics and Ecology, 27 (4), 395–406, 1999.

Zubairi, S. I., Othman, Z. S., Sarmidi, M. R., Aziz, R. A. Environmental friendly bio-pesticide Rotenone extracted from Derris sp.: A review on the extraction method, toxicity and field effectiveness. Jurnal Teknologi, 78(8), 2016.

17 Fungal Biopesticides and Their Uses for Control of Insect Pests and Diseases*

Deepak Kumar, M. K. Singh, Hemant Kumar Singh, and K. N. Singh

CONTENTS

DOI: 10.1201/9781003265139-20

17.1 INTRODUCTION

Agriculture has been facing the vicious activities of many pests like fungi, weeds, and insects from time immemorial, leading to drastic decline in yields. Pests are regularly being introduced to fresh areas either naturally or accidentally, or, in a few cases, organisms that are deliberately introduced become pests. Global employment has resulted in increased numbers of insidious nonnative pest species being introduced to fresh areas. Controlling these invasive species presents an unparalleled challenge worldwide. After many years of successful control by conventional agrochemical insecticides, a number of factors are challenging the effectiveness and continued use of these agents. These include the development of insecticide resistance and use-cancellation or deregistration of some insecticides due to human health and environmental issues. Therefore, an ecofriendly substitute is the need in the present scenario. Advances in pest control approaches are one method of generating elevated quality and a greater quantity of agricultural products. Therefore, there is a requirement to develop biopesticides that are efficient, biodegradable, and not harmful to the environment. Their shield against pests is a main concern, and due to the poor impact of chemical insecticides, use of biopesticides is escalating.

17.2 CONCEPT OF BIOPESTICIDES

According to the US Environmental Protection Agency (EPA), biopesticides are pesticides prepared from natural sources such as animals, plants, bacteria, and minerals. Biopesticides also include live organisms that obliterate agricultural pests. The EPA divides biopesticides into three main classes on the basis of the type of active ingredient used: namely, biochemical, plant-incorporated protectants, and microbial pesticides (USEPA, 2008). Biochemical pesticides are chemicals either isolated from natural source or manufactured to have a similar structure and function to naturally occurring chemicals. Biochemical pesticides differ from conventional pesticides in their construction (source) and mode of action (method by which they kill or control pests) (O'Brien et al., 2009). At the global level, there is a discrepancy in understanding the term *biopesticide* as the aforementioned definition of the term used by the EPA is not followed around the world. That is why the International Biocontrol Manufacturer's Association (IBMA) and the International Organization for Biological Control (IOBC, 2008) endorse the term *biocontrol agents* (BCAs) instead of *biopesticides* (Guillon, 2003). IBMA categorize BCAs into four groups: (1) macrobials, (2) microbials, (3) natural products, and (4) semiochemicals (insect behavior-modifying agents). Among the BCAs, the most important products are microbials (41%), followed by macrobials (33%), and, finally, other natural products (26%) (Guillon, 2003). This chapter spotlights microbe-based biopesticides, specifically fungal biopesticides.

Biopesticides or biological pesticides rely on pathogenic microorganisms specific to a target pest and present an ecologically effective solution to pest problems. They cause less threat to the environment and to human health. The most frequently used biopesticides are living organisms that are pathogenic for the pest of concern. These include biofungicides (*Trichoderma*), bioherbicides (*Phytopthora*), and bioinsecticides (*Beauvarria bassiana, Metarhizium anisopliae*). The benefits to agriculture and public health programs of the use of biopesticides are substantial. The interest in biopesticides is based on the rewards associated with such products, which include:

- Inherently fewer harmful effects and a lower environmental load
- Designed to affect only one specific pest or, in a few cases, some target organisms
- Often effective in very minute quantities and often decaying quickly, thereby resulting in less exposure and largely avoiding pollution problems
- Contribute much when used components of integrated pest management (IPM) programs

17.2.1 FUNGAL BIOPESTICIDES

Fungal pathogens have a major role in the development of diseases on various important field and horticultural crops, resulting in huge plant-yield losses (Khandelwal et al., 2012). Intensified use of fungicides has amassed toxic compounds potentially hazardous to humans and the environment, as well as the buildup of resistance to the pathogens. Fungal biopesticides, including other fungi, bacteria, nematodes, and weeds, can be used to manage insects and plant diseases. Their means of action are varied and depend on both the pesticidal fungus and the target pest. One benefit of fungal biopesticides, in contrast with many bacterial and all viral biopesticides, is that they do not need to be eaten to be effective. However, they are living organisms that often need a narrow range of situations, including moist soil and cool temperatures, to flourish. BCAs like *Trichoderma* are acclaimed as efficient, ecofriendly, and cheap, nullifying the ill effects of chemicals (Table 3.1). Therefore, of late, these BCAs are recognized as acting against an array of important soil-borne plant pathogens causing severe diseases of crops (Bailey and Gilligan, 2004). Fungal biopesticides used against plant pathogens include *T. harzianum*, which is an antagonist of *Rhizoctonia, Pythium, Fusarium*, and other soil- borne pathogens (Harman, 2005). *Trichoderma* is a fungal antagonist that enters the main tissue of a disease-causing fungus and exudes enzymes that degrade the cell walls of the other fungus and then eat the contents of the cells of the target fungus and multiply its own spores. *Trichoderma* is one of the general fungal BCAs being used worldwide for management of a variety of foliar and soil-borne plant pathogens like *Ceratobasidium, Fusarium, Rhizoctonia, Macrophomina, Sclerotium, Pythium*, and *Phytophthora* spp. (Dominguesa et al., 2000; Anand and Reddy, 2009). *Trichoderma viride* has shown to be very promising against soil-borne plant parasitic fungi (Khandelwal et al., 2012). A precise strain, *Muscodor albus* QST 20799, is a naturally occurring fungus initially isolated from the bark of a cinnamon tree in Honduras. The *M. albus* strain is reported to generate a number of volatiles, mainly alcohols, acids, and esters, that inhibit and kill specific bacteria and other organisms that

TABLE 17.1
Fungal Biopesticides Developed or Being Developed for the Biological Control of Pests

Product	Fungus	Target Pests	Producer Companies and Countries
BIO 1020	*Metarhizium anisopliae*	Vine weevils	Licensed to Taensa, US
Bio-Blast	*Metarhizium anisopliae*	Termites	EcoScience, US
Biogreen	*Metarhizium anisopliae*	Scarab larvae in pasture	Bio-care Technology, Australia
Bio-Path	*Metarhizium anisopliae*	Cockroaches	EcoScience, US
Boverin	*Beauveria bassiana*	Colorado beetles	Former USSR
Conidia	*Beauveria bassiana*	Coffee berry borers	Live Systems Technology, Colombia
Corn Guard	*Beauveria bassiana*	European corn borers	Mycotech, US
Engerlingspilz	*Beauveria brongniartii*	Cockchafers	Andermatt, Switzerland
Jas Bassi	*Beauveria bassiana*	Colorado beetles	Shri Ram Solvent Ext. Pvt., India
Jas Meta	*Metarhizium anisopliae*	Sugarcane spittle bugs, termites	Shri Ram Solvent Ext. Pvt., India
Jas Verti	*Verticillium lecanii*	Whiteflies and thrips	Shri Ram Solvent Ext. Pvt., India
Laginex	*Lagenidium giganteum*	Mosquito larvae	AgraQuest, US
Metaquino	*Metarhizium anisopliae*	Spittle bugs	Brazil
Mycotal	*Verticillium lecanii*	Whiteflies and thrips	Koppert, The Netherlands
Mycotrol WP	*Beauveria bassiana*	Whiteflies, aphids, thrips	Mycotech, US
Naturalis-L	*Beauveria bassiana*	Cotton pests including bollworms	Troy Biosciences, US
Ostrinil	*Beauveria bassiana*	Corn borers	Natural Plant Protection (NPP), France
Pae-Sin	*Paecilomyces fumosoroseus*	Whiteflies	Agrobionsa, Mexico
PFR-97	*Paecilomyces fumosoroseus*	Whiteflies	ECO-tek, US
Proecol	*Beauveria bassiana*	Army worms	Probioagro, Venezuela
Schweizer Beauveria	*Beauveria brongniartii*	Cockchafers	Eric Schweizer, Switzerland
Vertalec	*Verticillium lecanii*	Aphids	Koppert, The Netherlands

Source: Adapted and modified from Butt et al. (2001) and Rai et al. (2014)

generate soil-borne and postharvest diseases in plants. Products containing QST 20799 can be also used in fields, greenhouses, and warehouses (USEPA, 2008) for disease control. Other two fungi—namely, *B. bassiana* (Balsamo) Vuillemin and *Metarhizium anisopliae* (Metchnikoff) Sorokin—are naturally occurring entomopathogenic fungi that infect sucking pests, including *Nezara viridula*

(L) (green vegetable bug) and *Creontiades* sp. (green and brown mirids) (Sosa-Gómez and Moscardi, 1998). Fungi have the unique ability to attack insects by penetrating the cuticle, making them ideal for the control of sucking pests. *B. bassiana* is currently registered in the US as Mycotrol ES (Mycotech, Butte) and Naturalis L (Troy Biosciences). These biopesticide products are registered against sucking pests, such as whitefly, aphids, thrips, mealybugs, leafhoppers, and weevils. Studies also show that *B. bassiana* is potent against *Lygus hesperus* Knight (Hemiptera: Miridae), a major pest of alfalfa and cotton in the United States (Noma and Strickler, 2000).

17.2.2 *BEAUVARIA* SPP

The genus *Beauveria* includes at least 49 species of which approximately 22 are regarded as pathogenic (Kirk, 2003). *Beauveria bassiana*, a white muscardine fungus, is one of the most historically central fungus commonly used in this genus. It was originally recognized as *Tritirachium shiotae* and renamed after the Italian lawyer and scientist Agostino Bassi, who first implicated it as the causative agent of a white (later yellowish or occasionally reddish) muscardine disease in domestic silkworms (Furlong and Pell, 2005; Zimmermann, 2007). *Beauveria bassiana* is a fungus that grows naturally in soils all through the world and acts as a pathogen on various insect species belonging to the entomopathogenic fungi (Sandhu et al., 2004; Jain et al., 2008). When the microscopic spores of the fungus get in touch with the body of an insect host, they germinate, penetrate the cuticle, and grow inside, killing the insect within a matter of days. Later, a white mold appears on the cadaver and produces fresh spores. A typical isolate of *B. bassiana* can attack a broad range of insects; various isolates differ in their host range. An attractive attribute of *Beauveria* spp. is the high host specificity of several isolates. Hosts of agricultural and forest implication include the Colorado potato beetle, the codling moth, several genera of termites, and the American bollworm *Helicoverpa armigera* (Thakur et al., 2010). *Beauveria bassiana* can easily be isolated from insect cadavers or from soil in forested regions using media (Beilharz et al., 1982), as well as by baiting soil with insects.

17.2.3 *METARHIZIUM* SPP

The fungus *Metarhizium anisopliae*, first known by the name *Entomphthora anisopliae*, was first described near Odessa in Ukraine from infected larvae of the wheat cockchafer *Anisopliae austriaca* in 1879 and, later on, *Cleonus punctiventis* by Metschnikoff. It was later renamed *M. anisopliae* by Sorokin in 1883 (Tulloch, 1976). *Metarhizium* causes a disease in insect hosts known as "green muscardine" for the green color of its conidial cells. When these mitotic (asexual) spores (called conidia) of the fungus come in touch with the body of an insect host, they germinate, and the hyphae that emerge penetrate the cuticle. The fungus then expands within the body, ultimately killing the insect after a few days; this lethal outcome is very likely aided by the creation of insecticidal cyclic peptides (destruxins). If the ambient moisture is high enough, a white mould then develops on the cadaver that soon turns green as spores are produced. Several insects living near the soil have advanced natural defences against entomopathogenic fungi like *M. anisopliae*. The genus *Metarhizium* is pathogenic to a large number of insect species, many of which are agricultural and forest insects (Ferron, 1978). The taxonomy of *Metarhizium* is complex. Using the morphological characteristics of conidia and conidiogenous cells, Tulloch (1976) documented *M. flavoviride* and *M. anisopliae*, of which the latter was further subdivided into two variants: *majus* (short conidia up to 9 μm) and *anisopliae* (long conidia up to 18 μm). Currently, *M. anisopliae* consists of four varieties or genetic groups (Driver et al., 2000).

17.2.4 *VERTICILLIUM LECANII*

Verticillium lecanii is a broadly distributed fungus which can cause huge epizootics in tropical and subtropical regions, as well as in warm and humid environments (Nunez et al., 2008). It was reported by Kim et al. (2002) that *V. lecanii* is an effective BCA against *Trialeurodes vaporariorum* in South Korean greenhouses. The conidia (spores) of *V. lecanii* are slimy and connect to the cuticle of insects. The fungus infects the insects by producing hyphae from germinating spores that enter the insect's integument; the fungus then obliterates the internal contents, and the insect dies. The fungus eventually rises out through the cuticle and sporulates on the outside of the insect body. Infected insects emerge as white to yellowish cottony particles. Diseased insects typically appear in seven days. However, due to environmental conditions, there may be considerable lag time from infection to the death of insect. The fungus *V. lecanii* works best at temperatures of 15–25 °C and at a relative humidity of 85–90%. This fungus attacks nymphs and adults and sticks to the leaf underside by means of a filamentous mycelium (Nunez et al., 2008). In the 1970s, *V. lecanii* was developed to manage whitefly and several aphid species, including green peach aphids (*Myzus persicae*), in greenhouse chrysanthemums (Hamlen et al., 1979).

17.2.5 *PAECILOMYCES* SPP

The ungus *Paecilomyces fumosoroseus* is one of the most important natural enemies of whiteflies worldwide and causes a sickness called yellow muscardine (Nunez et al., 2008). A strong epizootic prospective against *Bemisia* and *Trialeurodes* spp. in both greenhouse and open field environments was reported with *Paecilomyces fumosoroseus*. The ability of this fungus to grow broadly over the leaf surface under humid conditions is a characteristic that certainly enhances its ability to extend rapidly through whitefly populations (Wraight et al., 2000). Kim et al. (2002) accounted that *P. fumosoroseus* is best for managing the nymphs of whitefly. These fungi cover the whitefly's body with mycelial threads and stick them to the bottom of the leaves. The nymphs show a "feathery" aspect and are surrounded by mycelia and conidia (Nunez et al., 2008). *Paecilomyces lilacinus* has been characterized as aggressive, and Dunn et al. (1982) stated, "The fungal egg parasites as group appear to be more promising to investigate as potential biological control agents of nematodes." Research results utilizing this fungus have been contradictory and erratic. In one experiment, the fungus caused a 71% decline in root galls and 90% decline in egg masses on root-knot nematode–infected corn (Ibrahim et al., 1987).

17.2.6 NOMURAEA SPP

Nomuraea rileyi, another potential entomopathogenic fungi, is a dimorphic hyphomycete that can cause epizootic death in various plant insects. It has been demonstrated that several insect species belonging to Lepidoptera, including *Spodoptera litura*, and few belonging to Coleoptera are susceptible to *N. rileyi* (Ignoffo, 1981). The host specificity of *N. rileyi* and its ecofriendly nature encourage its use in insect pest management. Its mode of infection and development have been reported for several insect hosts, such as *Trichoplusiani*, *Heliothis zea*, *Plathypena scabra*, *Bombyx mori*, *Pseudoplusia includes*, and *Anticarsia gemmatalis*.

17.3 PATHOGENICITY AND MODE OF ACTION OF ENTOMOPATHOGENIC FUNGAL BIOPESTICIDES

Insect pathogenic fungi are different in pathogenicity from bacteria and viruses in that they infect insects by breaching the host cuticle. The cuticle is composed of chitin fibrils embedded in a matrix of proteins, lipids, pigments, and N-acylcatecholamines (Richard et al., 2010). Fungi exude extracellular enzymes proteases, chitinases, and lipases to degrade the major constituents of the cuticle (i.e., protein, chitin, and lipids) and allow hyphal infiltration (Wang et al., 2005; Cho et al., 2006). The success of the infection was reliably proportional to the emission of exo-enzymes (Khachatourians, 1996). It is considered that both mechanical force and enzymatic action are involved in the penetration of fungus to the hemocoel of the insect. There are a large number of toxic compounds in the filtrate of entomopathogenic fungi, such as small secondary metabolites, cyclic peptides, and macromolecular proteins. *B. bassiana* is reported to secrete low molecular weight cyclic peptides and cyclosporins A and C with insecticidal properties such as beauvericin, enniatins, bassianolide (Roberts, 1981; Vey et al., 2001). Unlike other biopesticides such as bacteria and viruses, entomopathogenic fungi do not have to be ingested to cause infection, making them valuable as BCAs. Although little information suggests a mode of infection through the siphon tips or gut of insect larvae (Lacey et al., 1988), entomopathogenic fungi generally infect or penetrate their targets percutaneously (Charnley, 1989). This can occur by adhesion of spores to the insect integument, particularly the intersegmental folds, or by simple tarsal contact (Clarkson and Charnley, 1996).

Normally, germinated conidia manufacture an appressorium, which then forms a contagion peg (St. Leger et al., 1991). The penetration of conidia throughout the host's cuticle involves both mechanical pressure and enzymatic degradation (Bidochka et al., 1995; Clarkson and Charnley, 1996). Enzymatic degradation involves the production of large amounts of numerous cuticle-degrading enzymes, which differ according to the species and strains of the fungi. These enzymes show varying levels of pathogenicity toward their hosts. Following successful penetration of the cuticle, the fungus then produces blastospores or hyphae bodies, which are passively distributed in the hemolymph and the fat body (Hajek and Leger, 1994). So as to kill their host, fungal pathogens discharge a wide variety of secondary metabolic compounds, commonly called toxins, inside the insect host, mainly in the hemocoel. Two fungi, *M. anisopliae* and *B. bassiana*, secrete large amounts of a single extracellular protease called chymoelastase protease or Pr1 to degrade the host cuticle (St. Leger et al., 1992 and 1996). The endomoprotease Pr1, which is the major enzyme secreted by *Metarhi- zium* throughout the degradation process of the cuticle, also differs among strains in terms of biochemistry. However, the cuticle degrading-enzymes are manufactured sequentially, with the proteolytic enzymes and esterases first, followed by the chitinases; in other words, the proteins surrounding the cuticle must be degraded before the actions of the chitinases begin (Leger et al., 1996).

Time to fatality of an infected insect varies from 2 to 15 days post-infection, depending on the fungal strain and species and the characteristics of the host (Boucias and Pendland, 1998). When the infection procedure followed by the death of the host is complete, the fungus changes back to its hyphal mode. Under relatively humid conditions, the fungus then grows out of the cadaver surface to produce new, external, infective conidial saprophytic growth (Jianzhong et al., 2003; Mitsuaka, 2004). Under very dry conditions, the fungus may remain in the hyphal stage inside the cadaver, where the conidia are produced (Hong et al., 1997). Under positive conditions, sporulated cadavers can infect other individuals from the same target species through horizontal transmission (Meadow et al., 2000; Quesada-Moraga et al., 2004).

17.4 BIOFUNGICIDE OR MYCOFUNGICIDE

A range of fungal species can be used as biological control means and may offer effective activity against a variety of pathogenic microorganisms; these are known as biofungicides or mycofungicides (Table 17.2). Examples of these biofungicides or mycofungicides are *Trichoderma harzianum*, a species with biocontrol potential against *Botrytis cineria*, *Fusarium*, *Pythium*, and *Rhizoctonia* (Khetan, 2001); *Ampelomyces quisqualis*, a hyperparasite of powdery mildew (Liang et al., 2007; Viterbo et al., 2007); *Chaetomium globosum* and *C. cupreum*, which have biocontrol activity against root rot caused by *Fusarium*, *Phytophthora*, and *Pythium* (Soytong et al., 2001); and *Gliocladium virens*, an effective biocontrol of soil-borne pathogens (Viterbo et al., 2007). An efficient BCA should be genetically constant, effective in low concentrations, easy to mass produce in culture on inexpensive media, and be very effective against a wide range of pathogens (Wraight et al., 2001; Irtwange, 2006).

17.4.1 *Trichoderma*

Trichoderma species are common in the soil and root ecosystems of several plants, and they are easily isolated from soil, decaying wood, and other organic material (Howell, 2003; Zeilinger and Omann, 2007). *Trichoderma* species have been used

TABLE 17.2
Fungal Biofungicides/Mycofungicides Developed or Being Developed for the Biological Control of Diseases

Product	Fungus	Target Pests	Producer Companies and Countries
AQ10 Biofungicide	*Ampelomyces quisqualis*	Powdery mildews	Ecogen Inc., US
Aspire	*Candida oleophila*	*Botrytis* spp., *Penicillium* spp.	Ecogen Inc., US
Binab T	*Trichoderma harzianum*	Fungi causing wilt	Bio-Innovation, Sweden
Biofox C	*Fusarium oxysporum*	*Fusarium oxysporum, F. moniliforme*	SIAPA, Italy
Bio-Trek, RootShield	*Trichoderma harzianum*	*Rhizoctonia solani, Sclerotium rolfsii, Pythium*	BioWorks (TGT Inc.) Geneva, US
Cotans WG	*Coniothyrium minitans*	*Sclerotinia* species	Prophyta, Germany KONI, Germany
Fusaclean	*Fusarium oxysporum*	*Fusarium oxysporum*	Natural Plant Protection, France Thailand
Ketomium	*Chaetomium* sp.	*Botrytis cinerea, Didymella applanata, Fusarium oxysporum*, and *Rhizoctonia solani*	
Neemoderma	*Trichoderma harzianum, Trichoderma viride*	Wide range of fungal diseases	Shri Ram Solvent Ext. Pvt., India
Polygandron, Polyversum	*PPythium oligandrum*	*PPythium ultimum*	PPlant Protection Institute, Slovak Republic
Primastop	*Gliocladium catenulatum*	Several plant diseases	Kemira, Agro Oy, Finland
SoilGard (GlioGard)	ThermoTrilogy, USA	*Gliocladium virens*	Several plant diseases, damping off, and root pathogens
T-22 and T-22HB	*Trichoderma harzianum*	*Rhizoctonia solani, Sclerotium rolfsii, Pythium*	BioWorks (TGT Inc) Geneva, US
Trichoderma 2000	*Trichoderma harzianum*	*Rhizoctonia solani, Sclerotium rolfsii, Pythium*	Mycontrol (EfA1) Ltd, Israel
Trichodex	*Trichoderma harzianum*	Fungal diseases e.g., *Botrytis cinerea*	Makhteshim-Agan, several European companies e.g., DeCeuster, Belgium.
Trichodowels, Trichoject,	*Trichoderma viride*	*Chondrostereum purpureum* and other soil and foliar pathogens	Agrimms Biologicals, New Zealand
Trichopel	*Trichoderma harzianum*	Wide range of fungal diseases	Agrimm Technologies Ltd, New Zealand
YIELDPLUS	*Cryptococcus albidus*	*Botrytis* spp., *Penicillium* spp.	Anchor Yeast, S. Africa

Source: Adapted and modified from Butt et al. (2001) and Soytong et al. (2001)

as biological control means against a wide range of pathogenic fungi: for example, *Rhizoctonia* spp., *Pythium* spp., *Botrytis cinerea*, and *Fusarium* spp. *Phytophthora palmivora*, *P. parasitica*, and different species can be used: for example, *T. harzianum*, *T. viride*, *T. virens* (Sunantapongsuk et al., 2006; Zeilinger and Omann, 2007). Among them, *Trichoderma harzianum* is reported to be the most widely used as an efficient BCA (Szekeres et al., 2004; Abdel-Fattah et al., 2007). *Trichoderma* has a combination of biocontrol mechanisms (Benítez et al., 2004; Zeilinger and Omann, 2007). The main mechanisms are mycoparasitism and antibiosis (Vinale et al., 2008). Mycoparasitism relies on the recognition, binding, and enzymatic disturbance of the host fungus cell wall (Woo and Lorito, 2007). *Trichoderma* species have been very successfully used as mycofungicides as they are fast growing and have a high reproductive capability. *Trichoderma* also inhibit a broad spectrum of fungal diseases and have a variety of control mechanisms. *Trichoderma* have excellent competitors in the rhizosphere and the capacity to modify it and are tolerant of or resistant to soil fungicides. This fungus also has the ability to survive under unfavorable conditions, is efficient in utilizing soil nutrients, is strong against phytopathogenic fungi, and also promotes plant growth (Benítez et al., 2004; Vinale et al., 2006). Its ability to colonize and produce in association with plant roots is known as rhizosphere competence.

17.4.2 CHAETOMIUM

Chaetomium species are usually found in soil and organic compost (Soytong et al., 2001). The application of *Chaetomim* as a biological control for controlling of plant pathogens began in about 1954 when Martin Tviet and M. B. Moor found *C. globosum* and *C. cochliodes* on oat seeds and that these taxa provided some control of *Helminthosporium victoriae* (Tviet and Moor, 1954). *Chaetomium* species have been reported to be potential antagonists of various plant pathogens, especially soil-borne and seed-borne pathogens (Aggarwal et al., 2004; Park et al., 2005). Numerous species of *Chaetomium* with probable BCA capacity suppress the growth of bacteria and fungi through competition (for substrate and nutrients), mycoparasitism, antibiosis, or various combinations of these (Marwah et al., 2007; Zhang and Yang, 2007). *Chaetomium globosum* and *C. cupreum* have been extensively considered and successfully applied to control root rot in citrus, black pepper, and strawberries and have been reported to minimize damping off in sugar beets (Soytong et al., 2001; Tomilova and Shternshis, 2006). These taxa have been prepared in the form of powder and pellets as Ketomium, a broad spectrum mycofungicide. Ketomium has been also registered as a biofertilizer for decaying organic matter and to induce plant immunity and stimulate plant growth (Soytong et al., 2001).

17.4.3 GLIOCLADIUM

Gliocladium species are general soil saprobes, and numerous species have been reported to be parasites of several plant pathogens (Viterbo et al., 2007): for example, *Gliocladium catenulatum* parasites, *Sporidesmium sclerotiorum*, and *Fusarium* spp. These species demolish the fungal host by direct hyphal contact and form pseudoappressoria (Punja and Utkhede, 2004; Viterbo et al., 2007). *Gliocladium catenulatum* (Strain JI446) has also been employed as a wettable powder called Primastop by Kemira Agro Oy in Finland. This product can be used on soils, roots, and foliage to minimize the frequency of damping-off disease caused by *Pythium ultimum* and *Rhizoctonia solani* in the greenhouse (Paulitz and Belanger, 2001; Punja and Utkhede, 2004). *Gliocladium virens* produces antibiotic metabolites such as gliotoxin, which have important functions, including antibacterial, antifungal, antiviral, and antitumor. Recently, molecular confirmation indicated that *G. virens* is more strongly related to *Trichoderma* than those of *Gliocladium*. This chains suggest that this taxon should be referred to as *Trichoderma virens* (Hebbar and Lumsden, 1999; Punja and Utkhede, 2004).

17.4.4 AMPELOMYCES

Ampelomyces quisqualis is the mycoparasitic anamorphic ascomycete that decreases the growth of and kills powdery mildews. It can affect the pathogen through antibiosis and parasitism (Kiss, 2003; Viterbo et al., 2007). The fungus *A. quisqualis* was the initial organism reported to be a hyperparasite of powdery mildew, and it can be established simply in powdery mildew colonies (Paulitz and Belanger, 2001). Hyphae of *Ampelomyces* pierce the hyphae of powdery mildews and grow inside, then kill all the parasitized cells (Kiss, 2003). *Ampelomyces quisqualis* isolate M-10 has been prepared as AQ10 Biofungicide, developed by Ecogen, Inc. in the United States. This mycofungicide includes conidia of *A. quisqualis* and is formulated as water-soluble granules for the control of powdery mildew on carrots, cucumbers, and mangos (Kiss, 2003; Viterbo et al., 2007).

17.4.5 OTHER FUNGI AS BIOFUNGICIDES

Coniothyrium minitans is an anamorphic coelomycete which has been reported to be a mycoparasite of *Sclerotinia* species: for example, *Sclerotinia minor*, *S. sclerotiorum*, *S. trifoliorum*, and *S. cepivorum* (Viterbo et al., 2007; Whipps et al., 2008). It has been used effectively to control disease in numerous crops including lettuce, (Jones et al., 2004), rapeseed (Li et al., 2006), peanuts (Partridge et al., 2006), and alfalfa (Li et al., 2005). The utilization of nonpathogenic strains of *Fusarium oxysporum* to control Fusarium wilt has been reported for several crops, but there has been little commercial production due of a lack of understanding of their genetics, biology, and ecology (Fravel et al., 2003; Kvas et al., 2009). Nonpathogenic *F. oxysporum* strain Fo47 has been promoted as a liquid formulation called Fusaclean by Natural Plant Products, Nogueres, France, for soilless culture (Khetan, 2001; Paulitz and Belanger, 2001). *Pythium oligandrum* has proven its ability to control soil-borne pathogens both in the laboratory and in the field. *Pythium oligandrum* oospores have been used as seed treatments which minimize damping-off disease caused by *P. ultimum* in sugar beets (Khetan, 2001). *Pythium oligandrum* has been formulated as a granular or powder product called Polygangron by Vyskumny Ustav in the Slovak Republic (Khetan, 2001). This fungus has indirect effects by controlling pathogens in the rhizosphere and/or direct effects by inducing plant resistance after application. It also provokes plants to respond more quickly and efficiently to pathogen infections and enhances phosphorus uptake (Le Floch et al., 2003). Several other fungi that can be employed as mycofungicides are *Aspergillus* and *Penicillium* species. *Aspergillus* species are useful against white-rot basidiomycetes (Bruce and Highley, 1991). The fungal antagonists *Aureobasidium pullulans* and *Ulocladium atrum* have also been analyzed for control of *Botrytis aclada*, which causes onion neck rot in the field (Köhl et al., 1997).

17.5 MECHANISMS OF BIOLOGICAL CONTROL

Biological control may result from direct or indirect communication between the beneficial microorganisms and the pathogen. Direct communication may entail physical contact and synthesis of hydrolytic enzymes, toxic compounds, or antibiotics as well as competition. An indirect communication may result from induced resistance in the host plant or the use of organic soil modifications to advance the activity of antagonists against the pathogens (Benítez et al., 2004; Viterbo et al., 2007). The mechanisms of BCAs and reaction with the pathogen are numerous and complex interactions among the host and microorganisms. These mechanisms are influenced by soil type, temperature, pH, and the moisture of the plant and soil environment, as well as by the existence of other microorganisms (Howell, 2003). There are four principle mechanisms of biological control in plants: namely, antibiosis, competition, mycoparasitism or lysis, and induced resistance (Fravel et al., 2003; Viterbo et al., 2007).

17.5.1 ANTIBIOSIS

Antibiosis is described as the inhibition or destruction of the microorganism by substances such as specific or nonspecific metabolites or by the production of antibiotics that inhibit the growth of one or more microorganisms (Benítez et al., 2004; Haggag and Mohamed, 2007). Many BCA agents produce several types of antibiotics (Lewis et al., 1989; Handelsman and Stabb, 1996). Some antibiotics have been reported to play function in disease control (Lewis et al., 1989), either by impeding spore germination (fungistasis) or killing the cells (antibiosis) (Benítez et al., 2004; Haggag and Mohamed, 2007). *Gliocladium* and *Trichoderma* species are well known BCAs which create a wide range of antibiotics and control disease by various mechanisms (Whipps, 2001; Harman et al., 2004). Gliovirin metabolites produced by *Gliocladium virens* can kill *Pythium ultimum* by causing coagulation of the protoplasm of the cell (Whipps, 2001; Viterbo et al., 2007).

17.5.2 COMPETITION

Competition occurs between microorganisms when space and nutrients are limited (Lewis et al., 1989; Viterbo et al., 2007). The rhizosphere of a plant and the soil zone are where competition for space and nutrients mainly occurs (Viterbo et al., 2007). Competition between the BCA and the pathogen can result in displacement of the pathogen. BCAs can compete with other fungi for food and essential elements in the soil and around the rhizosphere (Chet et al., 1990; Irtwange, 2006) and can narrow the space or change the rhizosphere by acidifying the soil so that pathogens cannot breed (Benítez et al., 2004). For example, *Trichoderma harzianum* T-35's control of the *Fusarium* species on numerous crops arises via competition for nutrients and rhizosphere colonization (Viterbo et al., 2007).

17.5.3 MYCOPARASITISM

Mycoparasitism involves a complex process that includes the following steps: (1) the chemothophic growth of the antagonist to the host, (2) recognition of the host by the mycoparasite, (3) attachment, (4) excretion of extracellular enzymes, and (5) lysis and exploitation of the host (Whipps, 2001; Benítez et al., 2004; Viterbo et al., 2007). BCAs are able to lyse hyphae of pathogens by releasing lytic enzymes, and this is an important and powerful tool for control of plant disease (Chet et al., 1990, Viterbo et al., 2007) such as chitinases, proteases, and β-1, 3 glucanases (Whipps, 2001). For example, β-1, 3 glucanases produced by *Chaetomium* sp. can degrade the cell walls of plant pathogens including *Rhizoctonia solani*, *Gibberella zeae*, *Fusarium* sp., *Colletotrichum gloeosporioides*, and *Phoma* sp. (Sun et al., 2006). Proteases produced by *Trichoderma harzianum* T-39 are involved in the degradation of pathogen hyphal membranes and cell walls.

17.5.4 INDUCED RESISTANCE

Induced resistance is reported in large number of plants in response to infestation by phytopathogens (Harman et al., 2004). Induced resistance of host plants can be restricted and/or systematic, depending on the nature, source, and amount of stimuli of pathogens (Pal and Gardener, 2006). Induced resistance by BCAs entails a similar suite of genes and gene products engaged in plant response known as systematic acquired resistance (SAR) (Handelsman and Stabb, 1996; Whipps, 2001). *Trichoderma* strains are proficient at creating interaction-induced metabolic changes in plants that enhance resistance to a broad range of plant pathogenic fungi (Harman et al., 2004).

17.6 BIOHERBICIDES

Phytopathogenic microorganisms or microbial phytotoxins are valuable source of biological weed management and apply in related ways to conventional herbicides (Goeden, 1999; Boyetchko et al., 2002; Boyetchko and Peng, 2004). Bioherbicides serve an additional significant role as a complementary component in flourishing integrated management strategies (Hoagland et al., 2007) and not as a replacement for chemical herbicides and other weed management tactics (Singh et al., 2006). Several microbial components have been under evaluation for their prospects as bioherbicides for horticultural crops, turf, and forest trees, including obligate fungal parasites, soil-borne fungal pathogens, nonphytopathogenic fungi, pathogenic and nonpathogenic bacteria, and nematodes (Kremer, 2005). DeVine (Encore Technologies, Plymouth, MN, US) was one of the first herbicides registered with the active ingredient *Phytophthora palmivora*, which was developed to manage the strangler

TABLE 17.3
Fungal Bioherbicides/Mycoherbicides Developed or Being Developed for the Biological Control of Weeds

Sl. no.	Fungus	Product Name	Target Weed	Supplier and Country Where Registered
1.	*Cercospora rodmanii*	ABG 5003	Water hyacinth (*Eichhornia crassipes*)	Abbott Labs, US
2.	*Chondrosterium purpureum*	BioChon	Black cherry (*Prunus serotina*) in forestry	Koppert, The Netherlands
3.	*Colletotrichum gloeosporioides* f. sp. *Malvae*	Biomal	Mallow (*Malva pusilla*) in wheat and lentils	Canada
4.	*Alternaria cassiae*	Casst	Sicklepod (*Cassia obtusifolia*) and coffee senna (*C. occidentalis*) in soybeans and peanuts	US
5.	*Colletotrichum gloeosporioides* f. sp *aeschynomene*	Collego	Northern jointvetch (*Aeschynomene virginica*) in rice	Encore Technologies, US
6.	*Phytophthora palmivora*	Devine	Milkweed vine (*Morrenia odorata*) in Florida citrus	Sumitomo, Valent, US
7.	*Colletotrichum gloeosporioides* f. sp *cuscutae*	Luboa 2	*Cuscuta chinensis, C. australis* in soybeans	PR China
8.	*Colletotrichum coccodes*	Velgo	Velvetleaf (*Abutilon theophrasti*) in corn and soybeans	US, Canada

vine (*Morrenia odorata*) on citrus in Florida (Charudattan, 2005) (Table 3.3). Plant pathogens are used as BCAs and can cause severe damage to target weed species. In order to become appropriate pathogens, they have to be mass produced and their pathogenicity tested on weeds in a variety of environmental conditions, followed by field efficacy and host range tests (Ayres and Paul, 1990). A range of phytotoxins produced by plant pathogens can interfere with plant metabolism, ranging from delicate effects on gene expression to plant mortality (Walton, 1996). A few fungal pathogens are toxic to a broad range of weed species. The early mycoherbicides (DeVine; Collego, with the active ingredient *Colletotrichum gloeosporioides* f. sp. *aeschynomene*; Biomal, with the active ingredient *Colletotrichum gloeosporioides*) had extremely virulent fungal plant pathogens that could be mass cultured to produce large quantities of inoculum for inundative application to the weed host. These fungi infect the aerial portion of weed hosts, resulting in noticeable disease symptoms (Charudattan, 2005).

The rust fungus *Puccinia canaliculata* is a foliar pathogen of yellow nutsedge (*Cyperus esculentus*), and it can be mass cultured on the weed host in little field plots or in greenhouses (Phatak et al., 1983). Application of the fungal pathogen *Chonrotereum purpureum* to damaged branches or stumps of weedy tree species inhibited resprouting and corroded the woody tissues of the plant (Prasad, 1996). Weidemann et al. (1992) reported that the fungal pathogen *Microsphaeropsis amaranthi* controlled a few pigweed (*Amaranthus*) species, while *Phoma proboscis* controlled field bindweed (*Convolvulus arvensis*), and *Colletotrichum capsici* controlled morning glory (*Ipomoea* spp.). The naturally occurring fungus *Phoma macrostoma* has been tested as a control agent for dandelion (*Taraxacum officinale*), Canada thistle (*Cirsium arvense*), chickweed (*Stellaria media*), and scentless chamomile (*Matricaria perforata*), and its result is comparable to the industry standard synthetic herbicide pendimethalin (Bailey and Derby, 2001). A study by Héraux et al. (2005) on *Trichoderma virens* (*Gliocladium virens*) reported that colonization of composted chicken manure significantly reduced the emergence and growth of redroot pigweed (*Amaranthus retroflexus*) and broadleaf weeds in fields of horticulture crops.

17.7 LIMITATION IN SUCCESSFUL UTILIZATION OF FUNGAL BIOPESTICIDES

Billions of dollars have been invested by the biopesticide companies for the development of a variety of microbial products so as to eliminate crop diseases. It is impossible to accurately predict the market trends for biopesticides, and there is a substantial difference between predicting global sales and selecting categories of biopesticides. The agriculture market is observing an increase in demand for environmentally friendly, chemical residue–free organic products. However, growth in several regions is hindered by vigorous established chemical pesticide markets. However, there is lack of awareness of benefits of biopesticides and their uneven efficiency. The lack of awareness, knowledge, and confidence by farmers is one of the major reasons for the lagging of these ecofriendly pest control options. Commercial biological control of fungal biopesticides in the present global scenario is a high-tech venture, in terms of both safety and sustenance. The viability and virulence of fungal inoculum (conidia) after field application is the prerequisite threshold for their efficacy (Doust and Roberts, 1983). Diverse isolates of *B. bassiana* and *M. anisopliae* have been the most relied-on entomopathogens, have been extensively researched, and have considerable effect in field tests and their commercial usage for insect pest management (Easwaramoorthy, 2003). Upon field application, the entomopathogens are exposed to a variety of abiotic stresses like temperature (Rangel et al., 2005a), UV radiation (Rangel et al., 2006a), humidity-osmolarity (Lazzarini et al., 2006), edaphic factors, and nutrient sources (Shah et al., 2005) that negatively affect the field use of fungi as BCAs. Solar radiation, which includes visible light, UV radiation, infrared rays, and radio waves, has been the dominant source in which all organisms evolve and adapt.

In a biological context, UV radiation merits a special mention in terms of its impact on life (Bjorn, 2006). Soil temperature is a major factor which affects the success or failure in the establishment and production of fungal inoculum (Thomas and Jenkins, 1997). Entomopathogenic fungi have tolerance to soil temperature and also have the capacity to survive through the thermoregulatory defence response of the host insect (Ouedraogo et al., 2003). It has been confirmed that high/low temperatures change the vegetative growth among isolates of entomopathogenic fungi (Ouedraogo et al., 2004). Dry heat exposure causes DNA breakage through base loss, leading to depurination, and this may cause mutation in several cases (Nicholson et al., 2000). Wet heat (that is, heat in conjunction with high humidity) results in protein denaturation and membrane disorganization of the cell. It has been reported that *M. anisopliae* has a temperature tolerance upper limit of 37–40 °C (Thomas and Jenkins, 1997).

B. bassiana, on the other hand can survive up to a maximum temperature of 37 °C (Fargues et al., 1997). In fungi, the temperature range for germination and mycelial growth has been reported to have a similar pattern. Environmental factors also affect pathogenicity as well as mode of virulence of entomopathogenic fungi (Hasan, 2014).

The efficacy of fungal bioherbicides, often due to environmental factors, is the major limiting feature for their use. There are humidity needs for the establishment and spread of many foliar and stem fungal pathogens for weed control. There are also a few elements necessary for the development of special formulations to ensure the effectiveness of agents applied in the field. An extended dew period is needed by a few pathogens for infection on the aerial surfaces of target weeds (Auld et al., 2003). The bioherbicide application process should be judged by how it enhances the efficacy of the BCA. This includes attention to spray droplet size, droplet retention and distribution, spray application volume, and the equipment used (Charudattan, 2001). Other factors, such as the spectrum of the bioherbicide (whether broad or targeted to specific species), the type of formulation, and whether it involves amino acid–excreting strains, can significantly affect efficacy.

17.8 CONCLUSION

The relevance of fungal biopesticides in biological control is increasing largely because of superior environmental awareness, food safety concerns, and the breakdown of conventional chemicals due to an increasing number of insecticide-resistant species. Whether or not the use of fungal biopesticides has been thriving in pest and disease management, it is necessary to regard each case individually, and direct comparisons with chemical insecticides are usually unsuitable. Fungal biopesticides as components of an integrated system can offer significant and selective insect and disease control. In the near future, we should see synergistic combinations of microbial control agents with other technologies (in combination with semiochemicals, soft chemical pesticides, fungicides, other natural enemies, resistant plants, chemigation, remote sensing, etc.) that will augment the effectiveness and sustainability of integrated control strategies for plant disease.

REFERENCES

Abdel-Fattah, M. G.; Shabana, M. Y.; Ismail, E. A.; Rashad, M. Y. *Trichoderma harzianum*: A Biocontrol Agent against *Bipolaris oryzae*. *Mycopathologia* **2007**, *164*, 81–89.

Aggarwal, R.; Tewari, A. K.; Srivastava, K. D.; Singh, D. V. Role of Antibiosis in the Biological Control of Spot Blotch (*Cochliobolus sativus*) of Wheat by Chaetomium Globosum. *Mycopathologia* **2004**, *157*, 369–377.

Anand, S.; Reddy, J. Biocontrol Potential of *Trichoderma* Sp Against Plant Pathogens. *Inter. J. Agri. Sci.* **2009**, *2*, 30–39.

Auld, B. A.; Hethering, S. D.; Smith, H. E. Advances in Bioherbicide Formulation. *Weed Biol. Man.* **2003**, *3*, 61–67.

Ayres, P.; Paul, N. Weeding with Fungi. *New Sci.* **1990**, *732*, 36–39.

Bailey, D. J.; Gilligan, C. A. Modeling and Analysis of Disease Induced Host Growth in the Epidemiology of Take All. *Phytopathology* **2004**, *94*, 535–540.

Bailey, K. L.; Derby, J. Fungal Isolates and Biological Control Compositions for the Control of Weeds. U.S. Patent 60/294,475, May 20, 2001.

Beilharz, V. C.; Parbery, D. G.; Swart, H. J. Dodine: a selective agent for certain soil fungi. *Trans. British Mycol. Soc.* **1982**, *79*(3), 507-511.

Benítez, T.; Rincón, M. A.; Limón, M. C.; Codón, C. A. Biocontrol Mechanisms of *Trichoderma* Strains. *Int. Microbiol.* **2004**, *7*, 249–260.

Bidochka, M. J.; St Leger, R. J.; Joshi, L.; Roberts, D. W. An Inner Cell Wall Protein (Cwp1) From Conidia of the Entomopathogenic Fungus *Beauveria bassiana*. *Microbiology* **1995**, *141*, 1075–1080.

Bjorn, L. O. Stratospheric Ozone, Ultraviolet Radiation, and Cryptogams. *Biol. Conservation* **2006**, *135*, 326–333.

Boucias, D. R.; Pendland, J. C. Entomopathogenic Fungi: Fungi Imperfecti. In *Principles of Insect Pathology*; Boucias, D. R., Pendland, J. C., Eds; Kluwer Academic Publishers: Dordrecht, 1998, pp 321–359.

Boyetchko, S. M.; Peng, G. Challenges and Strategies for Development of Mycoherbicides. In *Fungal Biotechnology in Agricultural, Food, and Environmental Applications*; Arora, D. K., Ed., Marcel Dekker: New York, 2004; pp 11–121.

Boyetchko, S. M.; Rosskopf, E. N.; Caesar, A. J. Charudattan, R. Biological Weed Control with Pathogens: Search for Candidates to Applications. In *Applied Mycology and Biotechnology*, Vol. 2; Khachatourians, G. G., Arora, D. K. Eds.; Elsevier: Amsterdam, 2002; pp 239–274.

Bruce, A.; Highley, L. T. Control of Growth of Wood Decay Basidiomycetes by *Trichoderma* Spp. and Other Potentially Antagonistic Fungi. *Forest Products J.* **1991**, *41*, 63–67.

Butt, T. M.; Jackson, C. W.; Magan, N., Eds. *Fungal Biological Control Agents: Progress, Problems and Potential*. CABI International: Wallingford, Oxon, 2001.

Charnley, A. K. Mechanisms of Fungal Pathogenesis in Insects. In *The Biotechnology of Fungi for Improving Plant Growth*; Whipps, J. M., Lumsden, R. D., Eds.; Cambridge University: London, 1989; pp 85–125.

Charudattan, R. Biological Control of Weeds by Means of Plants Pathogens: Significance for Integrated Weed Management in Modern Agroecology. *Biocontrol* **2001**, *46*, 229–260.

Charudattan, R. Use of Plant Pathogens as Bioherbicides to Manage Weeds in Horticultural Crops. *Proc. Fla. State Hort. Soc.* **2005**, *118*, 208–214.

Chet, I.; Ordentlich, A.; Shapira, R.; Oppenheim, A. Mechanisms of Biocontrol of Soil-borne Plant Pathogens by *Rhizobacteria*. *Plant Soil* **1990**, *129*, 85–92.

Cho, E. M.; Boucias, D.; Keyhani, N. O. EST Analysis of CDNA Libraries from the Entomopathogenic Fungus *Beauveria* (Cordyceps) *Bassiana*. II. Fungal Cells Sporulating on Chitin and Producing Oosporein. *Microbiology* **2006**, *152*, 2855–2864.

Clarkson, J. M.; Charnley, A. K. New Insights into the Mechanisms of Fungal Pathogenesis in Insects. *Trends Microbiol.* **1996**, *4*, 197–203.

Dominguesa, F. C.; Queiroza, J. A.; Cabralb, J. M. S.; Fonsecab, L. P. The Influence of Culture Conditions on Mycelial Structure and Cellulose Production by *Trichoderma reesei* rut C-30. *Enz. Microb. Technol.* **2000**, *26*, 394–401.

Doust, R. A.; Roberts, D. W. Studies on the Prolonged Storage of *Metarhizium Anisopliae* Conidia: Effect of Temperature and Relative Humidity on Conidial Viability and Virulence Against Mosquitoes. *J. Invertebrate Pathol.* **1983**, *41*, 143–150.

Driver, F.; Milner, R. J.; Trueman, J. W. H. A Taxonomic Revision of *Metarhizium* Based on a Phylogenetic Analysis of RDNA Sequence Data. *Mycol. Res.* **2000**, *104*, 134–150.

Dunn, M. T.; Sayee, R. M.; Carrell, A.; Wergin, W. R. Colonization of Nematodes Eggs by *P. lilacinus* Samson of Observed with SEM. *SEM* **1982**, *3*, 1351–1357.

Easwaramoorthy, S. Entomopathogenic Fungi. In *Biopesticides and Bioagents in Integrated Pest Management of Agricultural Crops*; Srivastava, R. P. Ed., International Book Distributing Co.: Lucknow, 2003; pp 341–379.

Fargues, J.; Goettel, M. S.; Smits, N.; Ouedraogo, A.; Rougier, M. Effect of Temperature on Vegetative Growth of *Beauveria Bassiana* Isolates from Different Origins. *Mycologia* **1997**, *89*, 383–382.

Ferron, P. Biological Control of Insect Pests by Entomologenous Fungi. *Ann. Rev. Entomol.* **1978**, *23*, 409–442.

Fravel, D.; Olivain, C.; Alabouvette, C. *Fusarium oxysporum* and Its Biocontrol. *New Phytologist* **2003**, *157*, 493–502.

Furlong, M. J.; Pell, K. J. Interactions Between Entomopathogenic Fungi and Arthropod Natural Enemies. In *Insect—fungal Associations: Ecology and Evolution*; Vega, F. E.; Blackwell, M., Eds.; Oxford University Press: Oxford, 2005; pp 51–73.

Goeden, R. D. Projects on Biological Control of Russian Thistle and Milk Thistle in California: Failures That Contributed to the Science of Biological Weed Control. In *Abstracts of the 10th International Symposium on Biological Control of Weeds*; Spencer, N., Noweierski, R., Eds.; Montana State University: Bozeman, MT, 1999; p 27.

Guillon, M. L. Regulation of Biological Control Agents in Europe. In *International Symposium on Biopesticides for Developing Countries*; Roettger, U., Reinhold, M., Eds.; CATIE: Turrialba, 2003; pp 143–147.

Haggag, W. M.; Mohamed, H. A. A. Biotechnological Aspects of Microorganisms Used in Plant Biological Control. *Am. Eur. J. Sustainable Agric.* **2007**, *1*, 7–12.

Hajek, A. E.; St. Leger, R. J. Interactions Between Fungal Pathogens and Insects Hosts. *Ann. Rev. Entomol.* **1994**, *39*, 293–322.

Hamlen, R. A. Biological Control of Insects and Mites on European Greenhouse Crops: Research and Commercial Implementation. *Proc. Florida State Horticultural Soc.* **1979**, *92*, 367–368.

Handelsman, J.; Stabb, V. E. Biocontrol of Soilborne Plant Pathogens. *Plant Cell* **1996**, *8*, 1855–1869.

Harman, G. E. Overview of Mechanisms and Uses of *Trichoderma* spp. 648. *Phytopathology* **2005**, *96*, 190–194.

Harman, G. E.; Howell, C. R.; Viterbo, A.; Chet, I.; Lorito, M. *Trichoderma* Species Opportunistic, Avirulent Plant Symbionts. *Nat. Rev. Microbiol.* **2004**, *2*, 43–56.

Hasan, S. Entomopathogenic Fungi as Potent Agents of Biological Control. *Int. J. Eng. Technol. Res.* **2014**, *2*, 234–237.

Hebbar, P. K.; Lumsden, R. D. Biological Control of Seedling Diseases. In *Methods in Biotechnology Vol. 5: Biopesticides: Use and Delivery*; Frinklin, R. H., Julius, J. M., Eds.; Humana Press: New York, 1999; pp 103–116.

Héraux, F. M. G.; Hallett, S. G.; Ragothama, K. G.; Weller, S. C. Composted Chicken Manure as a Medium for the Production and Delivery of *Trichoderma virens* for Weed Control. *Hort. Sci.* **2005**, *40*, 1394–1397.

Hoagland, R. E.; Weaver, M. A.; Boyette, C. D. Myrothecium Verrucaria Fungus: A Bioherbicide and Strategies to Reduce Its Non-target Risks. *Allelopathy J.* **2007,** *19* (1), 179–192.

Hong, T. D.; Ellis, R. H.; Moore, D. Development of a Model to Predict the Effect of Temperature and Moisture on Fungal Spore Longevity. *Ann. Botany* **1997,** *79,* 121–128.

Howell, R. C. Mechanisms Employed by *Trichoderma* Species in the Biological Control of Plant Diseases: The History and Evolution of Current Concepts. *Plant Dis.* **2003,** *87,* 4–10.

Ibrahim, I. K. A.; Raza, M. A.; El-Saedy, M. A.; Ibrahim, A. A. M. Control of *Meloidogyne incognita* on Corn, Tomato and Okra with *P. lilacinus* and the Nematicide Aldecarb. *Nematologia Mediterria* **1987,** *15,* 265–268.

Ignoffo, C. M. The Fungus *Nomuraea rileyi* as a Microbial Insecticide. In *Microbial Control of Pests and Plant Diseases*; Burges, H. D., Ed., Academic Press: London, 1981, 513–538.

IOBC. International Organization for Biological Control. *IOBC Newslet.* **2008,** *84,* 5–7.

Irtwange, V. S. Application of Biological Control Agents in Pre- and Postharvest Operations. Agricultural Engineering International: The CIGR Ejournal. *Invited Overview.* **2006,** *3,* 1–12.

Jain, N.; Rana, I. S.; Kanojiya, A.; Sandhu, S. S. Characterization of *Beaveria bassiana* Strains Based on Protease and Lipase Activity and Their Role in Pathogenicity. *J. Basic Appl. Mycol.* **2008,** *I-II,* 18–22.

Jianzhong, S.; Fuxa, J. R.; Henderson, G. Effects of Virulence, Sporulation, and Temperature on *Metarhizium anisopliae* and *Beauveria bassiana* Laboratory Transmission in *Coptotermes formosanus*. *J. Invertebrate Pathol.* **2003,** *84,* 38–46.

Jones, E. E.; Mead, A.; Whipps, J. M. Effect of Inoculum Type and Timing of Application of *Coniothyrium Minitans* on *Sclerotinia Sclerotio46 Rum*: Control of Sclerotinia Disease in Glasshouse Lettuce. *Plant Pathol.* **2004,** *53,* 611–620.

Khachatourians, G. G. Biochemistry and Molecular Biology of Entomopathogenic Fungi. In *Human and Animal Relationships*; Howard, D. H.; Miller, J. D., Eds, *Mycota* VI, Springer: Heidelberg, 1996, pp 331–363.

Khandelwal, M.; Datta, S.; Mehta, J.; Naruka, R.; Makhijani, K.; Sharma, G.; Kumar, R.; Chandra, S. Isolation, Characterization and Biomass Production of *Trichoderma Viride* Using Various Agro Products—A Biocontrol Agent. *Adv. Appl. Sci. Res.* **2012,** *3,* 3950–3955.

Khetan, S. K. *Microbial Pest Control*. Marcel Dekker Inc.: New York, Basel, 2001; p 300.

Kim, J. J.; Lee, M. H.; Yoon, C. S.; Kim, H. S.; Yoo, J. K.; Kim, K. C. Control of Cotton Aphid and Greenhouse Whitefly with a Fungal Pathogen. *J. Nat. Inst. Agric. Sci. Technol.* **2002,** 7–14.

Kirk, P. M. *Indexfungorum*, 2003. www.indexfungorum.org (accesses Oct 28, 2009).

Kiss, L. A Review of Fungal Antagonists of Powdery Mildews and Their Potential as Biocontrol Agents. *Pest Management Sci.* **2003,** *59,* 475–483.

Köhl, J.; Bélanger, R. R.; Fokkema, N. J. Interaction of Four Antagonistic Fungi with *Botrytis Aclada* in Dead Onion Leaves: A Comparative Microscopic and Ultrastructural Study. *Phytopathology* **1997,** *87,* 634–642.

Kremer, R. J. The Role of Bioherbicides in Weed Management. *Biopestic. Int.* **2005,** *1,* 127–141.

Kvas, M.; Marasas, W. F. O.; Wingfield, B. D.; Wingfield, M. J.; Steenkamp, E. T. Diversity and Evolution of *Fusarium* Species in the Gibberella Fujikuroi Complex. *Fungal Diversity* **2009,** *34,* 1–21.

Lacey, C. M.; Lacey, L. M.; Roberts, D. R. Route of Invasion and Histopathology of *Metrahizium anisopliae* in *Culex quinquefasciatus*. *J. Invertebrate Pathol.* **1988,** *52,* 108–118.

Lazzarini, G. M. J.; Rocha, L. F. N.; Luz, C. Impact of Moisture on In Vitro Germination of *Metarhizium anisopliae* and *Beauveria bassiana* and Their Activity on *Triatoma infestans. Mycol. Res.* **2006**, *100*, 485–492.

Le Floch, G.; Rey, P.; Benizri, E.; Benhamou, N.; Tirilly, Y. Impact of Auxin-compounds Produced by the Antagonistic Fungus *Pythium Oligandrum* or the Minor Pathogen *Pythium* Group F on Plant Growth. *Plant Soil* **2003**, *257*, 459–470.

Leger, R. J.; Joshi, L.; Bidochka, M. J.; Roberts, D. W. Construction of an Improved Mycoinsecticide Overexpressing a Toxic Protease. *Proc. Natl. Acad. Sci. U. S. Am.* **1996**, *93*, 6349–6354.

Lewis, K.; Whipps, J. M.; Cooke, R. C. Mechanisms of Biological Disease Control with Special Reference to the Case Study of Pytium Oligandrum as an Antagonist. In *Biotechnology of Fungi for Improving Plant Growth*; Whipps, J. M., Lumsden, R. D., Eds.; Cambridge University Press: Cambridge, UK, 1989; pp 191–217.

Li, G. Q.; Huang, H. C.; Acharya, S. N.; Erickson, R. S. Effectiveness of *Coniothyrium minitans* and *Trichoderma atroviride* in Suppression of Sclerotinia Blossom Blight of Alfalfa. *Plant Pathol.* **2005**, *54*, 204–211.

Li, G. Q.; Huang, H. C.; Miao, H. J.; Erickson, R. S.; Jiang, D. H.; Xiao, Y. N. Biological Control of Sclerotinia Diseases of Rapeseed by Aerial Applications of the Mycoparasite *Coniothyrium minitans. Eur. J. Plant Pathol.* **2006**, *114*, 345–355.

Liang, C.; Yang, J.; Kovács, G. M.; Szentiványi, O.; Li, B.; Xu, X. M.; Kiss, L. Genetic Diversity of *Ampelomyces Mycoparasites* Isolated From Different Powdery Mildew Species in China Inferred from Analyses of RDNA ITS Sequences. *Fungal Diversity* **2007**, *24*, 225–240.

Marwah, R. G.; Fatope, M. O.; Deadman, M. L.; Al-Maqbali, Y. M.; Husband, J. Musanahol: A New Aureonitol-related Metabolite from a *Chaetomium* sp. *Tetrahedron* **2007**, *63*, 8174–8180.

Meadow, R.; Vandberg, J. D.; Shelton, A. M. Exchange of Inoculum of *Beauveria bassiana* (Bals) Vuil. (Hyphomycetes) Between Adult Flies of the Cabbage Maggot *Delia radicum* L. (Diptera: Anthomyiidae). *Biocontrol Sci. Technol.* **2000**, *10*, 479–485.

Mitsuaka, S. Effect of Temperature on Growth of *Beauveria Bassiana* F-263, A Strain Highly Virulent to the Japanese Pine Sawyer, *Monochamus alternatus*, Especially Tolerance to High Temperatures. *Appl. Entomol. Zool.* **2004**, *39*, 469–475.

Nicholson, W. L.; Munakata, N.; Horneck, G.; Melosh, H. J.; Setlow, P. Resistance of *Bacillus* Endospores to Extreme Terrestrial and Extraterrestrial Environments. *Microbiol. Mol. Biol. Rev.* **2000**, *64*, 548–572.

Noma, T.; Strickler, K. Effects of *Beauveria bassiana* on Lygus *Hesperus* (Hemiptera: Miridae) Feeding and Oviposition. *Environ, Entmol.* **2000**, *29*, 394–402.

Nunez, E. J.; Iannacone; Omez, H. G. Effect of Two Entomopathogenic Fungi in Controlling *Aleurodicus cocois* (Curtis, 1846) (Hemiptera: Aleyrodidae). *Chilean J. Agric. Res.* **2008**, *68*, 21–30.

O'Brien, K. P.; Franjevic, S.; Jones, J. Green Chemistry and Sustainable Agriculture: The Role of Biopesticides. *Advancing Green Chem.* **2009**. *http://advancinggreenchemistry*.org/wp-content/uploads/Green-Chemand-   Sus.-Ag.-the-Role-of-Biopesticides.pdf. (accessed May 7, 2016).

Ouedraogo, A.; Fargues, J.; Goettel, M. S.; Lomer, C. J. Effect of Temperature on Vegetative Growth Among Isolates of *Metarhizium anisopliae* and *Metarhizium flavoviride. Mycopathologia* **2004**, *137*, 37–43.

Ouedraogo, R. M.; Cusson, M.; Goettel, M. S.; Brodeur, J. Inhibition of Fungal Growth in Thermoregulating Locusts, *Locusta Migratoria*, Infected by the Fungus *Metarhizium anisopliae* var. *acridum. J. Invertebr. Pathol.* **2003**, *82*, 103–109.

Pal, K.; Gardener, B. M. Biological Control of Plant Pathogens. *Plant Health Instructor* **2006**, 1–25. doi: 10.1094/PHI-A-2006-1117-02. APSnet.

Park, J. H.; Choi, G. J.; Jang, S. K.; Lim, K. H.; Kim, T. H.; Cho, Y. K.; Kim, J. C. Antifungal Activity Against Plant Pathogenic Fungi of Chaetoviridins Isolated from *Chaetomium globosum*. *FEMS Microbiol. Lett.* **2005**, *252*, 309–313.

Partridge, D. E.; Sutton, T. B.; Jordan, D. L.; Curtis, V. L.; Bailey, J. E. Management of Sclerotinia Blight of Peanut with the Biological Control Agent *Coniothyrium minitans*. *Plant Dis.* **2006**, *90*, 957–963.

Paulitz, T. C.; Belanger, R. R. Biological Control in Greenhouse System. *Ann. Rev. Phytopathol.* **2001**, *39*, 103–133.

Phatak, S. C.; Summer, D. R.; Wells, H. D.; Bell, D. K.; Glaze, N. C. Biological Control of Yellow Nutsedge with the Indigenous Rust fungus *Puccinia canaliculata*. *Science* **1983**, *219*, 1446–1447.

Prasad, R. Development of Bioherbicides for Integrated Weed Management in Forestry. In *Proceedings of the 2nd International Weed Control Congress*; Brown, H., Ed., Department of Weed Control and Pesticide Ecology: Slagelse, 25–28 June 1996; pp 1197–1203.

Punja, Z. K.; Utkhede, R. S. Biological Control of Fungal Diseases on Vegetable Crops with Fungi and Yeasts. In *Fungal Biotechnology in Agricultural, Food, and Environmental Applications*; Arora, D. K., Ed.; Basel: New York, 2004; pp 157–171.

Quesada-Moraga, E.; Santos-Quirós, R.; Valverde-García, P.; Santiago, Á. C. Virulence, Horizontal Transmission, and Sublethal Reproductive Effects of *Metarhizium Anisopliae* (Anamorphic Fungi) on the German Cockroach (Blattodea: Blattellidae). *J. Invertebr. Pathol.* **2004**, *87*, 51–58.

Rai, D.; Updhyay, V.; Mehra, P.; Rana, M.; Pandey, A. K. Potential of Entomopathogenic Fungi as Biopesticides. *Ind. J. Sci. Res. Tech.* **2014**, *2* (5), 7–13.

Rangel, D. E. N.; Braga, G. U. L.; Anderson, A. J.; Roberts, D. W. Variability in Conidial Thermo Tolerance of *Metarhizium Anisopliae* Isolates from Different Geographic Origins. *J. Invertebr. Pathol.* **2005a**, *88*, 116–125.

Rangel, D. E. N.; Butler, M. J.; Torabinejad, J.; Anderson, A. J.; Braga, G. U. L.; Day, A. W.; Roberts, D. W. Mutants and Isolates of *Metarhizium Anisopliae* are Diverse in Their Relationships Between Conidial Pigmentation and Stress Tolerance. *J. Invertebr. Pathol* **2006a**, *93*, 170–182.

Richard, J. S.; Neal, T. D.; Karl, J. K.; Michael, R. K. Model Reactions For Insect Cuticle Sclerotization: Participation of Amino Groups in the Cross-Linking of *Manduca Sexta* Cuticle Protein MsCP36. *Insect Biochem. Mol. Biol.* **2010**, *40*, 252–258.

Roberts, D. W. Toxins of Entomopathogenic Fungi. In *Microbial Control of Pests and Plant Diseases*; Burges, H. D., Ed.; Academic Press: New York, 1981; pp 441–464.

Sandhu, S. S.; Vikrant, P. Myco-insecticides: Control of Insect Pests. In *Microbial Diversity: Opportunities & Challenges*; Gautam, S. P.; Sandhu, S. S., Sharma, A., Pandey, A. K., Eds.; Indica Publishers: New Delhi, 2004.

Shah, F. A.; Wang, C. S.; Butt, T. M. Nutrition Influences Growth and Virulence of the Insect Pathogenic Fungus *Metarhizium anisopliae*. *Microbiol. Lett.* **2005**, *251*, 259–266.

Singh, H. P.; Batish, D. R.; Kohli, R. K. *Handbook of Sustainable Weed Management;* Food Products Press: Binghamton, NY, 2006.

Sosa-Gómez, D. R.; Moscardi, F. Laboratory and Field Studies on the Infection of Stink Bugs, Nezara viridula, Piezodorus guildinii, and Euschistus heros (Hemiptera: Pentatomidae) with Metarhizium anisopliae and *Beauveria bassianain*. *Brazil J. Invertebr. Pathol.* **1998**, *2*, 115–120.

Soytong, K.; Kanokmadhakul, S.; Kukongviriyapa, V.; Isobe, M. Application of *Chaetomium* Species (Ketomium®) As a New Broad Spectrum Biological Fungicide for Plant Disease Control: A Review Article. *Fungal Diversity* **2001**, *7*, 1–15.

St Leger, R. J.; Bidochka, M. J. Staples, R. C. Preparation Events During Infection of Host Cuticle by *Metarhizium anisopliae*. *J. Invertebrate Pathol.* **1991**, *58*, 168–179.

St Leger, R. J.; Joshi, L.; Bidochka, M. J.; Roberts, D. W. Construction of an Improved Mycoinsecticide Overexpressing a Toxic Protease. *Proc. Natl. Acad. Sci. U. S. A.* **1996**, *93*, 6349–6354.

St Leger, R. J.; May, B.; Allee, L. L.; Frank, D. C.; Staples, R. C.; Roberts, D. W. Genetic Differences in Allozymes and in Formation of Infection Structures Among Isolates of the Entomopathogenic Fungus *Metarhizium anisopliae. J. Invertebrate Pathol.* **1992**, *60*, 89–101.

Sun, H.; Yang, J.; Lin, C.; Huang, X.; Xing, R.; Zhang, K. Q. Purification and Properties of a B-1,3-Glucanase from *Chaetomium* Sp. that is Involved in Mycoparasitism. *Biotechnol. Lett.* **2006**, *28*, 131–135.

Sunantapongsuk, V.; Nakapraves, P.; Piriyaprin, S.; Manoch, L. In *Protease Production and Phosphate Solubilization from Potential Biological Control Agents Trichoderma Viride and Azomonas Agilis from Vetiver Rhizosphere*, International Workshop on Sustained Management of Soil-Rhizosphere System for Efficient Crop Production and Fertilizer Use, Land Development Department, Bangkok, Thailand, 2006; pp 1–4.

Szekeres, A.; Kredics, L.; Antal, Z.; Kevei, F.; Manczinger, L. Isolation and Characterization of Protease Overproducing Mutants of *Trichoderma harzianum. Microbiol. Lett.* **2004**, *233*, 215–222.

Thakur, R.; Sandhu, S. S. Distribution, Occurrence and Natural Invertebrate Hosts of Indigenous Entomopathogenic Fungi of Central India. *Ind. J. Microbiol.* **2010**, *50*, 89–96.

Thomas, M. B.; Jenkins, N. E. Effect of Temperature on Growth of *Metarhizium Flavoviride* and Virulence to the Variegated Grasshopper, *Zonocerus variegatus. Mycol. Res.* **1997**, *101*, 1469–1474.

Tomilova, O. G.; Shternshis, M. V. The Effect of a Preparation from *Chaetomium* Fungi on the Growth of Phytopathogenic Fungi. *Appl. Biochem. Microbiol.* **2006**, *42*, 76–80.

Tulloch, M. The Genus *Metarhizium. Trans. Britannic Mycol. Soc.* **1976**, *66*, 407–411.

Tviet, M.; Moor, M. B. Isolates of Chaetomium that Protect Oats from *Helminthosporium victoriae. Phytopathology* **1954**, *44*, 686–689.

USEPA. What Are Biopesticides? 2008. www.epa.gov/pesticides/biopesticides/whatare-biopesticides.htm. (accessed May 7, 2016).

Vey, A.; Hoagland, R.; Butt, T. M. Toxic Metabolites of Fungal Biocontrol Agents. In *Fungi as Biocontrol Agents*; Butt, T. M., Jackson, C. W., Magan, N., Eds.; CAB International: Wallingford, 2001; pp 311–345.

Vinale, F.; Marra, R.; Scala, F.; Ghisalbert, E. L.; Lorito, M.; Sivasithamparam, K. Major Secondary Metabolotes Produced by Two Commercial *Trichoderma* Strains Active Against Different Phytopathogens. *Lett. Appl. Microbiol.* **2006**, *43*, 143–148.

Vinale, F.; Sivasithamparam, K.; Ghisalberti, E. L.; Marra, R.; Woo, S. L.; Lorito, M. Trichoderma Plant Pathogen Interactions. *Soil Biol. Biochem.* **2008**, *40*, 1–10.

Viterbo, A.; Inbar, J.; Hadar, Y.; Chet, I. Plant Disease Biocontrol and Induced Resistance via Fungal Mycoparasites. In *Environmental and Microbial Relationships*, 2nd edn, *The Mycota IV*; Kubicek, C. P., Druzhinina, I. S., Eds.; Springer-Verlag: Berlin, Heidelberg, 2007; pp 127–146.

Walton, J. D. Host-selective Toxins: Agents of Compatibility. *Plant Cell* **1996**, *8*, 1723–1733.

Wang, C.; Hu, G.; St Leger, R. J. Differential Gene Expression by *Metarhizium Anisopliae* Growing in Root Exudate and Host (*Manduca Sexta*) Cuticle or Hemolymph Reveals Mechanisms of Physiological Adaptation. *Fungal Genetic Biol.* **2005**, *42*, 704–718.

Weidemann, G. J.; TeBeest, D. O.; Templeton, G. E. Fungal Plant Pathogens Used for Biological Weed Control. *Ark. Farming Res.* **1992**, *41*, 6–7.

Whipps, J. M. Microbial Interactions and Biocontrol in the Rhizosphere. *J. Exp. Botany* **2001**, *52*, 487–511.

Whipps, J. M.; Sreenivasaprasad, S.; Muthumeenakshi, S.; Rogers, C. W.; Challen, M. P. Use of *Coniothyrium minitans* as a Biocontrol Agent and Some Molecular Aspects of Sclerotial mycoparasitism. *Eur. J. Plant Pathol.* **2008**, *121*, 323–330.

Woo, L. S.; Lorito, M. Exploiting the Interactions Between Fungal Antagonists, Pathogens and the Plant for Biocontrol. In *Novel Biotechnologies for Biocontrol Agent Enhancement and Management*; Vurro, M., Gressel, J., Eds.; Springer: Berline, 2007; pp 107–130.

Wraight, S. P.; Carruthers, R. I.; Jaronski, S. T.; Bradley, C. A.; Garza, C. J.; Wraight, S. G. Evaluation of the Entomopathogenic Fungi *Beauveria Bassiana* and *Paecilomyces Fumosoroseus* for Microbial Control of the Silver Leaf Whitefly, *Bemisia argentifolii. Biol. Control* **2000,** *17,* 203–217.

Wraight, S. P.; Jackson, M. A.; de Kock, S. L. Production, Stabilization and Formulation of Fungi Biocontrol Agents. In *Fungi as Biocontrol Agents Progress, Problem and Potential;* Butt, T. M., Jackson, C. W., Magan, N. Eds.; CABi Publishing: Wallingford, 2001; pp 253–288.

Zeilinger, S.; Omann, M. Trichoderma Biocontrol: Signal Transduction Pathways Involved In Host Sensing and Mycoparasitism. *Gene Regulation Syst. Biol.* **2007,** *1,* 227–234.

Zhang, H. Y.; Yang, Q. Expressed Sequence Tags-based Identification of Genes in the Biocontrol Agent *Chaetomium cupreum. Appl. Microbiol. Biotechnol.* **2007,** *74,* 650–658.

Zimmermann, G. Review on Safety of the Entomopathogenic Fungus *Beauveria bassiana* and *Beauveria brongniartii. Biocontrol Sci. Technol.* **2007,** *17,* 553–596.

18 Allelochemicals

Hyda Haroon, Irzam Haroon, Zareeka Haroon, Yawar Sadiq, Kavita Munjal, and Showkat Rasool Mir

CONTENTS

18.1 INTRODUCTION

Plants can interfere with neighbouring plants' establishment and growth through competition, allelopathy, or both. Allelopathy, as opposed to resource competition, involves the release of allelochemicals into the environment from living or dead plants [1]. As a result, identifying allelochemicals from plants and their environments is critical to understanding plant-plant allelopathic interactions. Numerous allelochemicals from various plant species have been investigated and identified thus far. Allelochemicals are chemically diverse, with phenolic compounds (simple phenolics, flavonoids, coumarins, and quinones), terpenoids (monoterpenes, sesquiterpenes, diterpenes, triterpenes, and steroids), alkaloids, nitrogen-containing chemicals (non-protein amino acids, benzoxazinoids, cyanogenic glycosides), and many other chemical families represented [2]. The first confirmed allelochemical was juglone, a 1,4-naphthoquinone derived from black walnut (*Juglans nigra*) [3]. Black walnut

DOI: 10.1201/9781003265139-21

is a woody species that wreaks havoc on understory plants. Although this allelopathic phenomenon has been documented for thousands of years, the allelochemical juglone was only isolated and identified a few decades ago [3, 4]. Juglone occurs in black walnut as a non-toxic naphthol O-glycoside. The naphthol O-glycoside biosynthesised in living tissues may be released into the environment through leaves, barks, and roots. By hydrolysis or soil microbial interactions, naphthol O-glycoside is rapidly transformed into aglycone, a less phytotoxic naphthol, and then oxidised to phytotoxic juglone [5].

18.2 MAJOR ALLELOCHEMICALS PRESENT IN VARIOUS PLANTS

Phytochemicals are classified into broad categories that begin with lipids and include simple and fictionalised hydrocarbons as well as terpenes. Natural plant products are classified into two types—primary constituents and secondary constituents—depending on whether they play an important role in plant metabolism and are found in all plants. Primary constituents include common sugars, proteins, amino acids, nucleic acid purines and pyrimidines, and chlorophyll. All remaining plant constituents are secondary constituents, and their distribution varies from plant to plant. Allelochemicals are secondary metabolites that plants produce. Allelochemical reactions come in a variety of forms.

18.2.1 MAJOR PHENOLIC COMPOUNDS

A hydroxyl group is directly attached to an aromatic ring in a phenolic compound. Aromatic phenol, tannins, flavonoids, cinnamic acid derivatives, hydroxyl and substituted benzoic acids, and quinones are all present. Benzoic acid and its derivatives are the most common plant-derived allelochemicals [6]. One study looked at the effect of aqueous extract of *Delonix regia* on the growth of lettuce (*Lactuca sativa*) and Chinese cabbage (*Brassica chinensis*) and discovered that the extract inhibited plant growth. The major compounds present in the aqueous extract responsible for its allelopathy effect were chlorogenic acid, protocatechuic acid (3,4-dihydroxybenzoic acid), gallic acid, 3,4-dihydroxybenzaldehyde, p-hydroxybenzoic acid, caffeic acid (3,4-dihydroxycinnamic acid), and 3,5-dinitrobenzoic acid [7]. A study found that phenolic compounds from *Chenopodium murale* influence chickpea and pea growth and macromolecule content. When phenolic allelochemicals in *Chenopodium murale* were analysed using HPLC, the four phenolic allelochemicals found were protocatechuic (12.85%), ferulic (30.4%), p-coumaric (20.2%), and syringic acid (33.6%) [8]. Two cis-cinnamic acid glucosides, 1-O-cis-cinnamoyl-D-glucopyranose and 6-O-(49-hydroxy-29-methylenebutyroyl)-1-O-ciscinnamoyl-D-glucopyranose, are found in the leaves of *Spiraea thunbergii* Sieb [9] and have plant growth inhibitory effects.

18.2.2 MAJOR ALKALOIDS OF PLANTS

Alkaloids are heterocyclic nitrogen-containing basic compounds of plant origin and are named accordingly, due to their alkaline chemical nature. Plant alkaloids predominate in four families of plants, including Asteraceae, Apocynaceae, Borag-inaceae, and Fabaceae [10]. They have been important since antiquity due to their pharmacological properties and are among the largest group of secondary metabolites with approximately 20,000 compounds identified to date, representing a great structural biosynthetic diversity [11]. Based on their biosynthetic origin, alkaloids are classified into different classes. Indole alkaloids are derived from tryptophan, pyrrolizidine alkaloids are derived from ornithine or arginine, and quinolizidine alkaloids are derived from lysine [12].

18.2.3 MAJOR TERPENOIDS PRESENT

Terpenoids have been used medicinally since ancient times, but they also contain a number of allelochemicals. 1,8-cineole and camphor are volatile monoterpenes that inhibit plant growth [13]. According to one study, the extraction procedure can improve *Cynara cardunculus*'s allelopathic activity. Water, methanol, ethanol, and ethyl acetate were used as extracting solvents, and all showed an allelopathic effect of more than 50% when tested against six weeds: *Amaranthus retroflexus*, *Portulaca oleracea*, *Stellaria media*, *Anagallis arvensis*, *Echinochloa crus-galli*, and *Lolium perenne*. From the ethyl acetate fraction of the aqueous extract of *Cynara carduncu-lus*, four sesquiterpene lactones (cynaropicrin, cynaratriol, desacylcynaropicrin, and 11,13-dihydrodesacylcynaropicrin) and a lignan (pinoresinol) were isolated. Allelopathic effects were observed in all of them, with cynaropicrin, desacylcynaropicrin, and pinoresinol having the highest activity [14].

18.2.4 DIFFERENT GLUCOSINOLATES AND ISOTHIOCYANATES

Glucosinolates are sulfur-containing compounds that, when hydrolysed, convert to isothiocyanates [15]. Isothiocyanates play an important role in the defence against insect/microorganism attack. They can also be used for soil fumigation because they are volatile [16]. According to one study, the allelopathic chemicals obtained from *Rorippa indica* Hiern. (Cruciferae) roots are hirsutin, arabin, and camelinin [17], as shown in Figure 18.1.

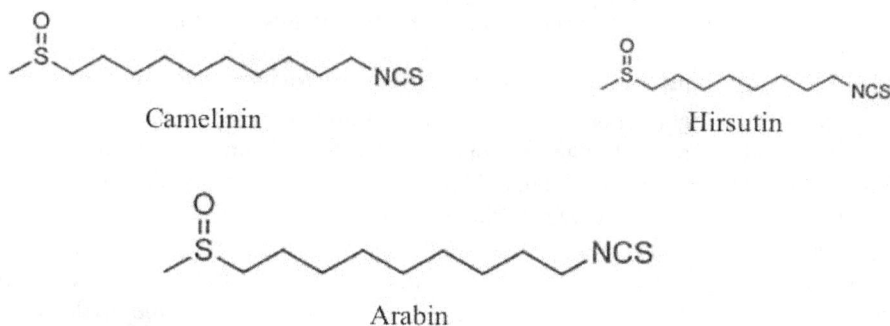

FIGURE 18.1   Camelinin, hirsutin, and arabin from *Rorippa indica*.

18.2.5 Various Benzoxazinoids Found in Plants

These are a type of allelochemical that is derived from indole. Natural insecticides, fungicides, and herbicides are produced by these allelochemicals. These allelochemicals protect *Zea mays* from a variety of pathogens and pests [18]. Plants in the Poaceae, Acanthaceae, Ranunculaceae, and Scrophulariaceae families produce benzoxazinoids, according to a study.

18.3 ALLELOCHEMICALS ACTING AS HERBICIDES

Allelochemicals, or natural plant products derived from higher plants/microbes, are becoming more popular as agrochemicals [19]. Initially, it was unclear why plants devote resources to the assembly of those compounds because they were thought to be useless waste products. However, it is now widely accepted that these compounds act as defensive agents against pathogens, insects, and neighbouring plants [20]. Many of these natural compounds have the potential to induce a wide range of biological effects and may be useful in agriculture and weed management [21]. There is evidence that higher plants release a variety of allelochemicals into their surroundings. Many factors, such as concentration, flux rate, plant age and metabolic state, and prevailing environmental conditions, confirm their toxicity [22]. According to Einhellig [23], both abiotic (temperature, nutrient quantity, and wetness deficit) and organic (disease and bug harm and plant interaction with herbivory) factors increase the number and biogenesis of allelochemicals in plants [24].

18.4 VARIATION IN THE ACTIVITY OF DIFFERENT ALLELOCHEMICALS

18.4.1 Changes in Activity during Natural Processes

Many biotic and abiotic factors must be considered in studies of allelochmical activity. Allelopathy is more intense in adverse or harsh environments where water, light, or nutrition are limited, according to some studies [25–27]. Allelopathy is frequently used by plants to increase their competitive ability and thus survival rates.

18.4.1.1 Temperature and Photoperiod Affect Allelochemicals' Activity

The intensity of light, photoperiod, and temperature all play important roles in allelochemical synthesis. Longer photoperiods and higher temperatures, in general, favour allelochmical activity. Pramanik et al. [28] discovered that the autotoxicity of *Cucumis sativus* root exudates varied with temperature and photoperiod. Under elevated temperatures and elongated photoperiods, the rate of root exudation in the vegetative and reproductive stages could be doubled. Lobon et al. [29] also demonstrated that a high temperature and a long photoperiod increased the allelopathic potential of *Cistus ladanifer* exudates. Chon et al. [30] discovered that extending the photoperiod can increase autotoxicity.

18.4.1.2 Effect of Water on the Activities of Various Allelochemicals

Water is necessary for life, and changes in water conditions have a significant impact on allelochemical activity. When there is a scarcity of water, the activity will generally increase. For example, Li [31] reported that a lack of water may increase the concentration of chlorogenic acid in some plants. Some terpenes, such as α-pinene, β-pinene, cineole, and camphor, increase their volatiles in dry conditions. Dias and Dias [32] discovered that drought can increase the allelopathic activity of *Datura stramonium*.

18.4.2 WITHIN-SPECIES VARIATIONS IN THE ACTIVITY OF ALLELOCHMICALS

Allelochemical activity varies by species and changes between tissues as well, as with plant maturity. Variation in allelochemical activity during the excrete process suggests that the plant itself is a key controlling factor. According to some studies [33], a specific gene controls the output of allelochemicals, and different allelochemicals are controlled by different genes at different times. This demonstrates the viability of using genes to investigate allelopathic mechanisms.

Allelochemicals are produced differently by different plants and even by different accessions of the same plant. Wu et al. [34] examined the phenolic acids in the root tissues of 58 wheat accessions and found that the concentrations of allelochemicals in different accessions varied greatly. Wu et al. [35] used the equal-compartment-agar method (ECMA) to assess allelopathy in seedlings of 453 wheat accessions against *Lolium rigidum*. He demonstrated that allelopathic activity in wheat germplasms varied significantly genetically. Tang and Sun [36] investigated the allelopathy of 700 rice accessions against vegetables and discovered that allelopathic intensity varied with accessions and that allelopathy was stronger among local varieties than among cultivated varieties. In a laboratory and field experiment, Kamara et al. [37] investigated the effects of extracts from levels and mulch of 14 trees on maize germination, growth, and yield, and all data showed significant variations in allelochemical activity.

The content and intensity of allelochemicals are also affected by the maturity of plant tissues. Some studies [38, 39] discovered that the amount and content of allelochemicals in soybean stubs varied with decomposing time and growth stage. Sharma et al. [40] discovered that the allelopathic intensity of *Populus deltoids* increased with age. Huang et al. [41] discovered that the total phenolic content of Chinese fir stump roots decreased with age.

18.5 CONCLUSION

Plant-derived allelochemicals have received increased attention in recent decades, as have their roles in agricultural pest management. Nonetheless, the chemical interactions between plants and organisms are a multidisciplinary science. The identification and detection of allelochemicals from plants and their environments are central to this science. Furthermore, in order to facilitate the design of new pesticides, a thorough understanding of the structural properties of allelochemicals is required. However, because most scientists in this field are not chemists, aspects of chemistry are overlooked. Unlike general phytochemicals, allelochemicals and signalling chemicals must be collected and determined in vivo, in situ, and in real time from intact and living plants. A better understanding of allelochemical production in relation to plant defence strategy may also enable us to better protect and manage developing crops, limit the spread of invasive weeds, conserve native plant stands, and develop strategies for allelochemical development and application as novel pesticides.

REFERENCES

1. Meiners, S.J.; Kong, C.H.; Ladwig, L.M.; Pisula, N.L.; Lang, K.A. Developing an ecological context for Allelopathy. Plant Ecol. 2012, 213, 1221–1227.
2. Macías, F.A.; Mejías, F.J.R.; Molinillo, J.M.G. Recent advances in allelopathy for weed control: from knowledge to applications. Pest Manag. Sci. 2019.
3. Soderquist, C.J. Juglone and allelopathy. J. Chem. Educ. 1973, 50, 782–783.
4. Willis, R.J. Juglans spp., juglone and allelopathy. Allelopathy. J. 2000, 7, 1–55.
5. Rietveld, W.J. Allelopathic effects of juglone on germination and growth of several herbaceous and woody Species. J. Chem. Ecol. 1983, 9, 295–308.
6. Zeng, R.S., Mallik, A.U., Luo, S.M. Allelopathy in sustainable agriculture and forestry. Springer Science+Business Media, LLC, New York, 2008.
7. Chou, C.H., Leu, L.L. Allelopathic substances and interactions of Delonix regia (BOJ) RAF. J Chem Ecol 1992, 18, 2285–2303.
8. Batish, D.R., Lavanya, K., Singh, H.P., Kohli, R.K. Phenolic allelochemicals released by Chenopodium murale affect the growth, nodulation and macromolecule content in chickpea and pea. Plant Growth Regul. 2007, 51, 119–128.
9. Hiradate, S., Morita, S., Sugie, H., Fujii, Y., Harada, J. Phytotoxic cis-cinnamoyl glucosides From Spiraea thunbergii. Phytochemistry 2004, 65, 731–739.
10. Haig, T. Allelochemicals in plants. Allelopathy in sustainable agriculture and forestry. Springer, NewYork, NY, 2008.
11. Yang, L., Stöckigt, J. Trends for diverse production strategies of plant medicinal Alkaloids. Nat Prod Rep. 2010, 27(10), 1469e1479.
12. Seigler, D.S. Pyrrolizidine, quinolizidine, and indolizidine alkaloids. Plant secondary Metabolism. Springer, 1998.
13. Romagni, J.G., Duke, S.O., Dayan, F.E. Inhibition of plant asparagine synthetase by Monoterpene cineoles. Plant Physiol 2000, 123, 725–732.
14. Scavo, A., Rial, C., Molinillo, J.M.G., Varela, R.M., Mauromicale, G., Macias, F.A. The extraction procedure improves the allelopathic activity of cardoon (Cynara cardunculus Var. altilis) leaf allelochemicals. Ind Crop Prod 2019, 128, 479–487.
15. Fenwick, G.R., Heaney, R.K., Mullin, W.J. Glucosinolates and their break down products in Food and food plants. Crit Rev Food Sci Nutr 1983, 18, 123–201.
16. Bangarwa, S.K., Norsworthy, J.K. Glucosinolate and isothiocyanate production for weed Control in plasticulture production system. In: Mérillon, J.M., Ramawat, K. (eds) Glucosinolates Reference series in phytochemistry. Springer, Cham, 2016, pp 1–35.
17. Yamane, A., Fujikura, J., Ogawa, H., Mizutani, J. Isothiocyanates as allelopathic compounds From Rorippa indica Hiern. (Cruciferae) roots. J Chem Ecol. 1992, 18, 1941–1954.
18. Zhou, S., Richter, A., Jander, G. Beyond defense: multiple functions of benzoxazinoids in Maize metabolism. Plant Cell Physiol. 2018, 59, 1528–1537.
19. Kohli, R., Batish, D., Singh, H. Allelopathy as a tool for weed and pest management. J Punjab Acad Sci. 1999, 1, 127–131.
20. Mattner, S.W. The impact of pathogens on plant interference and allelopathy In Inderjit, Mukerji, K.G. (eds) Allelochemicals: Biological control of plant pathogens and diseases. Springer, Dordrechts, 2006.

21. Macías, F.A., Chinchilla, N., Varela, R.M., Molinillo, J.M.G. Bioactive steroids from Oryza sativa L. Steroids. 2006, 71, 603–608.
22. Singh, H.P., Batish, D.R., Kohli, R.K. Autotoxicity: Concept, organisms, and ecological significance. Crit Rev Plant Sci. 1999, 18, 257–272.
23. Einhellig, F.A. Interactions involving allelopathy in cropping systems. Agronomy J. 1996, 88(6), 886–893.
24. Sodaeizadeh, H., Rafieiolhossaini, M., Van Damme, P. Herbicidal activity of a medicinal plant, *Peganum harmala* L., and decomposition dynamics of its phytotoxins in the soil. Indust Crops Prod. 2010, 31, 385–394.
25. Xu, T., Kong, C.-H., Hu, F. 1999. Allelopathy of *Ageratum conyzoides* III. Allelopathic effects of volatile oil from Ageratum on plants under different nutrient levels. Chin J Appl Ecol, 10, 748–750 (in Chinese with English abstract).
26. Zhou, Z.-H. Method of allelopathy bioassay and the affecting factors. Ecol Sci. 1999, 18, 35–38 (in Chinese with English Abstract).
27. Kong, C.H., Hu, F., Xu, X.H. Allelopathic potential and Chemical constituents of volatiles from *Ageratum conyzoides* Under stress. J Chem Ecol. 2002, 28, 1173–1182.
28. Pramanik, M.H.R., Nagai, M., Asao, T., Matsui, Y. Effects of Temperature and photoperiod on phytotoxic root exudates of Cucumber (*Cucumis sativus*) in hydroponic culture. J Chem Ecol. 2000, 26, 1953–1967.
29. Lobon, N.C., Gallego, J.C.A., Diaz, T.S., Garcia, J.C.E. Allelopathic potential of *Cistus ladanifer* chemicals in response to variations of light and temperature. Chemoecology. 2002, 12, 139–145.
30. Chon, S.U., Coutts, J.H., Nelson, C.J. Effects of light, growth Media, and seedling orientation on bioassays of alfalfa Autotoxicity. Agron J. 2000, 92, 715–720.
31. Li, S.-W. 2001. Ecological Biochemistry. Beijing University Press, Beijing, pp. 47–48 (in Chinese).
32. Dias, A.S., Dias, L.S. Effects of drought on allelopathic activity of *Datura stramonium* L. Allelopathy J. 2000, 7, 273–277.
33. Jensen, L.B., Courtois, B., Shen, L.S., Li, Z., Olofsdotter, M., Mauleon, R.P. Locating genes controlling allelopathic effects against Barnyard grass in upland rice. Agron J. 2001, 93, 21–26.
34. Wu, H., Pratley, J., Lemerle, D., Haig, T. Evaluation of seedling allelopathy in 453 wheat (*Triticum aestivum*) accessions against annual ryegrass (*Lolium rigidum*) by the equal-compartment agar method. Aust J Agr Res. 2000, 51, 937–944.
35. Wu, H.W., Haig, T., Pratley, J., Lemerle, D., An, M. Allelochemicals in wheat (*Triticum aestivum* L.): production and exudation of 2,4-dihydroxy-7-methoxy-1,4-benzoxazin-3-one. J Chem Ecol. 2001, 27, 1691–1700.
36. Tang, L.-H., Sun, J.-X. The allelopathy of rice accessions. Jiangsu Agr Sci. 2002, 1, 13–14.
37. Kamara, A.Y., Akobundu, I.O., Sanginga, N., Jutzi, S.C. Effect of mulch from selected multipurpose trees (MPTs) on growth, Nitrogen nutrition and yield of maize (Zea mays L.). J Agron Crop Sci. 2000, 184, 73–80.
38. Wang, P., Zhao, X.-Q. The effect of extracting condition on The analysis result of allelochemicals in wheat straw. Chin Bull Bot. 2001, 18, 735–738.
39. Hu, F., Kong, C.-H. Allelopathic potentials of Arachis Hypogaea on crops. J South China Agr Univ. 2002, 23, 9–12.
40. Sharma, N.K., Samra, J.S., Singh, H.P. Effect of leaf litter of Poplar on Phalaris minor weed. Allelopathy J. 2000, 7, 243–253.
41. Huang, Z.Q., Liao, L.P., Wang, S.L. Allelopathy of phenolics From decomposing stump-roots in replant Chinese fir Woodland. J Chem Ecol, 2000, 26, 2211–2219.

Section 4

Uses of Biopesticides

19 Biopesticides for Food and Medicinal Crops

*Irzam Haroon, Hyda Haroon, Yawar Sadiq,
Zareeka Haroon, and Showkat Rasool Mir*

CONTENTS

19.1 INTRODUCTION

Used to control agricultural pests and pathogens, biopesticides are naturally occurring chemicals or substances that are derived from animals, plants, and microorganisms like bacteria, cyanobacteria, and microalgae. According to the United States Environmental Protection Agency (EPA), biopesticides "are made from natural materials like animals, microorganisms, some minerals, and plants" [1]. Biopesticides are an unique class of chemically synthesised or naturally occurring active chemicals that protect plants [2]. The use of natural materials for pest management is a time-tested indigenous technical knowledge (ITK) that has proven to be very successful, yet many ITKs have been lost because of the development and widespread use of chemical pesticides. Less harm to the environment and human health is caused by biopesticides. They are typically less harmful than chemical pesticides, frequently target specific, have negligible to no aftereffects, and can be used in organic farming [3]. Many beneficial effects, such as a decline in pesticide residues in food, can result from the use of biopesticides in crop protection, lowering consumer risk. Biopesticides often only affect pest organisms and provide little risk to creatures that are not the target. Most of them break down fast, and some, like semiochemicals, are only

DOI: 10.1201/9781003265139-23

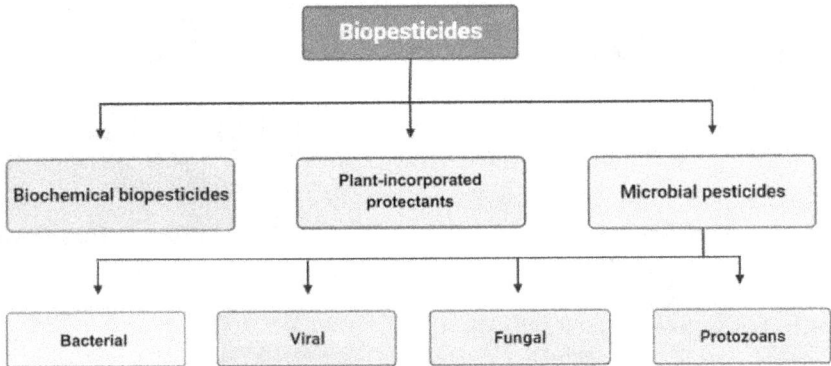

FIGURE 19.1   Classes of biopesticides.

utilised in extremely tiny dosages [2]. Based on the type of active ingredient utilised, the EPA divides biopesticides into three main classes: biochemical pesticides,
plant-incorporated protectants, and microbial pesticides [4] (Figure 19.1).

Compounds that have been manufactured or taken from natural sources with the
same structure and functionality as naturally occurring chemicals are known as biochemical pesticides [5].

Plants can be given genes that help the plants' bodies make substances to fight
pests. These self-made pesticides are called plant-incorporated protectants (PIPs).

Microorganisms (bacteria, fungus, viruses, or protozoans) are the active elements
in microbial pesticides, which have been employed effectively to manage insect
pests. The microbial pesticide *Bacillus thuringiensis*, also known as *Bt*, is one of the
most frequently used [3]. Numerous biopesticides have been discovered and developed over the past ten years, thanks to intensive research on microbial biopesticides,
which has also opened the door for their commercial viability [6]. Many new microbial species and strains, as well as their valuable toxins and virulence factors that
could be a boon for the biopesticide industry, have been discovered as a result of the
successful use of *Bacillus thuringiensis* and some other microbial species. Some of
these have also been turned into commercial products [6, 7].

19.2 VARIOUS BIOPESTICIDES

1. **Bacterial biopesticides**: The most prevalent type of microbial pesticides
   that work in a variety of ways are bacterial biopesticides. They can be used
   to limit the growth of bacteria and fungi that cause plant diseases, although
   they are mostly utilised as insecticides. As an insecticide, they typically
   target particular species of moths, butterflies, beetles, flies, and mosquitoes. They may need to be consumed in order to work, but they must come
   into contact with the pest that they are intended to control. By creating
   endotoxins that are frequently unique to the particular insect pest, bacteria in insects affect the digestive system. The bacterial biopesticide that is

applied to a plant to control pathogenic bacteria or fungus settles there and drives the pathogenic species out [5].

2. **Viral biopesticides**: Viral biopesticides like bacteria and fungi are extremely effective at combating diseases, particularly viruses that take the form of bacteriophages. Aside from that, viruses have negligible off-target effects since they only infect one or a small number of closely related species [8–11].

3. **Fungal biopesticides**: Many key field and horticultural crops are affected by fungal infections, which result in significant reductions in plant output [12]. Including other fungi, bacteria, nematodes, and weeds, fungus-based biopesticides can be used to manage insects and plant diseases. Depending on the pesticidal fungus and the target pest, different modes of action may be used. Fungal biopesticides have an advantage over many bacterial and all viral biopesticides in that they do not require ingestion to be effective. But because they are living things, they frequently need a specific set of circumstances in order to thrive, such as moist soil and moderate temperatures. *Trichoderma* is one of the most popular biocontrol agents since it is affordable, ecofriendly, and excellent at counteracting the negative effects of pesticides [13].

4. **Protozoans**: It is known that about 1,000 different protozoan species, mainly microsporidia, prey on invertebrates, including numerous insect species like grasshoppers and *Heliothine* moths. The two *Nosema* spp. and *Vairimorpha necatrix* species are the most notable insect pathogens in this category. The infectious stage in insects that are vulnerable is the protozoan spore. The host consumes *Nosema* spp. spores, which then germinate in the midgut. The germinating spore releases sporoplasm, which invades host target cells and causes a vast infection that kills organs and tissues. When released from infected tissues and consumed by a susceptible host, sporulation restarts, resulting in infection and an epizootic. In the natural world, parasiteoids and insect predators serve as disease vectors [14].

19.2.1 Role of Biopesticides in Food and Medicinal Crops

Biopesticides are safe to use on fresh fruits and vegetables since they have very short pre-harvest intervals [15]. Natural products are also environmentally beneficial because they are quickly biodegradable and do not harm the environment [16]. Because consumer tastes and preferences change over time and are influenced by the need for food grown organically, biopesticides are suitable substitutes for synthetic pesticides [17]. They work well in modest doses, and their usage encourages sustainable pest management, which supports sustainable agriculture [18]. Since they have no toxicity, biopesticides are safe for both the applicant and the consumer [19]. In order to reduce the number of chemical pesticides used to manage crop pests, biopesticides can therefore be effectively incorporated into integrated pest management (IPM) [20]. Because they break down fast, natural products are safer to use for the environment [12]. Additionally, biopesticides are utilised to clean up agricultural soils by introducing significant microbial populations [21].

19.2.1.1 Losses due to Pests

For all crops, the total loss from pests varies, probably being 50% for wheat, 80% for cotton, 31% for maize, 37% for rice, 40% for potatoes, and 26% for soybeans. The main production issue is weeds, which cause losses of up to 34%. Both infections and animal pests are issues in production, but their losses are lower than those caused by pests; pathogen losses are 16%, and animal pest losses are 18% [22].

19.2.1.2 Use of Pesticides

Pesticide insecticides are used to control agricultural pests such as weeds, insects, pests, and plant pathogens. The price of equipment, gasoline, and labour is decreased when eradicating the pests [23–25]. Pesticides have the benefits of decreased production costs, higher yields, and increased farmer income [25]. Pesticide use for crop production increased by roughly 20 times between 1960 and 2000 [22]. Pesticide use increased dramatically between 2002 and 2007, from 1.0 billion tonnes to 1.7 billion tonnes. China is a significant producer of crops and one of the world's top users of insecticides and pesticides [26, 27]. There are around 2,800 pesticide manufacturers in China. With a capacity of 5,000–10,000 tonnes annually, more than 20 significant industries produce pesticides. There are more than 600 registered active components, and there were 22,000 products (or formulations) in existence as of 2005. Pesticides were produced in excess of 1 million tonnes in 2018. In terms of active ingredients, 20 million hectares and 0.28 million tonnes of prepared products are used annually. It could be possible to avoid a yield loss of between 30 and 40%. More than 400 million farmers and 200,000 distributors exist [28].

19.2.1.3 Safety of Food in Field and Agricultural Practice

On agricultural farms, pesticides are employed to boost crop yield. But because they contain poison, pesticides also have certain negative effects on human health, including death. These residues work better on kids than adults. To keep the pest population below the economic threshold level, many agricultural techniques are applied. The method most widely employed by farmers to produce agricultural goods is the application of pesticides. But these chemical insecticides have a lot of negative effects. Food residues that stay in agricultural goods as a result of the indiscriminate application of pesticides to improve productivity lead to health issues in the people who consume these items [29].

19.2.1.4 Biopesticide Application and Food Safety

Food safety necessitates the use of a variety of biopesticides. An organism that feeds on another is referred to as that organism's natural enemy. Insects that prey on pests are considered beneficial insects. Spiders and mites, for instance, are advantageous to arthropods [30]. Natural fungus, protozoans, bacteria, algae, and fungi may produce microbial pesticides, as may genetically modified versions of these organisms. These are excellent substitutes for chemical insecticides. When a component of a biological toxin comes from a microorganism, such as a fungus or bacteria, it is known as a microbial toxin. Due of the extreme toxicity of these entomopathogens, these microorganisms may result in the pest's death or intestinal rupture. The development of insecticidal toxins by pathogens, which are crucial to pathogenesis, has been demonstrated by studies [31]. Antibiotics are the agents that inhibit the growth of or eradicate bacteria and fungus, among other microorganisms. When an antibiotic is "bactericidal," it kills the bacteria; when it is "bacteriostatic," it prevents the growth of the germs. The development of transgenic crops that are resistant to the main pests as well as their commercialisation are both results of biotechnology's success. Products from the first generation comprise plants with a single insecticidal *Bt* gene that exhibit resistance to the main cotton and corn pests [32]. Interest in plants and their chemo-biodiversity as a source of bioactive chemicals has grown as scientists look for alternate answers to crop protection issues. Secondary metabolites are tiny chemical compounds produced by plants that typically have extremely intricate and distinctive carbon skeleton structures [33]. These compounds have been used for many years to safeguard crops for the benefit of humanity [34]. Plants emit chemicals into the environment, and when they are cultivated in rotational sequences, employed as smother crops, green manures, intercrops, cover crops, or mulch, they can spread infectious illnesses and insect-borne pests, increasing crop yields [35]. Crude or semi-refined extracts, isolated or purified chemicals from various plant species, and commercial goods are all considered botanicals [33].

19.2.1.5 Efficacy of Different Types of Biopesticides

When it comes to controlling crop pests, synthetic pesticides are thought to be more successful than biopesticides [36]. Despite this, sometimes they are is not as significant in controlling a specific population of pests as biopesticides [37]. In some cases, biopesticides outperform synthetic pesticides when used in the proper regimens, concentrations, and frequency ranges [38]. Various plants, microbes, and predators have been highlighted in research reports from throughout the world as potential biopesticides. Insect pests are preyed upon by natural enemies, balancing their abundance in the ecosystem. These predators play a crucial role in agricultural systems [39, 40]. Scents and other attractants are two methods predators use to entice insects. Pheromones, which are some of these odours, have been commercialised and are used to control significant crop pests like *Tuta absoluta* [41]. Commercial pheromones are used as bait to help attract adult insects, which are either sterilised or starved to death to render them inactive [42]. In an effort to control various crop pests, researchers have looked into the substances that some plants possess to protect themselves from pests [43]. It has been discovered that some plants contain substances that are efficient against a variety of pests, including nematodes and fungi [44, 45]. Some microbe species are useful as biopesticides because they are antagonistic to other species [46].

Researchers have looked into the substances that some plants produce to shield themselves from pests in an effort to control a variety of agricultural pests [47]. A biopesticide mixture including ginger (*Zingiber officinale*) and onion (*Allium cepa*) was tested for effectiveness against the tomato fruit worm (*Helicoverpa armigera*), and it showed a 70% to 80% control rate [48]. In the study, plants treated with the formulation showed increased yield when compared to the untreated controls. Insect and microbial predators have been utilised successfully to control insect infestations. Thrips (*Frankliniella occidentalis* or *Scirtothrips dorsalis*) and spider mites (*Tetranychus urticae*) have all been managed with species of *Amblyseius swirskii* [49,

50]. Spider mites (*Tetranyshus* spp.) are effectively controlled by the predator mite *Phytoseiulus persimilis* [51]. Aphid predators (*Aphidius colemani*) were observed to be effective against *Aphis gossypii* on chrysanthemums (*Dendranthema grandiflora*) in a study by Vásquez et al. [52].

19.3 MERITS OF BIOPESTICIDES OVER CHEMICAL PESTICIDES

Compared to traditional chemical pesticides, biopesticides have a number of advantages. They are effective enough to substitute for synthetic pesticides in pest management since they are non-toxic to non-target organisms, ecologically benign, and target specific [53]. Because they can be effectively used in sustainable agriculture methods, the use of biopesticides has recently gained pace [54, 55]. Biopesticides can reduce the usage of conventional pesticides as an essential part of IPM programmes since they are extremely effective in small amounts and disintegrate quickly without leaving bothersome residues [56].

19.4 DEMERITS OF BIOPESTICIDES OVER CONVENTIONAL METHODS

The following factors explain why, despite the benefits of utilising biopesticides, their usage has not been as pervasive as anticipated:

- Due to the expenses associated with researching, testing, and obtaining regulatory approval for new biological agents, the cost of producing biopesticides is high.
- Biopesticides have a limited shelf life because they are sensitive to changes in temperature and humidity.
- Climatic/regional differences in temperature, humidity, soil conditions, etc. limit field efficacy.
- Farmers are uninterested in biopesticides because of their high specificity, which means that they only work against specific illnesses and pests. To control various infections and pests in the field, they must utilise a variety of biological agents. These substances are difficult to use, expensive, and inconvenient, and they are not suitable for all pests and pathogens.

19.5 CONCLUSION AND FUTURE PROSPECTS

Biopesticides continue to be viable alternatives to traditional pesticides despite the numerous obstacles facing their implementation. Because of their detrimental impacts on the environment, human health, natural enemies, and ecosystem equilibrium, the use of synthetic chemicals has given rise to a number of worries. Synthetic pesticides contain several active components that have been shown to cause cancer, endangering human life. Due to their low toxicity, biodegradability, and limited persistence in the environment, biopesticides provide a better option to synthetic pesticides. The building blocks of biopesticides are easily accessible and reasonably priced. To help with formulation and commercialisation, data on toxicity levels, chemistry, active chemicals, and their compatibility with other methods of managing pests and diseases are required. Studies on the efficacy of natural plant protection agents have been undertaken all over the world, with in vitro trials producing some of the most notable findings. Studies on the usefulness of biopesticides in field settings and controlled situations have also been conducted, with varying degrees of success. To fill the deficiencies in biopesticide formulation, more study is advised. The complete efficacy of biopesticides in crop pest management would be ensured by stable products under field settings. Therefore, in order to develop stable, long-lasting biopesticide formulations, researchers should collaborate with engineers in the public and private sectors and farmers.

REFERENCES

1. EPA. Ingredients used in pesticide products: Pesticides. What are biopesticides? www.epa.gov/Ingredients-used-pesticide-products/what-are-biopesticides (accessed on 10 May 2021).
2. Czaja K, Góralczyk, K, Struciński, P, Hernik, A, Korcz, W, Minorczyk, M, . . . Ludwicki, JK (2015) Biopesticides–towards increased consumer safety in the European Union. Pest Management Science, 71(1): 3–6.
3. Mazid S, Kalida JC, Rajkhowa RC (2011) A review on the use of biopesticides In insect pest management. International Journal of Science and Advanced Technology 1: 169–178.
4. USEPA (2008) What are biopesticides? www.epa.Gov/pesticides/biopesticides/whatare-biopesticides.htm
5. O'Brien KP, Franjevic S, Jones J (2009) Green chemistry and sustainable agriculture: the role of biopesticides. Advancing Green Chemistry. http://advancinggreenchemistry.org/wp-content/uploads/Green-Chemand-Sus.-Ag.-the-Role-of-Biopesticides.pdf
6. Ruiu L (2018) Microbial biopesticides in agroecosystems. Agronomy 8: 235.
7. Ujváry I (2001) Chapter 3—pest control agents from natural products, in Handbook of Pesticide Toxicology, 2nd ed., eds. Krieger RI, Krieger WC. Academic Press: San Diego, CA, pp. 109–179.
8. Cory JS, Myers JH (2003) The ecology and evolution of insect baculoviruses. Annual Review of Ecology, Evolution, and Systematics 34: 239–272.
9. England LS, Vincent ML, Trevors JT, Holmes SB (2004) Extraction, detection and persistence of extracellular DNA in forest litter microcosms. Molecular and Cellular Probes 18: 313–319.
10. Raymond B, Hartley SE, Cory JS, Hails RS (2005) The role of food plant and pathogen-induced behavior in the persistence of a nucleopolyhedrovirus. Journal of Invertebrate Pathology 88: 49–57.
11. Hewson I, Brown JM, Gitlin SA, Doud DF (2011) Nucleopolyhedrovirus detection and distribution in terrestrial, freshwater, and marine habitats of Appledore Island, Gulf of Maine. Microbial Ecology 62: 48–57.
12. Khandelwal M, Datta S, Mehta J, Naruka R, Makhijani K, Sharma G, Kumar R, Chandra S (2012) Isolation, characterization and biomass production of Trichoderma viride using various agro products: A biocontrol agent. Advances in Applied Science Research 3: 3950–3955.
13. Bailey DJ, Gilligan CA (2004) Modeling and analysis of disease induced host growth in the epidemiology of take all. Phytopathology 94: 535–540.

14. Brooks FM (1988) Entomogenous protozoa, in Handbook of Natural Pesticides, Vol. V, Microbial Insecticides, Part A, Entomogenous Protozoa and Fungi, ed. Ignoffo CM. CRC Press Inc: Boca Raton, FL, pp. 1–149.

15. Khater HF (2012) Prospects of botanical biopesticides in insect pest management. Pharmacologia 12: 641–656.

16. Leng P, Zhang Z, Pan G, Zhao M (2011) Applications and development trends in biopesticides. African Journal of Biotechnology 86: 19864–19873.

17. Okunlola AI, Akinrinnola O (2014) Effectiveness of botanical formulations in vegetable production and bio-diversity preservation in Ondo State, Nigeria. Journal of Horticulture and Forestry 1: 6–13.

18. Nawaz M, Mabubu JI, Hua H (2016) Current status and advancement of biopesticides: Microbial and botanical pesticides. Journal of Entomology and Zoology Studies 2: 241–246.

19. Ekwenye UN, Elegalam NN (2005) Antibacterial activity of ginger (Zingiber officinale Roscoe) and Garlic (*Allium sativum* L.) Extracts on *Escherichia coli* and *Salmonella typhi*. International Journal of Molecular Medicine and Advance Science 1: 411–416.

20. Sesan TE, Enache E, Iacomi M, Oprea M, Oancea F, Iacomi C (2015) Antifungal Activity of some Plant Extract against Botrytis cinerea Pers. in the Blackcurrant Crop (*Ribes nigrum* L). Acta Scientiarum Polonorum Technologia Alimentaria 1: 29–43.

21. Javaid MK, Ashiq M, Tahir M (2016) Potential of biological agents in decontamination of agricultural soil. https://doi.org/10.1155/2016/1598325

22. Oerke E-C (2006) Crop losses to pests. The Journal of Agricultural Science 144(1): 31–43.

23. Osteen CD, Szmedra PI (1989) Agricultural Pesticide Use Trends and Policy Issues. Agricultural Economics Report. US Department of Agriculture, Economic Research Service: Washinton, DC.

24. Fernandez-Cornejo J, Jans S, Smith M (1998) Issues in the economics of pesticide use in agriculture: A review of the empirical evidence. Review of Agricultural Economics 20(2): 462–488.

25. Fernandez-Cornejo J, Nehring R, Osteen C, Wechsler S, Martin A, Vialou A. (2014) Pesticide Use in US Agriculture: 21 Selected Crops (EIB-124). U.S. Department of Agriculture, Economic Research Service; 1960–2008: Washinton, DC.

26. Peshin R, Bandral RS, Zhang W, Wilson L, Dhawan AK (2009) Integrated pest management: A global overview of history, programs and adoption, in Integrated Pest Management: Innovation-Development Process. Springer, pp. 1–49. http://link.springer.com/10.1007/978-1-4020-8992-3_1

27. Zhang W, Jiang F, Ou J (2011) Global pesticide consumption and pollution: With China as a focus. Proceedings of the International Academy of Ecology and Environmental Sciences (IAEES) 1(2): 125.

28. Anastassiades M, Tasdelen B, Scherbaum E, Stajnbaher D, Ohkawa H, Miyagawa H, et al. (2007) Pesticide chemistry: Crop protection, in Public Health, Environmental Safety. Wiley-VCH Verlag GmbH Co. KGaA: Weinhei, pp. 439–458.

29 Abrol DP, Shankar U (2014) Pesticides, food safety and integrated pest management, in David Pimentel RP, ed. Integrated Pest Management: Pesticide Problems, 1st ed. Vol. 3. Springer, pp. 167–199.

30. Smith HA, Capinera JL (2017) Natural enemies and biological control, pp. 1–6. http://edis.ifas.ufl.edu/in120

31. Burges HD (1981) Safety, safety testing and quality control of microbial pesticides, in: Microbial Control of Pests and Plant Diseases 1970–1980, Academic Press Amsterdam Holland.

32. Ferry N, Edwards MG, Gatehouse J, Capell T, Christou P, Gatehouse AMR (2006) Transgenic plants for insect pest control: A forward looking scientific perspective. Transgenic Research 15(1): 13–19.

33. Cavoski I, Caboni P, Miano T (2011) Natural pesticides and future perspectives, in Pesticides in the Modern World-Pesticides Use and Management. InTech: Rijeka.
34. Rattan RS, Sharma A (2011) Plant secondary metabolites in the sustainable diamondback moth (*Plutella xylostella*) management. Indian Journal of Fundamental and Applied Life Sciences 1: 295–309.
35. Farooq M, Jabran K, Cheema ZA, Wahid A, Siddique KHM (2011) The role of allelopathy in agricultural pest management. Pest Management Science 67(5): 493–506.
36. Khan AI, Hussain S, Akbar R, Saeed M, Farid A, Ali I, Alam M, Shah B (2015) Efficacy of a biopesticide and synthetic pesticides against tobacco aphid, myzus persicae sulz. (Homoptera, Aphididae), on tobacco in Peshawar. Journal of Entomology and Zoology Studies 4: 371–373.
37. Ahmad S, Khan IA, Hussain Z, Shah SIA, Ahmad M (2007) Comparison of a biopesticide with some synthetic pesticides against aphids in rapeseed crop. Sahrad Journal of Agriculture 4: 1117–1120.
38. Shah JA, Inayatullah M, Sohail K, Shah SF, Shah S, Iqbal T, Usman M (2013) Efficacy of botanical extracts and a chemical pesticide against tomato fruit worm, Helicoverpa armigera. Sarhad Journal of Agriculture 1: 93–96.
39. Rao KS, Vishnupriya R, Ramaraju K (2017) Efficacy and safety studies on predatory mite, Neoseiulus longispinosus (Evans) against two-spotted spider mite, tetranychusurticae koch under laboratory and greenhouse conditions. Journal of Entomology and Zoology Studies 4: 835–839.
40. Kenis M, Hurley BP, Hajek AE, Cock MJW (2017) Classical biological control of insect pests of trees: facts and figures. Biological Invasions 19: 3401–3417.
41. Refki E, Sadok BM, Ali, BB, Faouzi, A, Jean VF, Rudy CM (2016) Effectiveness of pheromone traps against Tuta absoluta. Journal of Entomology and Zoology Studies 6: 841–844.
42. Galko J, Nikolov C, Kunca A, Vakula J, Gubka A, Zúbrik M, Rell S, Konopka B (2016) Effectiveness of pheromone traps for the European Spruce Bark Beetle: A comparative study of four commercial products and two new models. Lesnícky Casopis-Forestry Journal 62: 207–215. https://doi.org/10.1515/forj-2016-0027
43. Mendoza JLH, Pérez MIS, Prieto JMG, Velásquez JDQ, Olivares JGG, Langarica HRG (2015) Antibiosis of Trichoderma spp strains native to Northeastern Mexico against the pathogenic fungus *Macrophomina phaseolina*. Brazilian Journal of Microbiology 4: 1093–1101. https://doi.org/10.1590/S1517-838246420120177
44. Hussain F, Abid M, Shaukat S, Farzana S, Akbar M (2015) Anti-fungal activity of some medicinal plants on different pathogenic fungi. Pakistan Journal of Botany 5: 2009–2013.
45. Sidhu SH, Kumar V, Madhu MR (2017) Eco-friendly management of root-knot nematode, Meloidogyne javanica in Okra (*Abelmoschus esculentus*) Crop. International Journal of Pure and Applied Bioscience 1: 569–574. https://doi.org/10.18782/2320-7051.2507
46. Aw KMS and Hue SM (2017) Mode of infection of Metarhizium spp. fungus and their potential as biological control agents. Journal of Fungi 30: 1–20. https://doi.org/10.3390/jof3020030
47. Sumitra A, Kanojia AK, Kumar A, Mogha N, Sahu V (2014) Biopesticide formulation to control tomato lepidopteran pest menace. Current Science 7: 1051–1057.
48. Muzemu S, Mvumi BM, Nyirenda SPM, Sileshi GW, Sola P, Chikukura L, Kamanula JF, Belmain SR, Stevenson PC (2011) Pesticidal effects of indigenous plant extracts against rape aphids and tomato red spider mites. African Crop Science Conference Proceedings 10: 171–173.
49. Xu X, Enkegaard A (2010) Prey preference of the predatory mite, Amblyseius swirskii between first instar western flower thrips *Frankliniella occidentalis* and nymphs of the two-spotted spider mite *Tetranychus urticae*. Journal of Insect Science 149: 1–11. https://doi.org/10.1673/031.010.14109

50. Arthurs S, McKenzie CL, Chen J, Dogramaci M, Brennan M, Houben K, Osborne L (2009) Evaluation of Neoseiulus cucumeris and Amblyseius swirskii (Acari: Phytoseiidae) as biological control agents of chilli thrips, *Scirtothrips dorsalis* (Thysanoptera: Thripidae) on pepper. Biological Control 49: 91–96. https://doi.org/10.1016/j.biocontrol.2009.01.002

51. Fiedler Z (2012) Interaction between beneficial organisms in control of spider mite *Tetranychus urticae* (koch.). Journal of Plant Protection Research 2: 226–229. https://doi.org/10.2478/v10045-012-0035-2

52. Vá Squez GM, Orr DB, Baker JR (2006) Efficacy assessment of *Aphidius colemani* (Hymenoptera: Braconidae) for suppression of *Aphis gossypii* (Homoptera: Aphididae) in greenhouse-grown chrysanthemum. Journal of Economic Entomology 4: 1104–1111. https://doi.org/10.1093/jee/99.4.1104

53. Saberi F, Marzban R, Ardjmand M, Pajoum SF, Tavakoli O (2020) Optimization of culture media to enhance the ability of local Bacillus thuringiensis var. tenebrionis. Journal of the Saudi Society of Agricultural Sciences 19: 468–475.

54. Pathak J, Maurya PK, Singh SP, Häder DP, Sinha RP (2018) Cyanobacterial farming for environment friendly sustainable Agriculture practices: Innovations and perspectives. Frontiers in Environmental Science 6: 7.

55. Gonçalves AL (2021) The use of microalgae and cyanobacteria in the improvement of agricultural practices: A review on their biofertilising, biostimulating and biopesticide roles. Applied Sciences 11: 871.

56. Guerra F, Sepúlveda S (2021) Saponin production from Quillaja Genus species. An insight into its applications and biology. Scientia Agricola 78: 305.s

Index

Page numbers in *italics* Indicate figures; page numbers in **bold** indicate tables.

For Product Safety Concerns and Information please contact our EU
representative GPSR@taylorandfrancis.com
Taylor & Francis Verlag GmbH, Kaufingerstraße 24, 80331 München, Germany